669.0972 H826s

Hosler, Dorothy.

The sounds and colors of power

The Sounds and Colors of Power

Dorothy Hosler

The Sounds and Colors of Power

The Sacred Metallurgical Technology of
Ancient West Mexico

The MIT Press
Cambridge, Massachusetts
London, England

© 1994 Massachusetts Institute of Technology

All rights reserved. No part of this book may be reproduced in any form by any electronic or mechanical means (including photocopying, recording, or information storage and retrieval) without permission in writing from the publisher.

This book was set in Galliard by DEKR Corporation and was printed and bound in the United States of America.

Library of Congress Cataloging-in-Publication Data

Hosler, Dorothy.
    The sounds and colors of power : the sacred metallurgical technology of ancient West Mexico / Dorothy Hosler.
    p. cm.
    Includes bibliographical references and index.
    ISBN 0-262-08230-6
    1. Indian metal-work—Mexico—History. 2. Indians of Mexico—Antiquities. 3. Metallurgy—Mexico—History. 4. Founding—Mexico—History. 5. Mexico—Antiquities. I. Title.
F1219.3.M52H676 1994
669′.0972—dc20   94-25455
                                              CIP

```
669.0972 H826s

Hosler, Dorothy.

The sounds and colors of
  power
```

Contents

Acknowledgments — viii

1 The Perspective and the Region — 2
An Overview of Mesoamerican Prehistory — 6
The West Mexican Metalworking Zone — 13
South American Contacts in West Mexico — 15
The Research Agenda — 17

2 Resources, Metals, and Alloys — 20
Distribution of Native Metals and Ore Minerals in the Metalworking Zone — 21
Ore Types Used in West Mexican Metallurgy — 31
Documentary Evidence for Ancient Mining — 38

3 Period 1 of West Mexican Metalworking: A.D. 600 to A.D. 1200/1300 — 44
The Technological Chronology — 45
Archaeological Evidence for Period 1 Metallurgy — 46
The Metallurgical Technology of Period 1 — 51
Lost-Wax Casting: Bells — 52
Cold Working: Sumptuary and Ritual Objects — 59
Cold Working: Utilitarian Objects — 75
Summary and Observations — 83

4 Origins of Period 1 West Mexican Metallurgy — 86
Ecuador and West Mexico — 89
Colombia and Lower Central America — 98
The Introduction of the Technology to West Mexico — 99
The Evidence from Ecuador — 105
The Reinterpretation — 122

5 The Florescence of West Mexican Metallurgy: A.D. 1200/1300 to the Spanish Invasion — 126
Archaeological Evidence for Period 2 Metallurgy — 129
The Metallurgical Technology of Period 2: New Materials and New Designs — 131
Lost-Wax Casting: Bells — 132
Cold and Hot Work: Sumptuary Items — 139
Cold Work: Tools and Axe-monies — 156
The Focus of Period 2 Metallurgy — 168

6 Period 2: Origins and Transformations — 170
Alloying — 171
Alloys and Artifacts — 180
Mechanisms of Introduction — 184
The New Technology: West Mexican Alloys and Smelting Regimes — 186
The West Mexican Interpretation — 189

7 The Dissemination of West Mexican Metallurgy — 196
Western Morelos: Cuexcomate and Capilco — 202
Lamanai, Belize — 208
The Huastec Region: Vista Hermosa and Platanito — 216
Other Sites and Regions — 219
Discussion — 223

8 The Sounds and Colors of Power — 226
Color — 228
Sound — 233
Sound, Metal, and Creation — 246
The Social Context — 248

Appendix 1 Technical Studies: Data and Methods — 252
Measurements — 253
The Study Corpus — 254
Research Permissions — 257

Appendix 2 Quantitative Chemical Analyses of Artifacts in the RMG Collection — 260

Notes — 274

References — 282

Index — 302

I dedicate this book to two people. One is the late Cyril S. Smith, MIT Institute Professor of Metallurgy, whose intellect, craft, curiosity, play, and insight expanded and deepened my understanding of what it means to be fully and exuberantly human. I also dedicate this book to my father, who in his own style shares Cyril's intellect, creativity, imagination, and craft, and who showed me that work and play could be one and the same thing.

Acknowledgments

Everyone who writes a book depends on myriad people: I would like to thank those who were the most directly responsible for helping me produce this one. Professors Barbara Voorhies (UCSB), Michael Smith (SUNY Albany), and Heather Lechtman (MIT) had the patience and intellectual fortitude to read and comment on the whole text, sometimes through various drafts. Drs. Helen Pollard (Michigan State), Anabel Ford (UCSB), Professors R. Tom Zuidema (University of Illinois at Urbana), Maria Masucci (Drew), and Louise Burkhart (SUNY Albany), and my editor Larry Cohen made various organizational and substantive suggestions about particular chapters. Professor Kenneth Hale (MIT) assisted with Nahuatl and Professor Ulrich Petersen (Harvard) with geology and ore mineralogy. Professors Robert Rose and Regis Pelloux (MIT) helped with various questions concerning material-property relations; Professor David Parks (MIT) and Matt and Melissa Lewis (Los Alamos National Laboratories) refined certain problems concerning tweezer design and function. Matt and Melissa also helped with the Ecuadorian research while on their honeymoon in the irresistible metropolis of Guayaquil, Ecuador. Leslie Compton (then an MIT student) accompanied me on a trip to Mexico City during the eclipse of the sun to sample and later perform extended laboratory studies of material from the Huastec area. Patricia Capone (a Harvard graduate student) also helped with selected metallographic studies and with background research, Tona Hangen (an MIT student) with background research, Katherine Williams with some of the Belize material, and Gabriela Canalizo with sixteenth-century Spanish and chaos theory. MIT students Robin Huang and Nate Osgood saved the day with the data base management program, and Brett Snyder (an MIT student) capably assisted with the long task of assembly. In the final lap Jorge Calvo, sporting his green roller blades and purple shoelaces, combined his good humor and research skills, knowledge of English, Spanish, Nahuatl, and data base management programs to make vastly lighter what could have been a grueling task. Walter Correia performed the majority of the qualitative and quantitative chemical analyses, Carol Roberts offered patient good cheer and logistical support, Mike Aloisi inspired me by periodically reminding me that he couldn't retire until I finished my book, and Ann Lindemulder provided humor and help at an important moment.

The Undergraduate Research Opportunities Program at MIT funded some of the students who assisted me in the laboratory studies. MIT provided financial support for a one-semester leave for writing, and a HASS-D fund grant to help defray the cost of illustrations. ASARCO (the American Smelting and Refining Company) funded a second semester of leave, and ASARCO, the Wenner-Gren Foundation, and the National Science Foundation funded the original study and sampling. I am particularly grateful to ASARCO's Charles Barber and Juan Gallardo of Industrial Minera México (IMMSA), its Mexican affiliate, for their continued interest in this field.

I also owe many debts and profound thanks to my colleagues in Ecuador, Mexico, and elsewhere in the Americas who provided the initial encouragement that was critical to collecting the baseline chemical and metallographic data and helped me obtain the permits I needed to carry out that part of the work. In Mexico, Arq. Otto Schöndube, Ing. Frederico Solórzano, and the Consejo de Arqueología were crucial to the process; the French Archaeological Mission assisted with several subsequent permits. Dr. David Pendergast helped in Belize, in Ecuador the work was facilitated by Olaf Holm, Presley Norton, Ivan Cruz, and the Instituto del Patrimonio Nacional. Dra. Clemencia Plazas helped me gain access to collections in Colombia.

I thank Whitney Powell for her wonderful illustrations, and, at the MIT Press, Matthew Abbate for his copyediting job and Jean Wilcox for her book design. It has been a privilege and pleasure to work with my editor Larry Cohen.

Finally, I would like to thank Heather Lechtman, a mentor and colleague whose work initially interested me in this field and who, despite trying external circumstances, maintained her equanimity and supported this effort throughout. The training I received at CMRAE (the Center for Materials Research in Archaeology and Ethnology) at MIT made this work possible. I also thank Ralph Engle who adroitly helped me at various times with the less-than-joyous aspects of this enterprise. Despite those inevitable moments this project has been a thoroughly satisfying one from all perspectives, in part because I had to learn so much to do it, and in part because the answers—how world views, sensibilities, and values became expressed in this ancient New World technology—were so inherently fascinating.

The Sounds and Colors of Power

1

The Perspective and the Region

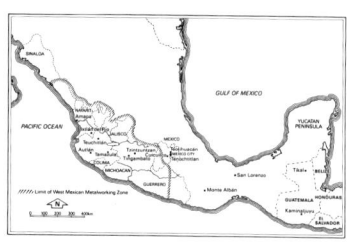

The sudden appearance of metallurgy between A.D. 600 and A.D. 700 in the western region of Mexico and its subsequent development there is one of the most interesting series of events in New World history. This technology came late, significantly after the emergence of civilization in ancient Mexico. These circumstances provide a singular opportunity to examine how a preindustrial people managed a completely new material, and one that appeared long after they had experimented with, and mastered, the properties of stone, bone, clay, fiber, and other materials to meet the necessities of social life. My goal is to examine technology as a cultural expression by investigating how West Mexican peoples chose to use the new material, metal. I identify the characteristics of the technology through laboratory studies of the artifacts, determine the raw materials used, and trace changes in the technology through time. I also establish the South and Central American origins of West Mexican metallurgy, and describe how West Mexican metalworkers incorporated certain facets of these external technical complexes, then transformed them in keeping with their own perceptions of this novel material. Finally, I explore the meaning of these transformations; why two properties of metal, its sound and its color, became so fundamentally important in West Mexico that they determined the course of an entire technology.

Embedded in this study is the premise that technologies and their products shape and are shaped by economic and political interests, social values, and other elements of culture. Staudenmaier puts it well:

Technical designs cannot be meaningfully interpreted in abstraction from their human context. The human fabric is not simply an envelope around a culturally neutral artifact. The values and world views, the intelligence and stupidity, the biases and vested interests of those who design, accept, and maintain a technology are embedded in the technology itself. (Staudenmaier 1985: 165)

Values, world views, intelligence and stupidity, biases and vested interests are all aspects of culture. What makes this study unique is that, by using the methods of materials science, I can determine *where* aspects of this particular ancient "human fabric" are visible in selection of raw materials and processing methods and in decisions about object design. This is possible because I can distinguish between those characteristics of the technology that result from technical *choices,* which reflect and express values, interests, and other social variables, and those arising from technical *requirements* imposed by the material's inherent physical and mechanical properties.

The documents I use to characterize this technology and to identify technical choices consist mostly of artifacts. Artifacts are the familiar domain of archaeologists. However, my documents are not simply the artifacts but their microstructures, chemistries, hardnesses, and other material property data. Through laboratory techniques, I collect and record the history they provide. This approach is still unusual even for the archaeologist. The data are liberating because artifact microstructures and artifact chemistries faithfully register the manufacturing history of an object. Microstructures and other material property data do not distort, lie, or opportunistically revise. They are immune to intellectual fads; they cannot be deconstructed.[1] They provide an extraordinarily rich and largely untapped source of information for historians of technology, archaeologists, cultural anthropologists—for anyone interested in how human beings have used objects (artifacts) to order, reflect, alter, and create their emotional, physical, and social world. The artifacts' manufacturing history is locked into these microstructures, and the history of a technology becomes visible in reading them. They reveal technical imagination, errors, experiment, inspirations, and business as usual.

The task I have set here is to identify, and subsequently explain, the choices that shaped West Mexican metallurgy by distinguishing these technical choices from technical requirements through such laboratory studies. Choices can occur at various points in a processing regime and are possible because technical alternatives usually exist in materials, design, and fabrication sequences. The most obvious choices have to do with the final product, that is with *what* to make. Metal items commonly found in preindustrial settings include body ornaments, hand tools, musical instruments, containers, armor, weapons, and agricultural tools. Metal serves well for weapons because it can be work-hardened and holds a sharp cutting edge, but properties such as fluidity and resonance make it useful for other objects. The choice of what to make takes place in a specific social context and is determined by an accommodation of social realities with the physical properties of materials. Local environmental circumstances impinge as well. A metal tool industry is unlikely if the raw materials for bronze or iron are inaccessible, but even when they are available the technology can develop along other lines.

Choices also occur in fabrication methods and raw materials, but these depend on the intended use or function of the particular item and on the specifics of its design. Making a choice in one domain, for example in design (thickness, length, and other dimensions), can impose corresponding technical requirements in others: in metals or alloys that can be used and in fabrication techniques. This is so for several reasons. First, properties (hardness, elasticity, toughness, color, etc.) vary by material and are affected by processing methods: cast bronze is harder than cast copper, but copper that has been work-hardened can approach the hardness of a low-tin bronze casting. In addition, the properties of any single material vary across certain ranges, and choices in developing particular properties are possible precisely because these ranges exist. Technical choices thus mutually influence one another. For example, if copper, which lacks the strength of bronze, is the only metal available, a copper work implement has to be made shorter and thicker than a bronze counterpart. Functionality or usability is also affected by fabrication methods. Some implements must deform or bend elastically in use; if they are shaped by cold work, they may become brittle and fail. All technical choices are conditioned by the social context; cumulatively they give form and definition to the technology.

Thus the possibility of choice among technical alternatives is what makes the technological enterprise a cultural activity. Yet, until relatively recently, archaeologists and historians of technology have considered technology as a kind of extracultural entity, and have confused it with its products, especially tools.[2] Technologies, defined implicitly as tools and acting apart from culture, have been assumed to behave as generic agents of social change through their progressive increase in efficiency and complexity. Historians of technology are now vigorously challenging this view[3] using data drawn primarily from western or industrial technologies.

The tendency to equate technology with tools has been especially strong among archaeologists studying ancient metallurgies. Metallurgy's purported "extracultural" nature has led some scholars to argue that, in their early and preindustrial forms, metallurgies developed invariably in a fixed, noncultural trajectory from simple to complex, guided by an inherent "technological logic" (Wertime 1973; see also Charles 1980). The idea is encapsulated in references to "ages" of stone, copper, bronze, and iron, associated invariably with increasingly complex, virtually inevitable cultural developments. The idea that technologies in general develop in linear sequences mirrors nineteenth-century concepts about social evolution, which was also held to move inexorably through progressive stages. Archaeologists generally agree that these ideas inhibit not only the study of social process (Johnson and Earle 1987; Wenke 1981) but the understanding of contemporary as well as nonindustrial technologies. Nonetheless, the view that technologies develop in a linear and sequential fashion is still very much in evidence in the archaeological literature.

This quasi-evolutionary and essentially nonanthropological model of metallurgical development has been refuted by Heather Lechtman's laboratory research on Andean metallurgy (see Lechtman 1977, 1984, 1988, 1991), presaged by Cyril Smith's pioneering work on the materials science of art and archaeological materials (see C. S. Smith 1965a, 1965b, 1968a, 1968b, 1972, 1975, 1978, 1981). Lechtman (1980) shows that Old World models claiming that people work native metals, like copper, before smelting ores do not apply in the Andean environment, where no evidence exists for such a primordial stage. There are good reasons for this. In the central Andes, native copper is uncommon and tends to be physically inaccessible; what were accessible and abundant were ore minerals. People used what was on hand, smelting copper ores in the earliest stages of metallurgical activity. Lechtman also shows that, in the Inka world, copper-tin bronze not only served utilitarian purposes but became a powerful symbol of Inka hegemony. The state disseminated the bronze alloy throughout its conquered northern territories. Tin bronze began to replace the indigenous copper-arsenic bronze not because of its technical superiority—it was technically equivalent—but because it was a marker of political power. Lechtman (1977) advocates laboratory-based case studies of indigenous and prehistoric technologies to identify similar culturally patterned technological behaviors; she calls these behaviors technological styles.

Mark Leone (1973, 1977), like Lechtman, argued for the symbolic role of technologies decades ago, although his work has not involved materials science approaches. Leone maintains that material culture and its associated technologies represent perceptions of social categories, and that utopian communities, such as the Mormons and the Shakers, are characterized by sacred technologies that express these basic ideological precepts (1973). One technological manifestation of Mormon ideology appeared in the way the community partitioned space, accomplished through town plans and fences that created closely spaced but separate compartments. The principle of mutually exclusive compartments reflected a basic tenet of Mormon philosophy while also solving the problem of ecological crowding resulting from life in an arid desert environment. Fences, houses, and yards enabled certain socially required behaviors while reinforcing ways of thinking about those behaviors.

Around the same time, Pierre Lemonnier (1976, 1986) raised the question of technical choice in artifact design from an ethnographic perspective. He recognized that the anthropologist relating technical procedures to characteristics of the societies that developed them must understand the range of technical alternatives known to a particular group. He did not deal with the critical issue of material properties, however, which is essential to understanding the difference between design decisions that derive from actual choices and those that cannot be avoided due to inherent material constraints.

Lechtman's 1977 call for laboratory-based case studies of indigenous technologies using materials science data has gone largely unheeded.[4] Researchers have carried out numerous technical studies on lithic materials, pottery, and textiles, using laboratory methods to document fabrication sequences,[5] although the focus has been quite narrow in some cases. A few are beginning to examine technical choice, usually as it optimizes certain material properties for specific functional objectives.[6] Nonetheless, until the present work, scholars generally have not used these approaches[7] systematically to identify the technical choices (or the material constraints and possibilities influencing such choices) that generate the broader patterns that distinguish particular indigenous or prehistoric technologies, or their industrial counterparts. Instead, archaeologists have been focusing on the products of technology—material culture—and its constitutive role in human society.

Thus research on technologies, whether undertaken by historians of technology, archaeologists, or anthropologists, for the most part has not examined production; and when it has, materials science approaches have been rare or the focus narrow. Materials science data, which can reveal choices among technical alternatives, allow the investigator to identify how, and in what aspects, attitudes, values, and interests have shaped the technological enterprise and are expressed by it. The objective of this work is to identify these technical choices and explore the patterns they represent.

My primary data for this investigation derive from laboratory studies of a large corpus of ancient Mexican metal artifacts, along with artifacts from other Mesoamerican regions and from several South American areas, most notably Ecuador. Some facets of the operational sequences I identify are predictable or unavoidable due to the immutable physical and mechanical properties and behaviors of materials; others are choices among technical alternatives. I interpret those choices using archaeological, ethnographic, historical, and ethnohistorical data. Lexical evidence from two indigenous Mesoamerican languages spoken in or near the metalworking zone, Tarascan and Nahuatl, also proves helpful. Let us begin with an outline of the prehistory of ancient Mesoamerica, then turn to its western margin where this technology developed.

An Overview of Mesoamerican Prehistory

The metallurgical technology I investigate here developed in the western region of ancient Mexico (figure 1.1) between A.D. 600 and the European invasion in A.D. 1521. Archaeologists refer to the region encompassing central and southern Mexico, Guatemala, Belize, western Honduras, and El Salvador as Mesoamerica; this was one of two major centers in the ancient Americas in which civilization emerged. At the time of the Spanish invasion the population of the area is estimated to have been as high as 25 million people. Mesoamerican peoples lived in settlements ranging from large multiethnic cities, such as the Aztec capital at Tenochtitlan (table 1.1), to towns and small agricultural hamlets. During some 3,500 years of sedentary life, the peoples of Mesoamerica developed mathematics, writing and notation systems, astronomy, philosophy, a wide range of agricultural strategies, monumental public architecture, and numerous other expressions of practical and abstract human endeavor associated with civilized existence. They did this without major outside influences other than intermittent contacts with other New World peoples. They also did so without metal, at least until very late in their history.

Mesoamerica was initially populated by small groups of foragers whose predecessors had moved into the Americas from Asia during the late Pleistocene (20,000 B.C. to 10,000 B.C.). We know for certain that humans were hunting species of Pleistocene mammals in the Valley of Mexico by around 8000 B.C. Following the end of the Pleistocene, nomadic, kin-based bands continued to form the basic Mesoamerican social unit for several mil-

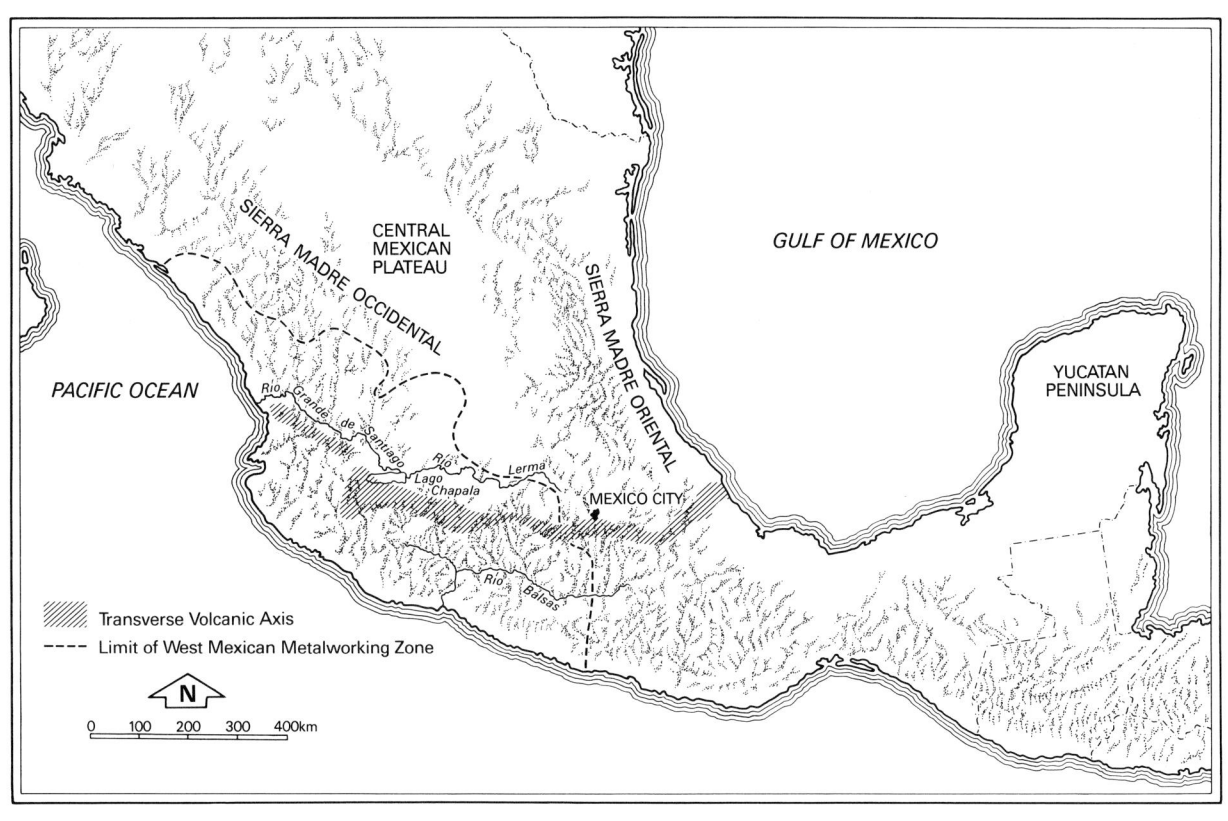

1.1

Topographic map of Mesoamerica, showing the limits of the West Mexican metalworking zone.

lennia. The archaeological record shows that by 4000 B.C., the complex events that led to plant domestication, food production, and sedentary village life were already under way.

Mesoamerica's rich and diverse physical environment certainly helped but does not entirely explain the emergence of civilization there. The region consists of a mountainous land mass that lies largely within tropical latitudes. Elevations range from sea level along the coastal plain and the Yucatán Peninsula to volcanic peaks, some of which reach to 6,000 meters. Abundant rainfall supports diverse floral and faunal communities. Among the flora were some of the wild progenitors of later domesticates, including corn, beans, peppers, and squash. Tertiary vulcanism further contributed to a hospitable environment for human life by creating a chain of volcanos that crosscut the central Mexican plateau (see figure 1.1). They blocked drainage to the west and to the east and formed a series of shallow, spring-fed highland lakes. These lakes offered plentiful fowl, including migratory

Table 1.1. Chronology of Mesoamerican Culture Periods, Selected Sites, and Cultural Groups

Relative Chronology	Central Highlands	Oaxaca	Southeastern Lowlands	West Mexico
Postclassic A.D. 900–A.D. 1521	Tenochtitlan (Aztec)			Amapa Autlán Ixtlán del Río Tamazula Tzintzuntzan (Tarascan)
Classic A.D. 150–A.D. 900	Teotihuacán	Monte Albán	Tikal (Maya)	Amapa Ixtlán del Río Teuchitlán Tingambato
Formative 2500 B.C.–A.D. 150	Cuicuilco		San Lorenzo (Olmec)	

Note: Cultural affiliations of sites, where known, appear in parentheses.

species of birds, freshwater fish, and other subsistence resources, and provided an environment for the eventual development of ingenious intensive agricultural techniques. The rich soils resulting from the decomposition of volcanic rock were also a boon to later agriculture. Mesoamerican peoples took advantage of one form of this volcanic rock, obsidian, for precise cutting tools and elegant ritual objects. They used *tezontle*, a porous volcanic stone, for construction, and basalt for grinding implements and other tools. Flint also was used for both tools and ritual items (figure 1.2). By 2000 B.C., Mesoamericans were harvesting corn, beans, squash, and other crops. At approximately this same time, unambiguous evidence begins to appear in some regions for settled life, social differentiation, and hierarchically organized society.

The transition from nomadic to settled life, and from foraging to farming, took place in several Mesoamerican areas during the same period. However, those conditions that allowed more complex social organization—cities, social classes, and full-time specialist occupations—were present in only some of these regions, and West Mexico was not one of them. The earliest significant Mesoamerican evidence for complex social forms appears in the Formative or Preclassic Period (2500 B.C. to A.D. 150) among the Olmec in the tropical lowlands of southern Veracruz and western Tabasco (see table 1.1). Archaeologists have excavated various sites that were centers of Olmec religious and ceremonial life. At San Lorenzo (1400 B.C. to 900 B.C.) (figure 1.3), the earliest of these, the population has been estimated at 1,000 persons. The

1.2

Flint eccentric. Maya (Late Classic Period, A.D. 600–900). Photograph from Willey 1974, figure 199.

material remains indicate not only that the Olmec were accomplished architects, sculptors, lapidaries, and potters, but that Olmec leaders were able to mobilize and manage a large labor force. Olmec managerial skills are most spectacularly manifest in huge blocks of basalt (some weighing to 20 metric tons) shaped into colossal heads and altars that were transported to the centers from as far as 60 kilometers away. Archaeologists have argued that an elite class—perhaps the first in ancient Mesoamerica—emerged at San Lorenzo through competition for the highly productive river levee soils that were deposited by seasonal flooding. Olmec artifacts recovered in highland Mexico, the Valley of Oaxaca, and elsewhere make it clear that the Olmec were also the first to widely disseminate elements of their material repertoire. The Olmec ceased to exist as an archaeologically known culture around 400 B.C., however, and the centers were abandoned.

Large religious and administrative centers began to appear to the east of the Olmec region during the Late Formative Period (circa 250 B.C.) in lowland areas of southeastern Mexico, Guatemala, and Belize. These areas became the heart of lowland Maya civilization during the subsequent Classic Period.

Complex societies also arose in the highlands during the Formative Period. Settlements in the Valley of Oaxaca southwest of San Lorenzo, and the Valley of Mexico where Cuicuilco flourished, provide early examples (figure 1.3). Archaeologists think that social differentiation in the highlands during the Formative Period was related to resource distribution: abrupt local variations in altitude resulted in closely juxtaposed but very different floral and faunal communities, particularly where the central Mexican plateau drops off toward the western and eastern tropical coastal plains. People exploited multiple resource zones in such areas and traded the products. The organizational necessities of such exchange relations required social mechanisms to facilitate them, and the presence of such mechanisms contributed to the formation of social hierarchies.

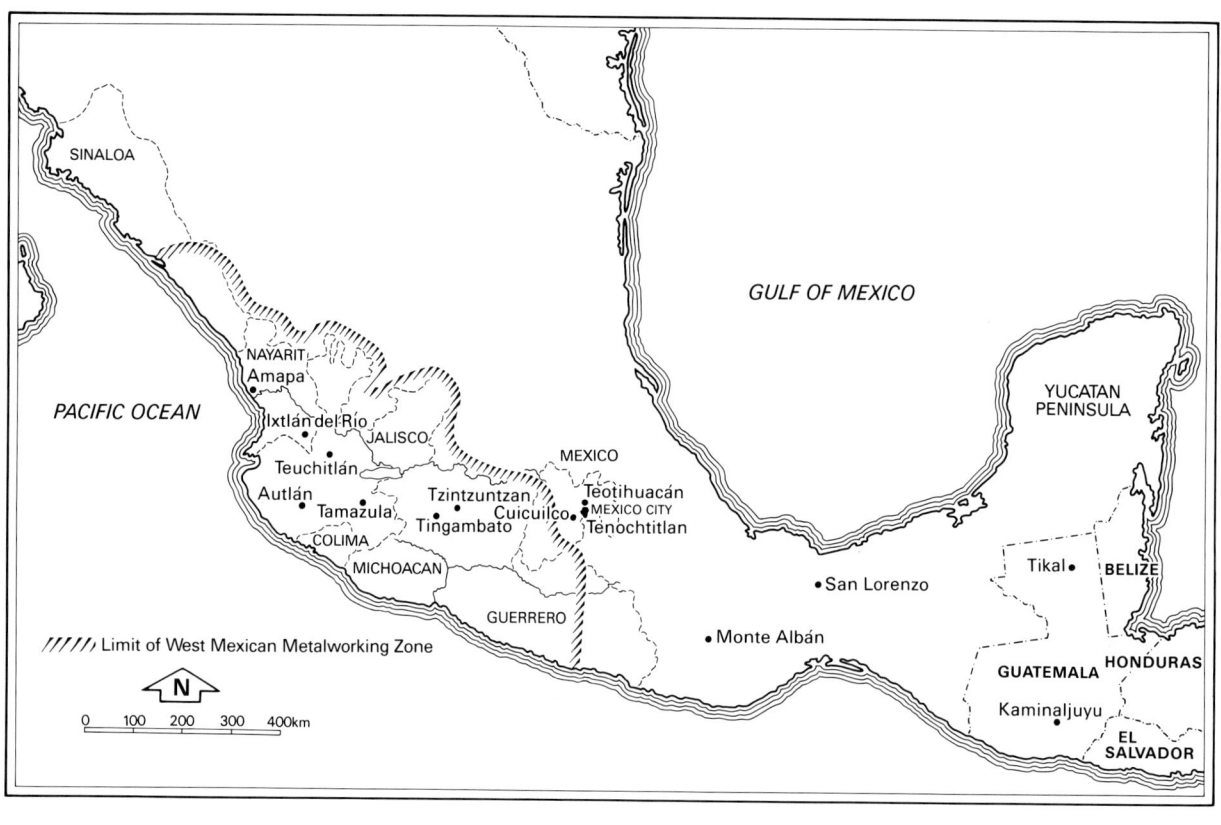

1.3

Mesoamerican archaeological sites and the West Mexican metalworking zone.

In the Valley of Oaxaca, public buildings began to appear around 1200 B.C. By 400 B.C., small polities were flourishing. Terracing and irrigation increased agricultural productivity and supported population increases. Cities also were developing in the Valley of Mexico. Cuicuilco, in the southern end of the valley, was the largest, inhabited by some 20,000 people. Cuicuilco was destroyed by a volcanic eruption about 150 B.C., and Teotihuacán, located some 50 kilometers to the north, subsequently became the valley's dominant political and economic power.

During the Classic Period (A.D. 150 to A.D. 900), urban centers—large planned cities, bureaucracies, and social classes—were well established in the Valley of Mexico, Oaxaca, and at some lowland centers in southeastern Mexico, Guatemala, and Belize. No metal objects have been recovered at any of these centers, apart from a few from the southeastern lowlands that were probably imported from lower Central America. The largest political entity at this time was Teotihuacán, followed by Monte Albán in the Valley of Oaxaca, the capital of the Zapotec state. Lowland Maya cities were smaller than Teotihuacán.

Complex societies also emerged from Formative Period beginnings in many other areas, for example highland Guatemala, the Pacific tropical coastal plain of Guatemala and Chiapas, and east of the Sierra Madre Oriental on the coastal plain of Veracruz. Many Mesoamerican peoples, representing distinct ethnic and linguistic groups, created their own iterations on basic Mesoamerican themes in monumental sacred and civic architecture, pottery traditions, settlement patterns, agricultural technologies, and other domains of human existence.

Teotihuacán (A.D. 100 to A.D. 750), the capital of the largest Mesoamerican Classic Period state, is believed to have been one of the largest cities in the world at the time. The city was dominated by the pyramid of the sun, a 71-meter-high structure located along a broad avenue lined by temples, elite residences, and administrative buildings. The city covered some 20 square kilometers at its apogee around A.D. 600, and was home to approximately 125,000 people, some of whom had come from the Maya area and Oaxaca as migrants, traders, or perhaps political emissaries. The high population densities resulted from a policy of concentrating rural peoples in the city. These 125,000 people could be supported because Teotihuacán exploited diverse and highly productive ecological niches: the valley (using a variety of irrigation techniques), the freshwater lakes, and the adjacent lower- and higher-altitude zones which furnished a wide spectrum of floral, faunal, and other resources. In addition, Teotihuacán enjoyed the economic benefits that accrued from control of large obsidian deposits. These deposits were mined and obsidian was processed in myriad city workshops; traders then moved obsidian throughout Mesoamerica. Teotihuacán exerted direct political control over the entire Valley of Mexico and parts of Puebla, and its influence was even more widespread. Teotihuacán-style architecture and artifacts appear in the Guatemalan highlands at Kaminaljuyu, in the Maya lowlands at Tikal, and at various sites in western and northern Mexico. Teotihuacán began to decline after A.D. 600, and it was burned around A.D. 750. But metal was unknown at Teotihuacán: not one of the varied subsistence endeavors, construction activities, military undertakings, or religious and ceremonial events at this great urban center was carried out using metal implements, metal weapons, or metal religious or ceremonial objects.

Classic Maya civilization also coalesced during this period (A.D. 150 to A.D. 900). The Maya developed a variety of technically complex materials-based technologies, but again metalworking did not figure among them. Maya-speaking peoples inhabited a large region from the Guatemalan highlands to the tip of the Yucatán Peninsula, although the elements that define what archaeologists refer to as Classic Maya were shared by peoples concentrated mainly in the tropical rain forests of the southern Yucatán Peninsula of Mexico, Guatemala, and Belize. Classic Maya traits include a calendar and hieroglyphic script and vibrant monumental architectural, sculptural (figure 1.4), and pottery technologies. Maya cities were not unified politically except through loose and shifting alliances. Tikal was the largest, with a population estimated at approximately 20,000 people. Maya polities tended to emerge where soil productivity was high, in well-drained upland areas. In less productive zones, Maya peoples developed a form of swamp reclamation, known as raised fields, that allowed for intensive agriculture. The characteristics that define Classic Maya culture—the architectural style, the hieroglyphic script, the calendar, and

1.4

Lintel 25, Yaxchilan, Chiapas, Mexico. Late Classic Period (A.D. 725). Limestone. Photograph from Schele and Miller 1986, plate 63a.

the polychrome pottery—begin to disappear from the archaeological record around A.D. 800 to A.D. 900. During this period some cities were abandoned or experienced dramatic population declines. Maya peoples continued to live in the lowland areas, but population densities were far lower than during the Classic Period.

By the beginning of the Postclassic Period (A.D. 900 to A.D. 1521) several of Mesoamerica's most powerful and ideologically charismatic civilizations had collapsed. Teotihuacán and numerous lowland Classic Maya cities had been abandoned. In Oaxaca, the highly aggregated population at Monte Albán also dispersed. Small city-states came to comprise the predominant political units throughout that region. New but smaller capitals or city-states also arose near the Basin of Mexico.

Metal was virtually unknown in these primary centers of Mesoamerican civilization until the Postclassic Period. The only area where metallurgy developed prior to the Postclassic Period was in West Mexico, and West Mexico did not foster early development of complex, stratified societies, such as those in the Valley of Mexico, Oaxaca, and the Maya area. Nonetheless, metallurgy's appearance in the west during the period A.D. 600 to A.D. 800 coincided with, and I believe was triggered by, events in those primary areas, a topic I treat in the final chapter.

Not surprisingly, West Mexico contains abundant deposits of native metals and ores. Native metals and ore minerals do not occur in or near the Valley of Mexico or the southeastern lowlands. Deposits of metallic minerals do appear in Oaxaca, the Guatemalan highlands, and other regions that supported early cultural developments, but these deposits do not compare in variety and abundance to those of the west. Ore deposits are also abundant in some areas of northern Mexico, but the aridity of those areas generally has kept population densities low.

West Mexico is also the only region in Mesoamerica where persistent evidence exists for contacts with northern South America. Although the details of these relations elude us so far, these contacts clearly sparked metallurgy's florescence in West Mexico. The metallurgical technology that took shape in this area spread to other regions of Mesoamerica during the late Postclassic Period (A.D. 1200 to A.D. 1521). During that time, West Mexican peoples, or their intermediaries, exported metal artifacts and certain techniques to Oaxaca,[8] to the Maya area, and to the region encompassed by the present state of Tamaulipas, where the Huastec peoples lived (chapter 7). West Mexican metal artifacts also appear in areas surrounding the Valley of Mexico during and prior to the time these areas were controlled by the Aztec, whose empire the Spaniards first invaded in A.D. 1519 and subsequently decimated.

THE WEST MEXICAN METALWORKING ZONE

The first West Mexican metal objects appear around A.D. 600 in the area I have designated the West Mexican metalworking zone (see figure 1.1). The zone encompasses the contiguous present-day states of Jalisco, Michoacán, Nayarit, Colima, and southern Sinaloa, as well as northern Guerrero and portions of the state of Mexico. This area differs from West Mexico as archaeologists usually conceive it: West Mexico ordinarily excludes Guerrero and the state of Mexico. The metalworking zone crosscuts several physiographic regions. It includes the western edge of the central Mexican plateau: the broad flat highland basins of Jalisco, Michoacán, and parts of the state of Mexico that offer relatively easy communication and ample space for urban centers. It also encompasses the southern portion of the Sierra Madre Occidental, the north-south-trending mountain range that borders the central Mexican plateau in areas of Nayarit and Jalisco. The Río Balsas drainage in Guerrero and Michoacán, an arid east-west structural depression, and the lowland "Tierra Caliente" (Hot Lands) regions of Nayarit, southern Sinaloa, Jalisco, Colima, Michoacán, and Guerrero also lie in this zone. The Balsas and the Lerma-Santiago river system are the two major river drainages in this metalworking area. The Lerma-Santiago system, originating in the Toluca basin, passing through Lago Chapala in Jalisco, and emptying into the Pacific Ocean in Nayarit, provided a natural corridor between the central Mexican plateau and the west coast.

The rich deposits of metallic minerals in the metalworking zone include native silver and silver ore minerals (including argentite, proustite, and others), native copper, copper oxide and sulfide ores, gold, arsenopyrite (the sulfide ore of iron and arsenic), and lead, among others.

The topographically dissected area that makes up West Mexico is much larger and far more diverse culturally than the other regions I have discussed. No single West Mexican "culture" existed at any time during its history. The area also has received far less archaeological attention than Oaxaca, the Maya lowlands, the Valley of Mexico, and other regions that exhibit spectacular features such as writing systems and extensive monumental architecture. This lack of archaeological attention makes the task of reconstructing the culture history particularly difficult. I will describe here in general terms those areas and settlements about which we know the most; subsequent chapters focus in more detail on sites where metal artifacts have been excavated.

During the Formative Period, small villages, hamlets, and scattered homesteads probably characterized most of the area that was suitable for either irrigation or rainfall agriculture. Because surface features including habitation sites and ceremonial structures are rare, we know little about the size and social organization of these communities. Archaeologists generally identify these "cultures" through their distinctive regional ceramic styles and mortuary practices. Polities with conspicuous ceremonial structures begin to appear in the Early Classic Period (A.D. 150 to A.D. 500), but they do not compare in complexity and scale to those in the Valley of Mexico, Oaxaca, and the Maya lowlands. State-level society did not take shape until very late in West Mexico. Archaeologists currently think that West Mexico actually consists of several independent subareas—the highlands of Michoacán, the lake region of Jalisco, and the lowland coastal plain[9]—and that researchers can better define these and other regions once they have clarified temporal sequences and cultural relations.

Metal objects appear first along the coast in all but a few cases. The technology moved gradually inland along the river drainages. By the Postclassic Period, archaeologists report metal objects at highland sites, and by the Late Postclassic, as I have already mentioned, elements of West Mexican metallurgy spread to other Mesoamerican regions.

During the earlier part of the period in which metallurgy was developing (A.D. 600 to A.D. 1000), hierarchically organized polities were emerging from a local base in West Mexico. Teuchitlán, in the lake region of Jalisco, provides a good example of them (Weigand 1985). Teotihuacán also influenced the West Mexican metalworking zone, however, and contacts with that city, visible in West Mexican architecture and ceramics, probably intensified local processes of social differentiation. Large settlements showing evidence of Teotihuacán's influence appear in highland Jalisco and Michoacán and along the coastal plain at this time. Amapa, on the coast (Meighan 1974, 1976), and Ixtlán del Río, which lies inland, are two examples. Both exhibit monumental public architecture with temple mounds and plaza-type structures oriented in cardinal directions, features typical of Central Mexican sites. In highland Michoacán, a number of widely dispersed centers such as Tingambato also display characteristic Central Mexican features. Some have ball courts, plazas, pyramids, and the *talud tablero* architectural motif, a hallmark of the Teotihuacán style. Pollard (1993) believes that the architectural features and community pattern at these highland sites may indicate elite contacts with Teotihuacán, or possibly migration out of that city following the dissolution of the empire. Apart from Amapa, where some 200 metal objects were recovered, metal is reported only very occasionally from these other sites.

The Teuchitlán tradition, characterized by a series of large centers with a unique non–Central Mexican architectural pattern, flourished in highland Jalisco between A.D. 200 and A.D. 800 to A.D. 900. No metal objects have been recovered there. Teuchitlán centers have residential compounds, evidence for craft specialization, and other markers of complex social organization (Weigand 1992), but the unusual feature is the public architecture, which consists of a concentric circular design unknown elsewhere in Mesoamerica. Weigand argues that Teuchitlán provides the uniquely West Mexican contribution to the mosaic comprising Mesoamerican civilization, a composite of elements from Oaxaca, the southern lowlands, the central highlands, and other regional hearths (Mountjoy and Weigand 1974; Weigand 1985, 1992, 1993).

State-level societies and expansionist empires did not form in the West Mexican region until the Late Postclassic Period, when the Tarascan state emerged (Pollard 1993). At the time of the Spanish invasion, the Tarascan state was the largest political unit in Mesoamerica apart from the Aztec, and it had successfully warded off various attempts at Aztec domination. Apart from the Tarascan capital at Tzintzuntzan and the cities of Tamazula and Autlán in Jalisco, Spanish chroniclers characterized all other West Mexican settlements as only "hamlets" or "villages" (Schöndube B. 1980a: 246), providing further evidence that developments in West Mexico were not equivalent in scale to those in the other regions I have described here.

Although the West Mexican metalworking zone is extraordinarily varied topographically and culturally, the element that distinguishes it as a whole is the copper-based metallurgy that I examine in this book. Another distinguishing feature, and one related to metallurgy, is that, over a very long period, concrete and convincing archaeological evidence exists for intermittent contacts with South America. This evidence, which I describe subsequently, does not appear elsewhere in Mesoamerica, and suggests that West Mexican societies especially encouraged such contacts for economic, political, religious, or other reasons.

Some Mesoamerican archaeologists argue that West Mexico is so diverse and shares so little with the great civilizations of the central highlands or the southeastern lowlands—urbanism defined by elaborate, extensive civic-ceremonial architecture, distinctive regional iconographies, social stratification, and large population concentrations—that they exclude West Mexico from Mesoamerica entirely until the Late Postclassic Period. They think of West Mexico as a sort of cultural backwater. Others, myself included, consider that West Mexico, or certainly some areas within it, comprises one of the multiple hearths of Mesoamerican civilization, contributing technically sophisticated regional pottery forms and styles, unusual architecture, and other unique traits—the most significant of which, as this work should make clear, is a multifaceted and inventive metallurgy.

South American Contacts in West Mexico

Strong evidence exists for contacts between West Mexico and northern South America in the prehispanic era and is highly relevant to the development of West Mexican metallurgy. In summarizing the archaeology of the state of Jalisco, Schöndube observes that:

muchos investigadores han tendido a buscar el origen, o al menos una explicación, de las manifestaciones culturales del Occidente de México en el arte y las costumbres de los pueblos instalados hacia la vertiente del Pacífico en Sudamérica. Pocos dudan actualmente que haya habido contacto entre ambas regiones, y las preguntas por contestar ahora mas bien se refieren a temporalidad, a la calidad e impacto de dichos acercamientos, o sobre las direcciones del movimiento de la difusión de rasgos. [Many researchers have tended to look for the origin, or at least an explanation, of the cultural manifestations of West Mexico in the art and customs of the peoples living along the Pacific coast of South America. Few currently doubt that there had been contact between the two regions, and the questions to answer have more to do with the time periods, the quality of the impact of such contacts, and the direction of the movement of traits.] (Schöndube B. 1980b: 211)[10]

The most convincing evidence for contact appears in a Formative Period West Mexican pottery style, in an unusual tomb form—shaft tombs—dating generally to between 200 B.C. and A.D. 400, and in the copper-based metallurgy that is the subject of this work. These traits do not occur in other areas of Mesoamerica nor further to the south, in Central America. Archaeologists have long argued that Pacific maritime contacts between northern South America and West Mexico best explain these features' unusual and discontinuous geographical distribution (Meighan 1969; Mountjoy 1969).

Ceramics and Shaft Tombs. Coastal Colima exhibits very early evidence for contacts with South America. Pottery from the Formative Period Capacha phase (1500 B.C.) looks remarkably like ceramics from northern South America. In fact, it resembles ceramics in Ecuador's Machalilla phase (1500 B.C. to 500 B.C.) far more closely than it resembles contemporaneous Mesoamerican pottery. Kelly, in fact, finds virtually all of Capacha's defining characteristics also present in northern South American pottery assemblages (1980: 37). Dating problems make the direction of the influences, however, whether north to south or south to north, unclear.

West Mexican shaft tombs point to ties with northern South America in later periods (Meighan and Nicholson 1989; Schöndube B. 1980b; M. E. Smith 1978). Shaft tombs are designed with a highly distinctive vertical or slightly inclined shaft leading to one or more chambers that open out from it. The shafts vary in length between 2 and 17 meters. Certain West Mexican tombs date to approximately 1500 B.C.,[11] but most fall between 200 B.C. and A.D. 400. They are widely distributed in Peru, Ecuador, Colombia, and Venezuela. In West Mexico they occur in an arc across Jalisco, Nayarit, and Colima. In South America, the earliest shaft tombs are found in Colombia and date to 545 B.C. (Duque G. 1964, cited in M. E. Smith 1978). Kelly (1980) reports shaft tombs in Colima associated with Capacha ceramics; other early West Mexican shaft tombs (also dating to around 1500 B.C.) appear in northwestern Michoacán, but many archaeologists think that their forms vary too greatly to group them with the later types linked to northern South America.

Unfortunately many of the tombs have been looted and others do not seem to be located in habitation areas, so we know little about the people who practiced these burial rites. However, in northern South America as well as West Mexico the dead are buried with hollow, hand-crafted ceramic figurines. The ceramics from the West Mexican shaft tombs exhibit a technically sophisticated, often naturalistic style that constitutes one of the defining characteristics of certain regional cultures of the period. The tombs and the figurines have been interpreted as one element in a shamanistic complex (Furst 1965a) imported from northern South America. Furst maintains that they could represent a body of esoteric and religious knowledge introduced from the northern Andes into West Mexico prior to approximately A.D. 400. Whether or not these tombs and associated ceramics specifically represent such a complex, the physical similarities between the northern South American and West Mexican tomb types are unmistakable. The subsequent appearance of metallurgy in West Mexico probably reflects the long-standing contacts between these two areas.

Metallurgy. Archaeologists generally have agreed that West Mexican metallurgy did not develop autochthonously but had its roots in the metallurgies of Central or South America. This work is the first comprehensive study of the question. I show here that well-defined technical complexes (identical artifact classes, fabrication methods, and metal and alloy types) were indeed introduced to West Mexico from the south, but from two distinct regions: one the Andean area that includes Ecuador, Peru, and Bolivia, and the other the area encompassing lower Central America and Colombia.

The earliest evidence for metal anywhere in the Americas occurs in the more southerly region, at about 1500 B.C. in highland Peru. By A.D. 600, when metallurgy had begun to take shape in West Mexico, Andean metallurgy was diverse and technically complex. Andean smiths took advantage of the wide range of nonferrous metals available in the region, using metal for tools, which they crafted from copper-tin and copper-arsenic bronze. They also fashioned ritual and status objects from gold, silver, and their alloys with copper, showing a particular technical enthusiasm for metallic colors, one of the hallmarks of that technology. By elaborating complex surface enrichment techniques, Andean smiths produced silvery and golden surfaces on objects hammered from copper-silver, copper-gold, and copper-silver-gold alloys (Lechtman 1970, 1979, 1984). In fact, shaping metal by hammering was so pronounced a preference that Andean smiths usually rendered three-dimensional forms by joining pieces of sheet metal rather than by casting solid or hollow shapes. The technical preference for hammering was totally appropriate for peoples interested in metallic color. Surface enrichment effects, which alter color, result inevitably from the annealing sequences required to produce such metal sheet.

In lower Central America and Colombia, metallurgy arose somewhat later, around 500 B.C. to 200 B.C. (Colombia) and A.D. 200 to A.D. 300 (lower Central America). The mineral resources in this area are far less diverse than in the Andean region. The primary deposits are of copper and gold, metals that lack optimal mechanical properties to elaborate fine, hard, cutting tools. Metalworkers did sometimes make small hand tools but from *tumbaga*, the copper-gold alloy. These people handled metal differently from their counterparts to the south as well. They most often shaped hollow and solid forms by casting them, using the lost-wax technique. The vast majority of objects were for ritual and status purposes; most were made from gold or from copper-gold alloys.

We do not yet know whether these two metalworking traditions, the Andean and the lower Central American and Colombian, developed independently or arose from a common source. What we do know is that they are not related to other world metallurgies. It is also clear that West Mexican metallurgy did not simply replicate either of these two technologies, although it did develop from them. The technology was shaped by technical choices and experiment arising from the social experience of the peoples of the metalworking zone, but grounded in physical realities: in the kinds of raw materials on hand, their distribution and relative abundance, and their physical and mechanical properties.

The Research Agenda

To collect the data to characterize West Mexican metallurgy, I carried out laboratory-analytical studies on a large assemblage of West Mexican objects made prior to the European invasion. The studies described the formal design attributes of the objects, and identified metal or alloy types, fabrication methods, alloy-property relations, and, where possible, how the objects were used. These analyses and examinations had to be performed before addressing the question of technical choices. A full description of the study sample, analytic methods, and sampling strategy appears in appendix 1.

The metal collections of the Regional Museum of Guadalajara (RMG), in Guadalajara, Jalisco, Mexico, comprised the basic study material. They contain approximately 3,200 prehispanic metal artifacts from the states of Nayarit, Colima, Michoacán, and Jalisco. Some artifacts are attributed to specific sites. I examined the entire corpus macroscopically, then performed laboratory-analytical studies on a sample of approximately 10% of these (see appendix 1) chosen from the major artifact classes. These included needles, tweezers, axes, and bells.

The results of the technical work on this RMG reference collection were amplified by studies of approximately 400 artifacts excavated from the sites of Tomatlán (Jalisco), Amapa (Nayarit), Urichu (Michoacán), Milpillas (Michoacán), Cuexcomate and Capilco (Morelos), Lamanai (Belize), and Platanito and Villa Hermosa in the Huastec region. I carried out technical studies of all objects from Cuexcomate, Capilco, Lamanai, Milpillas, Platanito, and Villa Hermosa, and examined the Urichu objects macroscopically. Objects from Amapa (Root in Meighan 1976: 116–118) and Tomatlán (Mountjoy and Torres M. 1985) had been analyzed for chemical composition. Those data, coupled with other published information on dated metal objects and my own technical evidence, provided basic temporal and associational information.[12]

No dated assemblages of West Mexican metal objects exist comparable in size and variety to the RMG collection. In appendix 1, I discuss how I confirmed assemblage representativeness and the authenticity of the objects. I was able to date certain RMG artifact types and the use of certain methods and materials (see chapter 3) when dated objects with similar design characteristics were reported in the literature. West Mexican smiths invariably made certain designs (artifact types) from alloys—copper-tin and copper-arsenic bronze—because, as the laboratory studies show, their design characteristics required specific alloy properties for functional success: hardness, elasticity, or other physical and mechanical properties. I plotted the chronological development of major elements of this metallurgy by establishing such patterns of association within the RMG collection and then comparing these with the design (and chemical composition, where available) of objects in assemblages with good temporal controls.

Documentation of fabrication techniques and metals and alloys used, and studies of alloy-property relations, set the stage for evaluating technical choices. However, to identify such choices I needed to establish how West Mexican peoples used these metal objects. To do this, I relied on data from the laboratory studies and used mechanical engineering criteria in conjunction with information from ethnographic, sixteenth-century documentary, and archaeological sources. From the perspective of a mechanical engineer, two characteristics determine an object's ability to perform a particular task: its design and its mechanical properties. Design is what we visually perceive as shape; it is expressed in dimensions such as length, width, and thickness. Mechanical properties govern the behavior of a material when force is applied. Mechanical properties (ductility, toughness, hardness, elasticity, and others) are a function of the inherent properties of the metal or alloy, and of fabrication methods, since certain properties, like hardness, are altered through manufacturing techniques. For example, a serviceable bronze axe must conform to certain design requirements, maintaining particular ratios between length, width, and thickness.

It must also meet certain mechanical property criteria: the blade must be hard enough to cut, but the interior must be tough and resilient to absorb the stress at impact.

I evaluated the design and mechanical properties of each artifact type against engineering criteria for a serviceable object with the same apparent function. The criteria were developed from mechanical engineering data and, in one case, through a computer simulation study of artifact design. Once I determined the mechanical properties normally required for functional success, I measured them directly where possible (for example, in the case of hardness) or derived them through standard calculations using experimentally determined data.

Two other classes of information also helped determine how the objects were used. One was microstructural evidence, which is especially apparent in tools, where the deformation from use is often visible. Sixteenth-century manuscripts were also helpful because they sometimes describe metal objects or illustrate them in use.

Artisans also made choices in raw materials and smelting technologies. I established the native metals and ores that these metalworkers may have used by compiling information on the ore geology of West Mexico and adjacent regions. By combining this information with data from artifact chemical analyses, and documentary information on prehispanic mining, I was able to determine the native metals, ores, and smelting regimes most probably used by these ancient smiths.

To examine the external origins of certain elements of the technology, I compared fabrication techniques, metals, alloy systems, and object classes in West Mexico with those from Ecuador, Peru, Colombia, and lower Central America. Most South American data derive from my macroscopic and laboratory studies of the metal objects in the collections of the Museo Antropológico del Banco Central in Guayaquil, Ecuador (MAG), and from

Ecuadorian sites (see chapter 4). Approximately 4,000 MAG objects were available for study; chemical analyses were performed on some 5% of them.[13] Macroscopic and laboratory-analytical studies were also carried out on objects from excavated assemblages from the Ecuadorian coastal sites of Salango, Loma de los Cangrejitos, El Azúcar, and elsewhere.

The evidence makes clear that specific metalworking and processing techniques and artifact types were introduced to West Mexico from lower Central America, Colombia, and the northern Andes, stimulating the development of a local Mesoamerican technology. However, the circumstances surrounding the technology's introduction, why it may have occurred during the period in question, who was directly involved, and what items may have been exchanged for technical know-how from the south have had to be reconstructed inferentially; for the most part, there is no direct archaeological evidence (for example, a West Mexican port with remains of South American watercraft, or South American metal objects found in West Mexico) bearing on these issues. Nonetheless, taken as a whole, these data furnish reliable evidence of many sorts about where the technology came from, how it was transmitted, and how it was transformed once it became established in the region.

In the final chapter, I address the most intriguing issue raised by the distinctive character of this ancient metallurgy. What was it about sound and metallic color that so captured the attention of these metalsmiths that they chose to orient an entire technology around them? The geological and laboratory-analytical evidence shows that they had many other technical options, and that they had, in fact, developed some of these. I examine their choices by seeking the meaning of metallic sound and metallic color for these ancient peoples, bringing to bear the variety of data described previously. I also consider the temporal and social factors that might have encouraged such a pronounced ritual and elite orientation to this technology when a wide range of utilitarian alternatives was clearly available.

2

Resources, Metals, and Alloys

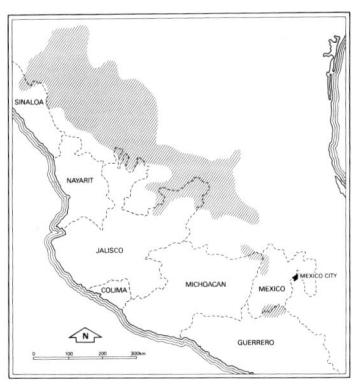

What native metals and ore minerals occur in the metalworking zone, and which did these ancient metalworkers use? Little archaeological evidence exists for either mining or extractive metallurgy in West Mexico or in Mesoamerica as a whole. (Extractive metallurgy involves winning metal from an ore through physical and chemical means. Workers crush the ore to separate the rocky gangue material from the metal-bearing minerals and then smelt these minerals in a furnace or a crucible to eliminate all remaining nonmetals, usually by forming a slag.) Not a single Mesoamerican site has been excavated where metal production was a major activity, and few artifacts associated with processing have been recovered.[1] To document mining and extraction, we need to excavate mining and ore-processing sites, smelting sites, and metalworking sites. We also need to study the ores from zones where mining took place, and to analyze by-products of intermediate stages of metal production—slags, fluxes, and partially and fully smelted material. Information from such studies is essential to identification of the ores, smelting regimes, and metals and alloys that characterized this ancient technology.[2]

Fortunately, the ore mineralogy of West Mexico has been sufficiently well investigated that it is possible to specify the native metals and ores that could have been used in the prehispanic era and to determine their geographical distribution. We can also reconstruct the ore and mineral types ancient metalworkers probably used by the presence and concentration of certain diagnostic elements in the artifact metals. In a few cases it is even possible to determine where metalworkers obtained certain raw materials, once we have circumscribed the distribution of a particular ore mineral. By combining these sources of information, we can identify the metals and alloys that characterized this technology, identify the ore types that were smelted to produce them, plot the geographical distribution of these ore types, and propose probable smelting regimes.

All the native metals and ore minerals that West Mexican smiths processed and used are abundant in the metalworking zone with the exception of cassiterite, an oxide ore mineral and the only source of tin in Mexico. These metals and ores include native copper, copper oxides and sulfides, native arsenic, arsenopyrite, sulfarsenides, and many minerals of silver including the native metal, silver sulfides (such as argentite), and silver sulfosalts (including polybasite and others). Tin deposits are rare. West Mexican smiths produced copper, tin, lead, silver, and gold, and also a number of common as well as exotic alloys: the two bronzes (copper-arsenic and copper-tin), a copper-arsenic-tin alloy, and alloys of copper-silver and copper-arsenic-silver. The most unusual alloy is one of copper-arsenic with arsenic concentrations reaching 23 weight percent.

Distribution of Native Metals and Ore Minerals in the Metalworking Zone

The metalworking zone lies within the precious and base metal province of Mexico (Ruvalcaba-Ruiz and Thompson 1988), an ore belt that extends from Sonora and Chihuahua in the north to Oaxaca in the south. Using maps produced by the Instituto Geológico of the Universidad Nacional Autónoma de México (UNAM), we can locate the deposits of economically significant metals in this region. These maps do not identify the ore mineralogy. However, they do show which metals were potentially available to the ancient metalsmiths and thus allow us to plot the overall geographical distribution of the metals that were actually used.[3]

The maps show that deposits of copper and of silver are plentiful in the West Mexican metalworking zone (figures 2.1 and 2.2). Both were used extensively. For each metal, two distinct zones of mineralization exist, separated by an area in western Michoacán and northeastern Jalisco that contains few or no deposits. The geological map for copper lists 53 deposits in the more southerly (actually southeast) region and 34 in the more northerly (northwest) region. The distribution of silver deposits generally coincides with that of copper; the UNAM map identifies 56 silver deposits in the southern zone and some 54 deposits to the north.

Only a few tin deposits appear in West Mexico proper. Nearly all Mexican tin is found in a zone known as the Zacatecas tin province (figure 2.3), a 500-kilometer-long belt extending along the eastern border of the Sierra Madre Occidental from southern Durango to northern Jalisco (Terrones 1984). This province contains some 1,000 scattered deposits, all extremely small. Modern exploitation of the metal has not been commercially viable, yet the tin bronze alloy was widely used prior to the Spanish invasion. Apart from the deposits lying in the Zacatecas tin province or on the periphery, cassiterite is virtually absent in Mexico, as it is in Central America and northern South America. In fact, the only significant deposits of tin in the Americas are in the tin-rich zones of south highland Peru, Bolivia, and northwest Argentina.

Arsenic, the other important metal in ancient West Mexican metallurgy, was used as the alloying component of copper-arsenic bronze. In Mexico arsenic appears primarily combined with other elements in diverse minerals; the native metal is unusual. The UNAM map for arsenic shows only a few deposits and lacks the detail available in the copper, silver, and tin maps. I was, however, able to collect data from other sources on the distribution of arsenic-bearing ore minerals.

2.1

Zones of copper deposits in the West Mexican metalworking zone.

Copper, silver, arsenic, and tin appear in many forms. Copper, for example, can occur as native copper, as copper oxides and carbonates, such as malachite, or as copper sulfides, such as chalcopyrite. The mineral form in which the metal appears is key to reconstructing ancient processing technologies, because the methods used to extract metal from oxide and carbonate ores differ from those used to extract it from a sulfide ore. The parent material must be identified to determine whether smiths were employing a native metal or an ore mineral, and what the ore mineral was. The associations of metal and/or ore types in a particular geological deposit also matter, because these associations suggest how certain alloys—for example, copper with arsenic—may have been made.

Information concerning how these metals present themselves in the metalworking zone, their relative abundance, and the metals and ore types with which they co-occur comes from several sources. The *Mapa Metalogénico de México* (Salas 1980) locates major metallic ore deposits, identifies the primary metals associated with them, and describes the surrounding geological environment. Specific information regarding the ore mineralogy of particular deposits can (but does not always) appear in geological reports, which often also describe the relative abundance and associations of particular ore minerals.

The *Mapa Metalogénico* locates the principal Mexican ore deposits but, like the UNAM maps, does not specify ore mineralogy. In the case of copper, for example, we do not know whether the ore mineral is chalcopyrite, malachite, or something else. (The documentation accompanying the map does indicate whether native metals or oxide or sulfide ores predominate in any deposit.) The

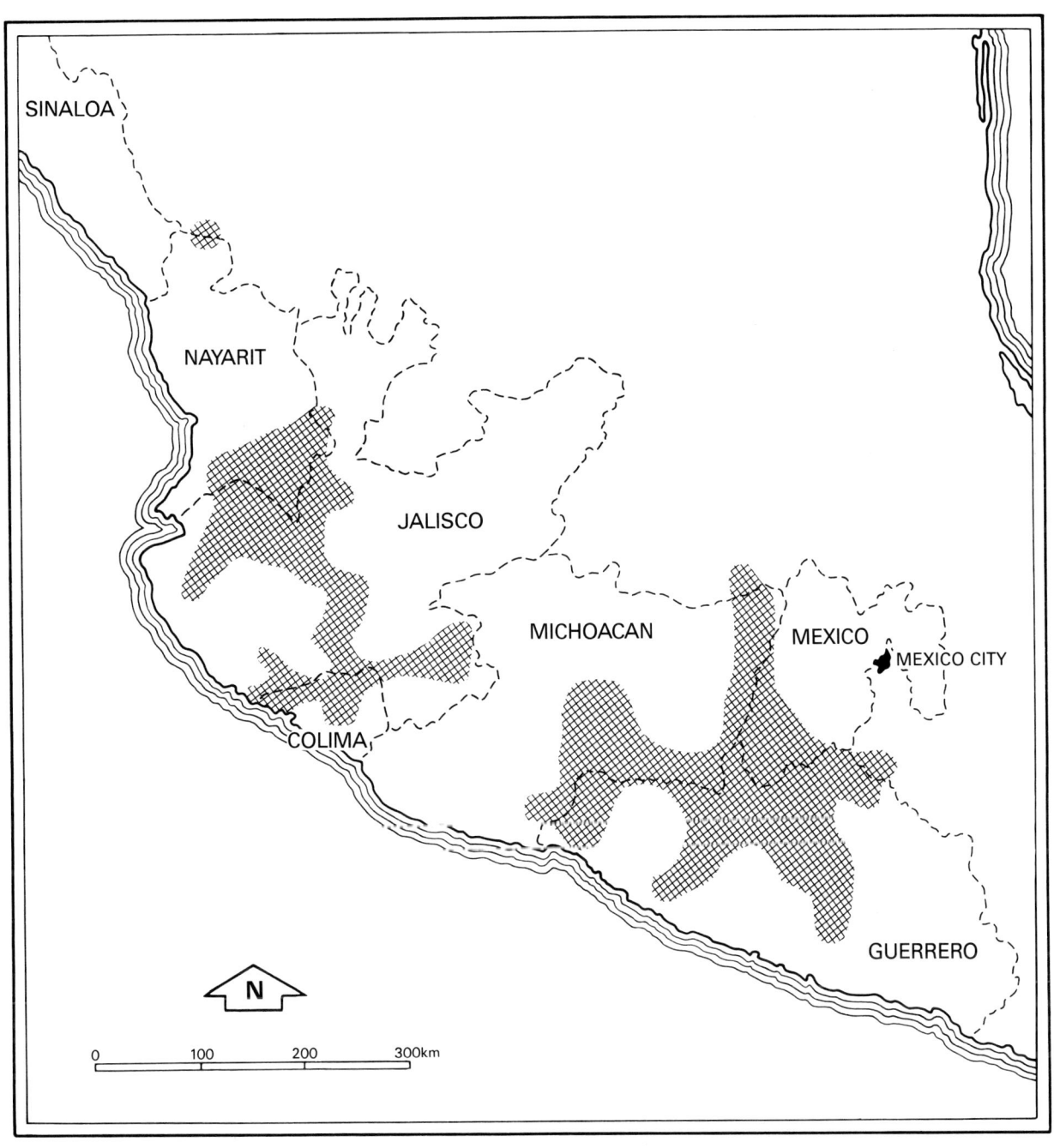

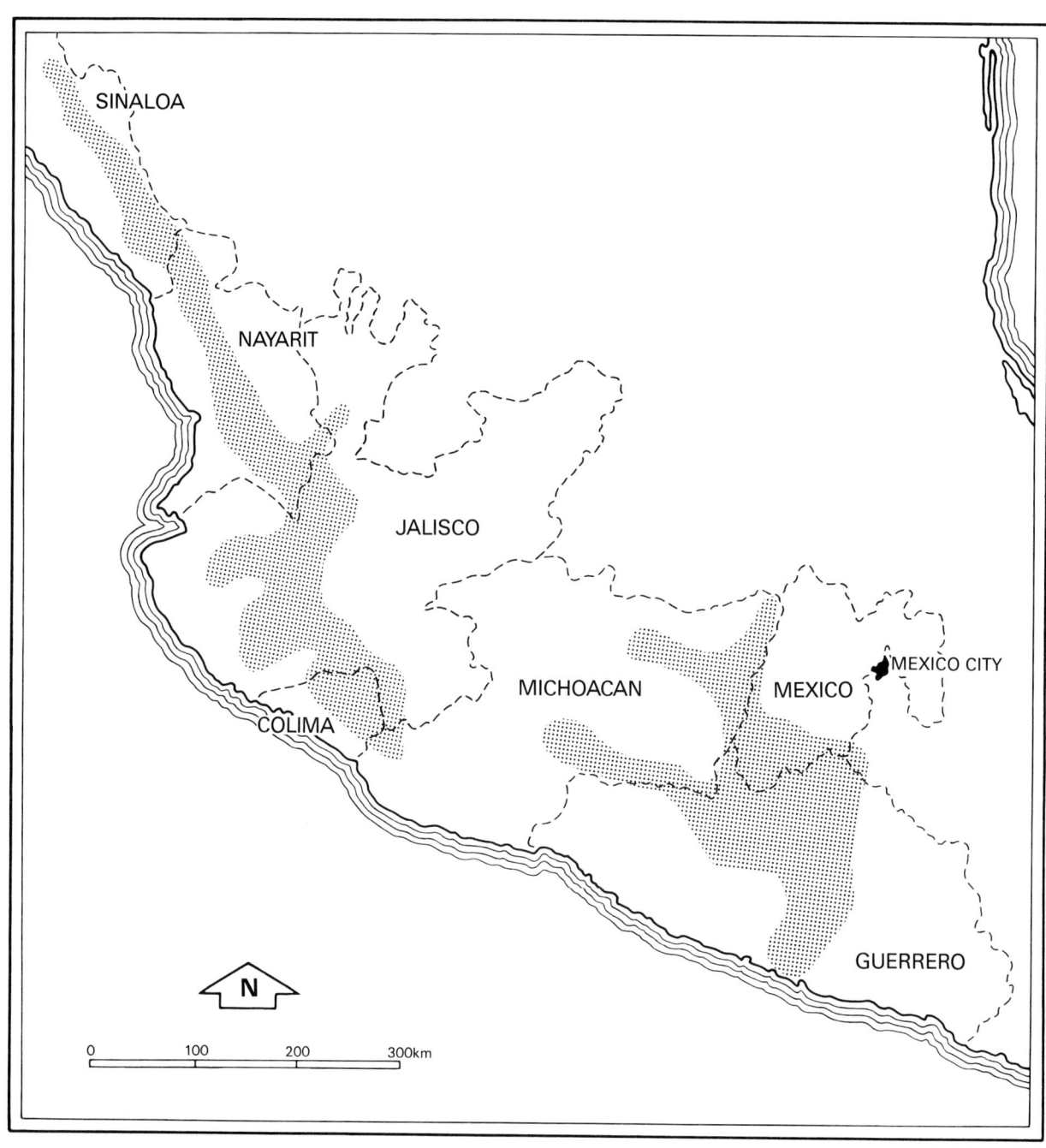

2.2

Zones of silver deposits in the West Mexican metalworking zone.

map is helpful nonetheless because it is often possible to infer ore mineralogy and ore type from the nature of the geological environment. That environment is described in detail on the map. Such inferences are especially important in this case, because for some West Mexican regions or deposits, descriptions of ore mineralogy are unavailable.

The ore minerals that may be expected to occur in the geological environments described on the *Mapa Metalogénico* were inferred by Ulrich Petersen, Professor of Geology at Harvard University, on the basis of his considerable experience with Mexican and Andean ores. The metals found at deposits shown on the map are reproduced in figure 2.4. The ore minerals typical of these deposit types and inferred from the map are listed in table 2.1.

The most useful information for our purposes comes from identifying specific ore minerals in particular deposits on the basis of analytic data. Such information, available in geological reports, makes concrete the inferences drawn from the *Mapa Metalogénico*. I compiled information on the ore minerals reported in the metalworking zone from these reports (Barrera 1931; Berrocal L. and Querol S. 1991; Flores 1946; Lorinczi and Miranda V. 1978; Lyons 1988; Parga P. and Rodríguez S. 1991; Salas 1991a, 1991b; Santillán 1929; Scheubel, Clark, and Porter 1988) and from extensive archival information made available to me by Industrial Minera México (IMMSA, formerly the American Smelting and Refining Company). IMMSA, a major mining company, has been prospecting and mining in the West Mexican region since 1900.

Figure 2.5 is constructed from these field surveys and reports. It identifies the native metals, ore minerals, and mineral associations that characterize each site or deposit. (Each number on the map locates the deposit or mine and identifies it in the legend by name, if the data are published, and by the acronym IMMSA if those data derive from IMMSA company archives. The symbols following each number identify the native metals and ore minerals found at the deposit.) Some deposits are or have been large working mines; others are prospects. Some important ore deposits shown on the *Mapa Metalogénico* do not have their mineralogy described in figure 2.5 because information on the specifics of their mineralization is unavailable. On the other hand, various small deposits and prospects, which are not included on the more general UNAM geological maps, appear in figure 2.5. Probably not all minerals present in some deposits were reported because of the nature of the survey or because the geologists were only interested in the presence and abundance of economically important ores. The value of this map lies in its specificity. It documents the presence and associations of particular ore minerals that could have served as parent materials for metal objects made and used in the prehispanic era. Not surprisingly, the mines on this map that also appear on the *Mapa Metalogénico* show the same ore minerals that Petersen has inferred for them (see table 2.1).

The five maps—three based on the UNAM distribution data, one based on the *Mapa Metalogénico,* and one constructed from field reports—reaffirm that two geographically distinct zones of mineralization characterize the metalworking zone. Deposits cluster in a southern area including eastern Michoacán, western Guerrero, and the state of Mexico, and in a northern zone comprising western Jalisco, Colima, and Nayarit. In the more southerly region the most abundant copper mineral is chalcopyrite, which appears at all of the deposits in Guerrero and southwestern Michoacán where copper is shown as an important metal. Malachite, which is frequently a

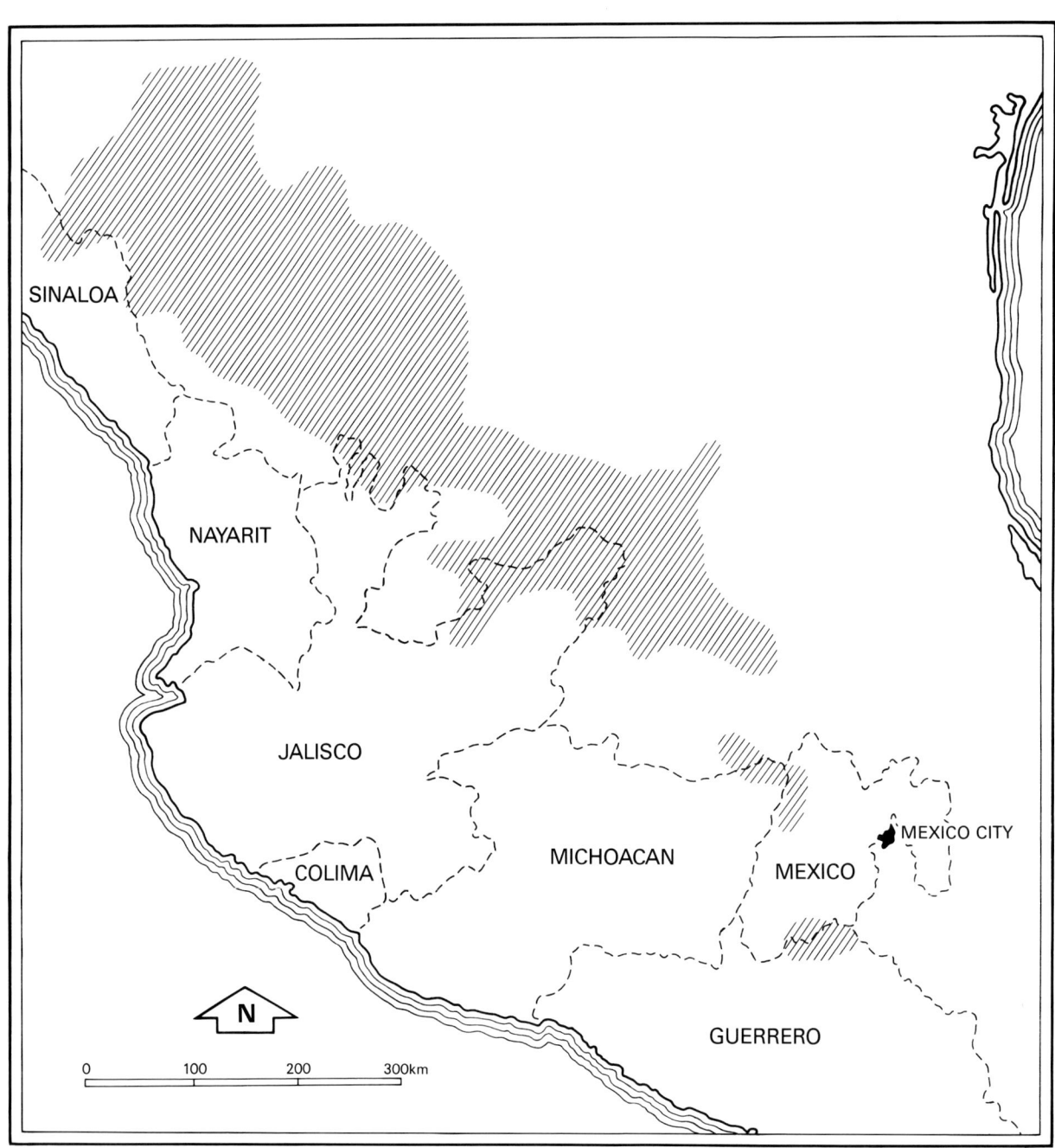

Resources, Metals, and Alloys

2.3

The Zacatecas tin province, and tin-bearing areas to the south.

weathering product of chalcopyrite, also appears often at these mines. Some of these deposits contain chalcocite, which like chalcopyrite is a copper sulfide, and several show the copper oxides, malachite and azurite.

Copper ore from Churumuco and La Verde, two mines in Michoacán in the southern area (La Verde appears in figure 2.4), have been analyzed by Grinberg (1989) using atomic absorption methods. Her results furnish element concentrations for the ores but do not identify the minerals. Since the samples were not analyzed for sulfur, mineral identification is made more difficult, but the data do allow educated guesses concerning ore minerals. The samples from La Verde contain high levels of iron and copper; other elements are present only in trace concentrations, suggesting that the ore is probably chalcopyrite or bornite. All but one of the samples from Churumuco, in contrast, lack iron or show its presence in trace amounts; there, the ore is probably malachite or chalcocite.

The primary ore in which arsenic appears in the southern zone is arsenopyrite (table 2.1). In most cases where this ore occurs, it is associated with chalcopyrite. Deposits of tetrahedrite, a sulfide ore of copper that always contains antimony but that can contain arsenic as well (see table 2.2), also appear in this zone. In tetrahedrites, arsenic always occurs in concentrations lower than that of antimony. Tetrahedrite forms a solid solution series with tennantite, which contains arsenic in concentrations higher than antimony, but it is not typical of these geological environments. Figure 2.5 shows that arsenopyrite occurs in thirteen mines or sites in this zone, and tetrahedrite at four. A deposit of native arsenic is reported in the Sierra de Tlatlaya (no. 45 on map) in the state of Mexico. Neither tennantite nor enargite is reported.

The most common ore of silver in the southern area is the silver-arsenic-antimony sulfosalt, proustite. Data inferred from the *Mapa Metalogénico* show proustite probably present at seven mines and the silver sulfide, argentite, at six of the seven. Figure 2.5 from the survey data shows the sulfosalts (proustite, polybasite, and pyrargyrite) present at seven deposits, native silver appearing in six mines, and argentite at nine.

All sources indicate that tin is unusual in this region. One deposit is shown on the *Mapa Metalogénico;* the mineralogical data in figure 2.5 show two deposits in the state of Mexico.

The most common ore minerals in the more northerly region are the same as those in the south. Data from the *Mapa Metalogénico* (see table 2.1) show that the most usual ore of copper is chalcopyrite, which appears at ten of the eleven mines on the map. The survey data in figure 2.5 also show that chalcopyrite is common; chalcocite occurs less frequently, and native copper appears once. As for the arsenic-bearing ores, arsenopyrite is inferred as present at all ten of the mines where chalcopyrite is shown; tetrahedrite at seven. The deposits shown in figure 2.5 indicate that arsenic occurs as arsenopyrite in two cases and could be present as tennantite in one. Silver is identified as argentite at nine deposits and is accompanied by proustite at seven of these nine mines. Proustite is also present by itself once. Figure 2.5 indicates that argentite is the most abundant silver ore mineral. Native or metallic silver occurs only once. The tin ore reported is cassiterite, but it is shown at only two deposits.

Copper-tin bronze alloys became central to West Mexican metallurgy in spite of the fact that cassiterite is rare in the metalworking zone. Nearly all cassiterite deposits appear in the Zacatecas tin province, apart from a few isolated examples along the northern border of

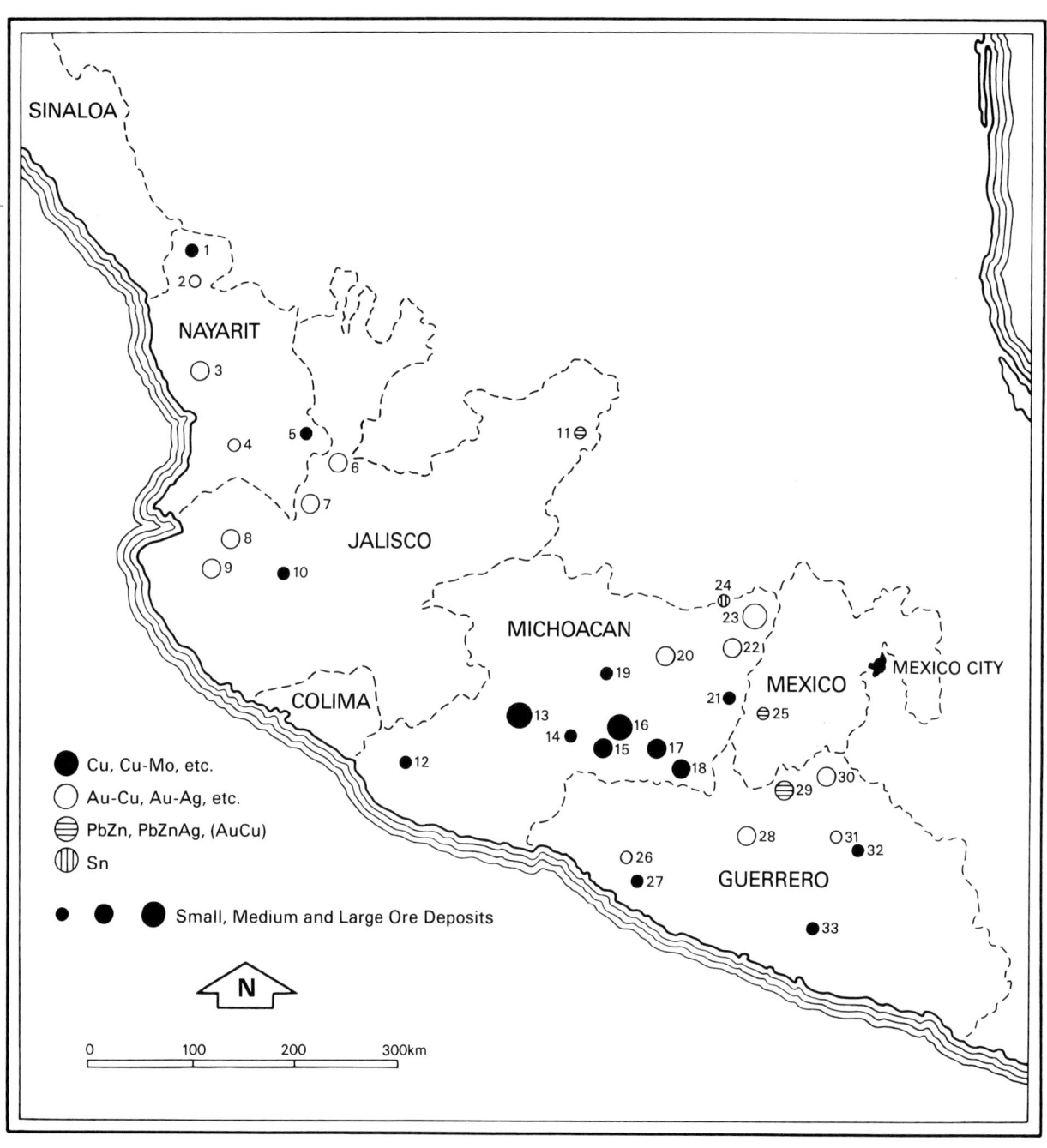

2.4

The distribution of primary metals in the West Mexican metalworking zone, from the Mapa Metalogénico de México (Salas 1980).

Nayarit:
1. Cucharas
2. El Tigre
3. Santiago
4. Compostela
5. La Yesca

Jalisco:
6. Cinco Minas
7. Etzatlán
8. Mascota
9. Talpa de Allende
10. Ayutla
11. Comanja de Corona

Michoacán:
12. Coalcomán
13. La Verde
14. San Isidro
15. Manga de Cuimbo
16. Inguarán
17. Bastan
18. Huétamo
19. Caltzontzin
20. Otzumatlán
21. Juárez
22. Angangueo
23. Tlalpujahua
24. Los Cabires

México:
25. Temascaltepec

Guerrero:
26. Real de Guadalupe
27. Copper King
28. Cuetzala
29. Xitinga
30. Taxco
31. Mezcala
32. La Unión
33. La Dicha

Michoacán and Jalisco and in the state of Mexico. Cassiterite occurs in vein deposits, which in some cases give rise to alluvial placers. The ore-bearing vein deposits are so small that contemporary miners recover the mineral with hand tools. The alluvial deposits themselves contain little tin because the supply of cassiterite at the source is so small. Also, the tin content of the ore is quite low, so that the amount of tin recovered during the smelting process is also extremely low (Foshag and Fries 1942). Mexican tin deposits are so small and widely dispersed that most mining in the present and recent past has been carried out by *gambusinos*, miners working individually or in small groups who collect the ore to sell it to the smelters.

One reason smelting operations yield so little tin is that Mexican cassiterites occur in close association with specularite, an oxide mineral of iron. In fact, specularite is generally more abundant than cassiterite. The two minerals are intimately intergrown, and in some cases specularite replaces cassiterite. Their association is so close that they cannot be completely separated even with fine mesh screens, so that smelting operations allow recovery of only small amounts of tin. In some cases as little as 12% of the tin in the ore is recovered as metal (Foshag and Fries 1942).

A potentially useful characteristic of cassiterite is that it contains the trace element indium. In Mexican cassiterites indium is present in unusually high concentrations, making it easier to detect when the ore, or metal smelted from the ore, is analyzed. This may provide an opportunity to characterize cassiterite deposits by their tin-to-indium ratios.

The observations I have made about the relative abundance of certain ore minerals in the metalworking zone are corroborated by geologists familiar with the region and in published treatments of Mexican geology. Copper generally appears as a sulfide in Mexico, and chalcopyrite is the copper ore that serves as the source for most copper metal produced in that nation. The primary ores of silver are argentite, native silver, proustite, and other silver sulfides, such as pyrargyrite, stephanite, and polybasite (González R. 1956). In Mexico, arsenic is present most commonly as arsenopyrite (González R. 1956).[4] Enargite is not characteristic of these geological environments.[5] Finally, the only known ore mineral of tin is cassiterite.

In summary, all of the native metals and ore minerals most likely to have been used by ancient metalsmiths, except for cassiterite, are found in the West Mexican

Table 2.1. West Mexico: Inferred Mineralogy of Principal Ore Deposits

Inferred Mineral	Principal Deposits				
	Guerrero	Michoacán	Jalisco	Nayarit	Estado de México
Chalcopyrite, arsenopyrite	Copper King La Dicha La Unión	Bastan Caltzontzin Coalcomán Huétamo Inguarán Juárez Manga de Cuimbo San Isidro La Verde			
Chalcopyrite, arsenopyrite, tetrahedrite, argentite, sphalerite			Ayutla	La Yesca	
Chalcopyrite, arsenopyrite, gold				Cucharas	
Chalcopyrite, arsenopyrite, argentite, proustite, gold, galena		Otzumatlán		Compostela Santiago El Tigre	
Chalcopyrite, argentite, proustite, gold, galena	Cuetzala Mezcala Real de Guadalupe	Tlalpujahua			
Chalcopyrite, arsenopyrite, tetrahedrite, argentite, proustite, sphalerite, galena	Taxco	Angangueo	Mascota		

continued

Table 2.1 (continued)

Inferred Mineral	Principal Deposits				
	Guerrero	Michoacán	Jalisco	Nayarit	Estado de México
Chalcopyrite, arsenopyrite, tetrahedrite, argentite, proustite, sphalerite	Real de Guadalupe		Cinco Minas Etzatlán Talpa de Allende		
Tetrahedrite, proustite, galena	Xitinga		Comanja de Corona		Temascaltepec
Cassiterite		Los Cabires			

metalworking zone. The most common ore of copper, chalcopyrite, is a sulfide, although copper also appears as the native metal, as copper oxides (especially malachite), and as other sulfide ores (including chalcocite). Arsenic appears in the native form at one site in the state of Mexico, but its primary occurrence is as the complex arsenical iron sulfide, arsenopyrite. Arsenopyrite is very frequently associated with chalcopyrite in the metalworking zone. This particular association is especially important in determining how copper-arsenic alloys were made. Arsenic can also occur in copper sulfarsenide ores, such as tennantite-tetrahedrite. Tennantite is rare in West Mexico, but tetrahedrites are not unusual. Tetrahedrite itself always contains high concentrations of antimony, levels that I have rarely detected in the artifact metal, making it unlikely that this ore mineral was customarily used. On the other hand, as the tetrahedrite series grades into tennantite, arsenic concentration increases as arsenic replaces antimony. Thus we must leave open the possibility that some arsenic-rich metal (such as copper-arsenic bronzes) could have derived from a tetrahedrite ore of tennantite-like composition. Silver appears most commonly as the sulfide proustite in the more southerly region and as the sulfide argentite to the north, although proustite is also found there. Native silver and silver sulfosalts, including polybasite and others, also occur but less frequently.

ORE TYPES USED IN WEST MEXICAN METALLURGY

Which of these native metals and ore minerals did the smiths of ancient West Mexico use, and how did they process them? Thus far, the clearest indication comes from qualitative and quantitative analyses of the composition of metal artifacts. These data do not provide direct evidence for the ores that were smelted, but they do identify the metals and alloys that were used. On the basis of that information, specifically the presence and concentration of diagnostic elements in the artifact metal, and knowledge of the presence, relative abundance, and associations of various ore minerals in the metalworking

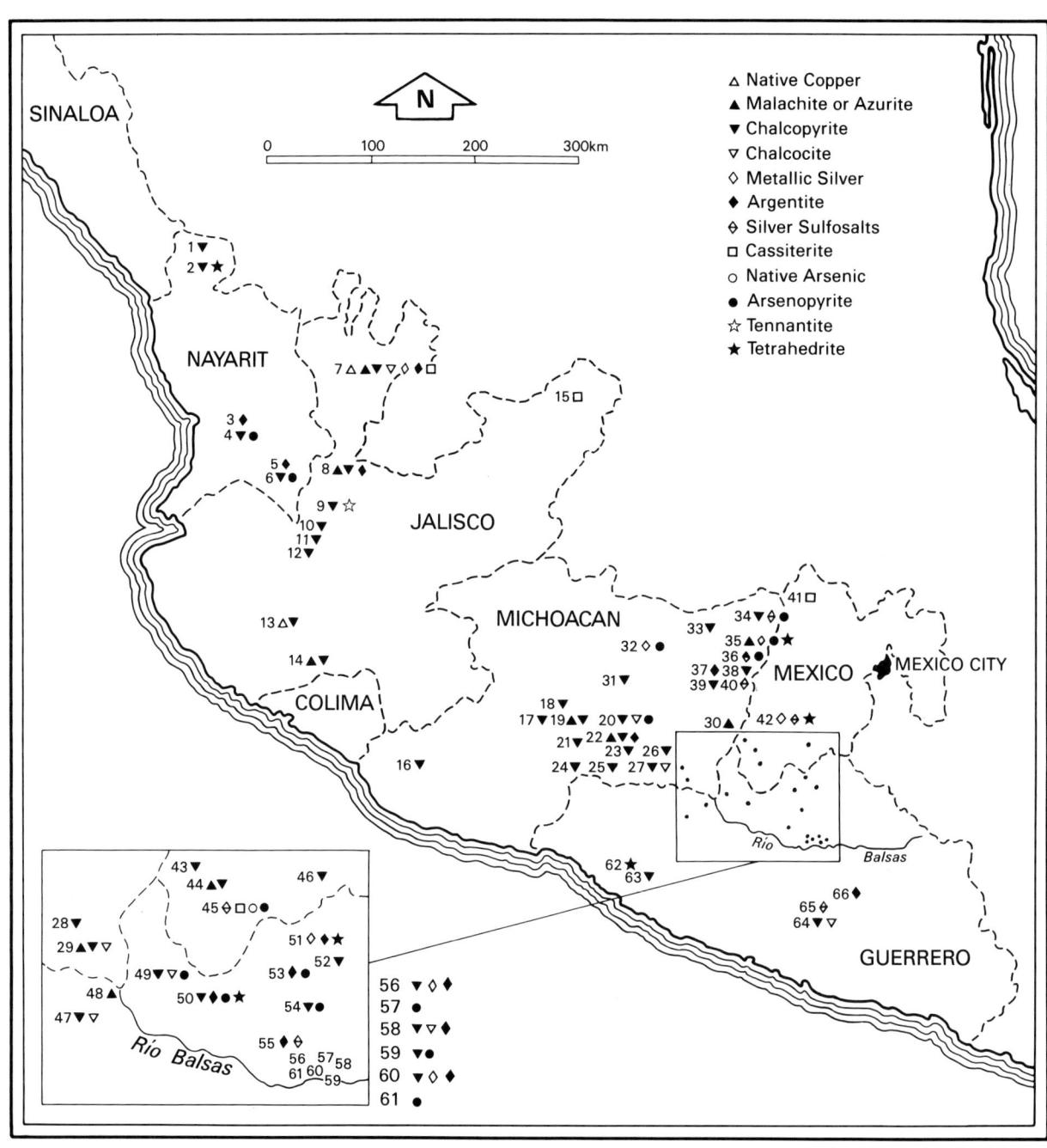

2.5

Native metals and ore mineral associations in the metalworking zone.

Nayarit:
1. IMMSA
2. La Paloma
3. IMMSA
4. IMMSA
5. Ixtlán
6. Barranca del Oro

Jalisco:
7. Bolaños
8. Cinco Minas
9. La Mazata
10. Magistral
11. IMMSA
12. IMMSA
13. Ayutla
14. IMMSA
15. IMMSA

Michoacán:
16. IMMSA
17. La Verde
18. IMMSA
19. IMMSA
20. Inguarán
21. IMMSA
22. Los Olivos
23. Manga de Cuimbo
24. IMMSA
25. Churumuco
26. La Libertad
27. IMMSA
28. Huétamo
29. IMMSA
30. IMMSA
31. Caltzontzin
32. IMMSA
33. Zinapécuaro
34. Maravatío
35. IMMSA
36. IMMSA
37. IMMSA
38. Los Cerros
39. IMMSA
40. Zitácuaro

México:
41. IMMSA
42. Temascaltepec
43. IMMSA
44. IMMSA
45. IMMSA
46. IMMSA

Guerrero:
47. Los Placeres
48. IMMSA
49. IMMSA
50. IMMSA
51. Teololoapán
52. Río de la Plata
53. San Jerónimo
54. IMMSA
55. IMMSA
56. IMMSA
57. IMMSA
58. Chichihualco
59. Perú
60. Puerto Hondo
61. Tepantitlán
62. Cruz de Oro
63. Peñon Blanco
64. Balsas
65. San Nicolás del Oro
66. Pezuapa

zone, reasonable inferences are possible about the most likely parent materials that were employed. It is also possible to make plausible reconstructions of smelting technologies.

To identify particular ore minerals I used qualitative chemical analytical data from 374 RMG artifacts and quantitative analyses from 263 of them.[6] The methods and sampling criteria are outlined in appendix 1. As noted there, the objects were selected from the seven most common RMG artifact types: bells, axe-monies, rings, tweezers, needles, axes, and awls.

I sorted the qualitative analytical data on these objects by searching for suites of elements that would normally characterize those native metals and minerals that were the objects' most likely parent material. The elements used to profile or represent each of these native metals and minerals were identified by Ulrich Petersen and are listed in table 2.2.

I identified two primary compositional groups on the basis of these analyses: one contains objects made from copper, with other elements present only in trace amounts; the other consists of copper alloys. I was able to identify the native metals and/or minerals that had likely served as the parent materials for the objects in the two groups by evaluating these data in conjunction with information from the quantitative analyses, which, by indicating the concentration of various elements, make clear which artifacts were probably intentional alloys. The two primary compositional groups are given in figure 2.6, along with the number of artifacts belonging to each.

Objects in Group I are made of "pure" copper, but the copper derives from a variety of sources—from native copper and very pure copper oxides and carbonates, on the one hand, and from minerals such as chalcopyrite and impure oxides and carbonates on the other. Group II consists of objects made from alloys: binaries such as copper-silver, copper-arsenic, and copper-tin, and ternaries like copper-silver-arsenic, copper-arsenic-antimony, and copper-arsenic-tin.

In theory, each of the distinct compositional subgroups can be derived in a variety of ways. Which of those alternatives was the most likely, given the ore resources available in the metalworking zone and the data from the chemical compositions of the artifacts? Let us consider the objects in Group I first.

Unalloyed copper objects can be made from native copper or from one of several copper ore minerals that we know were common in the metalworking zone: chalcopyrite or its weathered products, normally oxides, carbonates, and sulfates such as malachite and azurite. Group I contains 71 unalloyed copper objects and is divided into two subgroups. Each subgroup contains examples of all of the artifact types present in the RMG collection. None of the artifacts in Group I contains arsenic, antimony, or indium. Also, none contains tin or silver in larger than trace amounts. As table 2.2 indicates, trace amounts of tin and silver can be present in certain copper ores.

Group IA contains objects that are made either from native copper or from oxide or carbonate ores that are relatively pure. None of these 17 artifacts contains tin, bismuth, or nickel. The characteristic microstructure of native copper, confirmed in the metallographic studies of some of these 17 artifacts, identifies them as having been made from the native metal. Others were made from metal smelted from relatively pure weathered ores of copper. Trace amounts of silver, iron, and manganese are present, as is common in such ores (C. S. Smith 1968b; Maddin, Wheeler, and Muhly 1980; Rapp 1982).

The metal for some of the 54 Group IB objects was probably smelted from chalcopyrite. Metal for others apparently derived from relatively impure oxides or carbon-

Table 2.2 Elements in Artifact Metal Used to Identify Ore Types

Ore	Mineral Composition	Ore Contains	Ore Does Not Contain	Ore Can Contain
Native copper		Cu	As*, In, Ni, Sb	Ag
Chalcopyrite	$CuFeS_2$	Cu, Fe	As, In, Sb	Ag, Bi, Ni, Sn**
Tetrahedrite-tennantite	$(Cu, Ag)_{10}(Fe, Zn)_2Sb_4S_{13}$–$(Cu, Ag)_{10}(Fe, Zn)_2As_4S_{13}$	As, Cu, Sb	In, Ni	Ag, Bi
Enargite	Cu_3AsS_4	As, Cu	In, Ni	Ag, (Sb, Zn)***
Arsenopyrite	FeAsS	As, Fe	In, Ni, Sb	Ag
Cassiterite	SnO_2	In, Sn	Ni	As, Bi, Pb, Sb
Freibergite (high silver tetrahedrite)	$(Cu, Ag)_{10}(Fe, Zn)_2(Sb, As)_4S_{13}$	Ag, As, Cu	In, Ni	Bi, Sb

* Arsenic has been excluded as a diagnostic element for native copper. Native copper sometimes does contain arsenic, but the artifacts in this assemblage that metallographic analysis showed to be made from native copper exhibited no arsenic upon chemical analysis.

** When tin is found present at a minor level (m) in the qualitative analyses, I have identified the ore type as chalcopyrite; this corresponds, in quantitative analyses, to less than approximately 0.45 weight percent.

*** Enargite usually contains antimony and zinc.

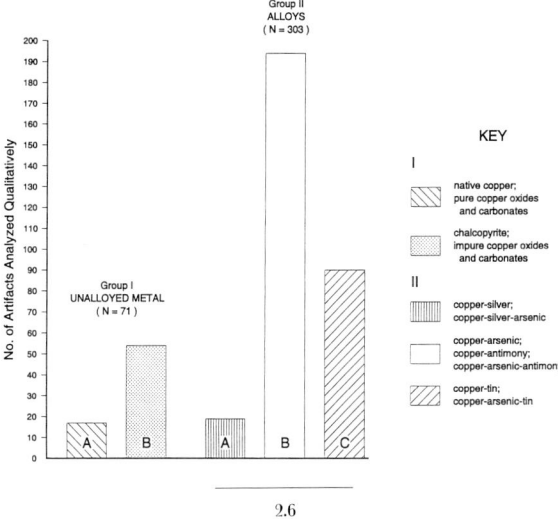

2.6

Compositional groups: West Mexican metal artifacts (RMG collection).

ates. All 54 artifacts contain either tin, nickel, bismuth, or combinations of these three elements in trace amounts. (All also contain traces of silver, iron, and manganese.) The three elements tin, nickel, and bismuth are particularly characteristic of those chalcopyrite ores in which tin occurs in solid solution with the copper.[7] Group IB objects that do not contain tin in trace amounts were smelted from copper ores associated with ores containing nickel and bismuth.

The Group II objects are made from alloys and fall into three main subgroups: IIA, alloys of copper and silver; IIB, alloys of copper with arsenic, antimony, or both of these elements; and IIC, alloys of copper with tin or with arsenic and tin. The 19 artifacts that are alloys of copper and silver contain silver in concentrations that range from 14% to 99%. (For this group, I use the term "alloy" when the element in question is present in con-

centrations high enough to affect the working properties of the metal.) These copper-silver alloys can be derived in several ways. One is to direct-smelt an ore mineral such as freibergite, which contains both copper and silver (and usually antimony and/or arsenic) and which might produce a copper alloy containing as much as 20% silver. Such minerals are not abundant in the metalworking zone, and the compositions of only three objects indicate freibergite as a possible parent material. (These objects are not represented in figure 2.6.) Generally, however, copper-silver alloys are produced by smelting copper ores to win copper metal, smelting silver ores to produce metallic silver, then melting the two metals together to form the alloy. (In this discussion, the term "copper-silver alloy" refers to alloys in which copper is the matrix metal; "silver-copper alloy" refers to alloys in which silver is the matrix metal.) The Group IIA alloys had to be produced in this way. There are no ores that contain both copper and silver in sufficiently high concentrations to yield the range of alloy compositions found in these objects. The copper ore mineral for these alloys appears to have been chalcopyrite, because no object in Group IIA contains arsenic or antimony but all contain tin and/or bismuth in trace concentrations. With only 19 objects of copper and silver, it was not possible to evaluate the likely sources of the silver from chemical analyses. Nevertheless three different silver ores may be represented. One set of artifacts contains minor amounts of lead and gold; another lacks gold. A few objects of silver-copper alloy also lack gold. In the one object made of pure silver no trace elements but gallium are present, and this may represent a native silver.

Copper-arsenic alloys (Group IIB) can be produced by smelting copper sulfarsenide ore minerals such as enargite or tennantite after initial roasting to remove the sulfur.[8] These ore minerals contain copper and arsenic (and, in the case of tennantite, antimony at low levels), and the metal won from them will be an immediate alloy of copper and arsenic. Another method of achieving the alloy is to smelt mixed sulfide ore minerals: minerals of copper, such as chalcopyrite, and those containing arsenic, such as arsenopyrite, or the weathered products of these two co-occurring minerals,[9] which may be malachite with copper arsenate or chenevixite[10] (Lechtman 1985; Rostoker and Dvorak 1991). In the rare case where native arsenic metal is available (as the geological data indicate it was in one deposit in the state of Mexico), it can be added to molten copper. Instead of adding native arsenic to the molten copper, one can add arsenic-containing minerals such as enargite, tennantite, or arsenopyrite to form the alloy.

The 194 objects made of copper-arsenic alloy contain arsenic in concentrations between 0.02% and 23%. When arsenic is present in concentrations greater than 0.5%, it affects the working properties of the metal. The issue that concerns us here, for all 194 objects, is whether they represent the smelting of mixed ores such as chalcopyrite and arsenopyrite, which we know occur in extremely close association in West Mexican deposits, or result from smelting a copper sulfarsenide ore such as enargite or tennantite, which is the method that may have been used in the Andean region. One group of objects contains arsenic alone as the alloying element with copper, another contains arsenic and antimony, and a third group contains antimony alone. The concentration of antimony is always less than that of arsenic when both elements are present. All the objects contain silver in trace amounts, between 0.03% and 0.17%. The majority also contain traces of bismuth, nickel, and tin.

There are two ways to explain the suite of elements appearing in these objects. The first is that the metal came from the smelting of a mixed ore of chalcopyrite and arsenopyrite associated with bismuthinite, accounting for the bismuth in the metal. The metallogenic association of these three ore minerals is quite common elsewhere in the world. Tin in trace concentrations can occur in chalcopyrite; nickel minerals such as pentlandite can associate with arsenopyrite. The presence of antimony may be explained by the mining of an ore body that could contain arsenopyrite near the center and a mineral of antimony, such as stibnite, at the surface.[11] Stibnite is common in the metalworking zone.

By contrast, these alloys could have been produced by smelting copper sulfarsenide ores such as enargite, which would contribute arsenic to the metal, or an ore of tennantite, which would contribute arsenic and antimony, with the arsenic present in concentrations higher than the antimony. However, copper sulfarsenides generally contain no nickel or tin and in many cases contain silver in high concentration, whereas most of the objects do contain nickel and tin, and their silver content is low. Thus, the sulfarsenide ores alone do not account for the typical elemental compositions of the objects. What is more, enargite and tennantite are uncommon in this region.

Based on the compositional data, the likelihood is high that the metal for these Group IIB objects was won through the smelting of mixed-ore minerals of chalcopyrite and arsenopyrite that might have been associated with minerals bearing nickel and bismuth.[12] It is also possible that either the weathered forms of chalcopyrite and arsenopyrite, such as chenevixite and limonite, or malachite and a copper arsenate could have served as the parent materials for these alloys, although these minerals have not been reported in the studies of West Mexican ore mineralogy.

Copper-tin alloys (Group IIC) can derive from the direct smelting of copper ore minerals that contain tin, such as stannite, but there is no geological evidence for stannite ore in Mexico. The copper-tin alloy therefore could have been produced in one of two ways: either by first smelting cassiterite to win metallic tin, then adding the tin to molten copper, or by co-smelting cassiterite with copper ore minerals. The chemical compositions of the Group IIC objects indicate that the metal for the 90 artifacts made from copper-tin bronze alloys resulted from intentionally alloying the two metals. The presence of indium in all artifacts indicates unequivocally that the tin metal for these objects derived from cassiterite. Indium was detected only in artifacts that contain tin in concentrations greater than about 1%. Since indium is a trace element characteristic of cassiterite, the indium detected in the copper-tin alloys was introduced through the smelting of cassiterite ore minerals. Nearly all tin bronze alloy objects contain nickel and bismuth as well.

To investigate the possibility of characterizing deposits on the basis of indium-to-tin ratios, I examined the proportion of indium to tin through a neutron activation study of 90 of the copper-tin bronze objects. (Neutron activation detects indium when present even in very low concentrations.) The results are presented in figure 2.7. The plot of indium-to-tin concentrations shows that indium concentration generally increases proportionately with the concentration of tin, although there are striking exceptions to this pattern. That these two elements generally associate in a linear fashion may suggest that the relation derives from the ore body. The fact that there are striking deviations from the linear pattern suggests that indium-to-tin ratios may vary between ore deposits, and that objects were made using different cassiterite sources. The topic is currently under investigation.

In Group IIC, 66 of the 90 objects contain arsenic and are alloys of copper, arsenic, and tin. All contain tin in concentrations higher than 1%, and arsenic is present in concentrations consistently lower than the tin, sometimes only in trace amounts. Antimony appears in all of these objects in trace amounts. The copper-arsenic-tin alloys may result from any of the permutations discussed for the production of copper-arsenic and copper-tin alloys. The copper metal for these copper-tin and copper-arsenic-tin alloys was smelted from a copper ore that is likely to have had associated with it ore minerals such as arsenopyrite and perhaps bismuthinite and pentlandite, which served to introduce the traces of arsenic, bismuth, and nickel found in the analyses. Arsenic occurs in greater than trace amounts in some cases; a few objects contain between 1% and 3% arsenic. It is very likely that the copper-tin alloys in Group IIC were made by melting together tin smelted from cassiterite and copper smelted from mixed chalcopyrite/arsenopyrite ore minerals.

Summary. We can now identify the likely source materials for the artifact metal on the basis of these chemical analytical data and the information concerning the presence, abundance, and associations of ore minerals in the metalworking zone. Unalloyed copper derived primarily

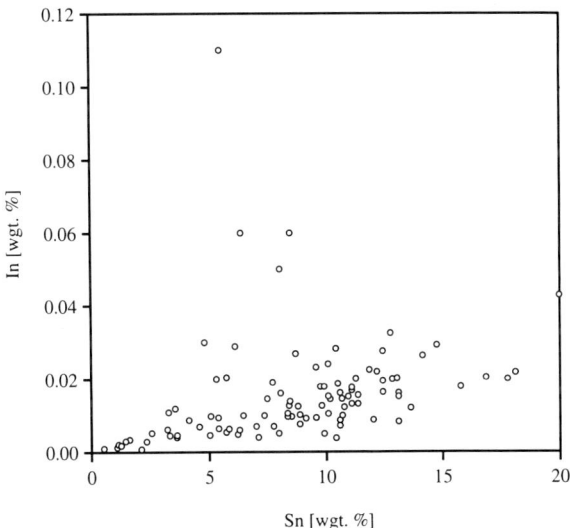

Ratio of indium to tin in RMG collection artifacts.

from smelting chalcopyrite (a sulfide) or impure copper oxides. Metalsmiths employed native copper or very pure copper oxide ores less commonly. They made copper-arsenic alloys either from chalcopyrite associated with arsenopyrite or by co-smelting arsenopyrite with a non-sulfide ore of copper. These low-arsenic copper-arsenic alloys may also have been smelted directly from an arsenic-bearing copper ore mineral, such as enargite or tennantite, but both the geological evidence concerning relative ore abundance and artifact chemical compositions point far more directly to the use of chalcopyrite and arsenopyrite or their weathered products. The copper-tin alloys were made through deliberate alloying of copper smelted from chalcopyrite with tin smelted from cassiterite. The copper and silver or silver and copper alloys likewise represent the intentional alloying of the separate metals.

From the geological information just reviewed, coupled with studies of artifact chemical compositions, we have a clear picture of the native metals and ore minerals West Mexican smiths used. Sixteenth-century documents provide another significant source of evidence concerning the location and exploitation of certain mines. I will cite major published sources; unpublished manuscripts in historical archives offer another useful but largely unexplored corpus of data.

Documentary Evidence for Ancient Mining

A key document concerning the exploitation of the mineral-rich region of eastern Michoacán and western Guerrero is the *Lienzo de Jucutacato,* a manuscript that illustrates the migration of the Tarascan people to what today is the state of Michoacán. According to one interpretation (Jiménez M. 1948), the *Lienzo* represents the migration of a metallurgical guild from the Gulf coast to the region around Uruapan, Michoacán (figure 2.8). Jiménez argues that the *Lienzo* also depicts subsequent migratory expeditions: one to Pátzcuaro, to establish a metalworkers' guild, and others to La Huacana, Churumuco, Arcelia, and Coalcomán to look for mines (figure 2.8). Four of these areas are rich in metallic minerals, and two (La Huacana/Inguarán and Coalcomán) appear on the *Mapa Metalogénico* as major copper mines.

The expedition that moved toward La Huacana, a community located in the region of the Inguarán mine, continued to a place identified in the *Lienzo* as Tepulan. Inguarán is one of the largest copper mines in Mexico. Barrett (1981) argues that Tepulan was its ancient name. The *Lienzo* follows the migration to Churumuco, in the same general area where a large mine containing chalcopyrite is currently being exploited. Barrett cites documentary evidence that copper was being mined in the prehispanic era at Churumuco and Guaraxo (Inguarán), both subject to the town of Sinagua, and at Cocían located to the northeast.

The *Relaciones Geográficas* provide complementary evidence. The *Relaciones Geográficas* are the replies by local officials to a 50-item questionnaire issued in 1577 by the Spanish throne. The questions deal with political geography, environment and terrain, town boundaries, languages, native government, mineral resources, and economic life. The *Relaciones* report tribute in gold, silver, and copper ingots from many communities throughout this region (Hosler 1986; Pollard 1987), but more germane to this particular discussion are entries that allow inferences about smelting technologies and ore types. The *Relación de Sinagua* suggests, for example, that ores exploited in that region may, when smelted, have produced copper-arsenic bronze metal. The *Relación* mentions that there is a mine from which copper is extracted that serves for tools "como el hierro, con que ellos trabajan y labran sus sementeras" [like iron, that they use for work and to work their fields] (Acuña 1987: 254). This copper is described as hard, like iron. Data from artifact analyses and from the reports describing ore mineralogy suggest that the copper is a copper-arsenic bronze alloy, smelted from mixed ores of arsenopyrite and chalcopyrite or their weathered products. The data presented here describing the ore mineralogy of the two mines, Churumuco and Inguarán, indicate that chalcopyrite and arsenopyrite are present at both.

J. Benedict Warren, in 1968, published *Legajo 1204,* a sixteenth-century manuscript dealing with copper mines in Michoacán. The document describes a visit in 1533 to Michoacán by Vasco de Quiroga, the purpose of which was to investigate the existence, location, and extent of the copper deposits in the region. The manuscript verifies that several of the mines shown on the maps in figures 2.4 and 2.5—Inguarán, Churumuco, Huétamo, Bastan— were exploited prior to the Spanish invasion. The *Legajo* names other mines—Tancítaro and Coyuca—that do not appear on the maps but that reportedly were sources of copper metal. The *Relaciones Geográficas,* in addition, note that there are mines of both silver and copper in the Sinagua and Guayameo areas, at Tacámbaro, and at Turicato (near La Huacana), as well as copper mines at La Huacana. The *Relaciones* also indicate that copper and silver metal were both tribute items in Michoacán: thirteen of the towns listed rendered tribute in copper and twelve in silver (Hosler 1986; Pollard 1987).

One of the more tantalizing pieces of information offered by the *Legajo* published by Warren is that the ore gathered from some of these mines was extremely difficult to smelt (suggesting that the metalsmiths were working with copper sulfides like bornite or chalcopyrite), whereas the metal from others was softer (suggesting that the ore mineral consisted of a copper oxide). At Sinagua, for example, "porque es muy recio de coger y después tardan en soplar y fundirlo cada uno dos o tres días" [some ore is very difficult to mine and later they take a long time to smelt it, each one takes the ore two or three days] (Warren 1989: 50).[13] By contrast, "la mina del dicho pueblo de Cocían es blanda la tierra . . . y habiendo allí gente . . . se sacaría mucha cantidad" [At the mine at the village of Cocían the earth is soft; if there were enough people it would be possible to obtain a large quantity of copper] (Warren 1989: 46). The repeated smelting reported at Sinagua may suggest that, after breaking up the ore, the first step in processing was to roast it to remove the sulfur before the ore could be smelted to extract the copper.

Returning to the *Lienzo de Jucutacato,* Jiménez (1948) argues that another migration proceeded to the east to the region of Arcelia, Guerrero, near the border with the state of Mexico. Arcelia lies approximately 30 kilometers south of a deposit in the Sierra de Tlatlaya where arsenopyrite, silver, cassiterite, and native arsenic are reported. The *Relaciones* from Sultepec, the tribute town closest to Tlatlaya, relate that there were many tin mines in that region (Paso y Troncoso 1905–1906 vol. 7: 13). The expedition in search of mines that went to that area thus may have exploited native arsenic, arsenopyrite, tin, and silver; the region was probably a key source for these raw materials.

The UNAM geological map showing the distribution of tin locates a deposit near Taxco, Guerrero. The existence of this deposit may corroborate Hernán Cortez's statement that after the Spanish invasion of Tenochtitlan he sent his men to Taxco for tin to make bronze cannons (Caley and Easby 1964). No cassiterite is mined at the modern mine of Taxco itself, but the tin ore mineral may have been obtained near there.

Also in Guerrero, the *Relaciones* report copper mines in the province of Tepecoacuilco and Tetela del Río. The *Relación de Tetela,* for example, states, "dijeron que tenían dos mynas de cobre . . . y questas mynas las

2.8

Population centers and mines mentioned in sixteenth-century Spanish and native documents.

labrauan en tiempo antiguo y agora no las labran" [they said that they used to have two copper mines and that they had worked these mines in the past but they are not working them now] (Paso y Troncoso 1905–1906 vol. 6: 136). Silver mines also are reported at Taxco, Zumpango, and Iguala (figure 2.8).

A silver mine at Temascaltepec (figure 2.8) was shown to the Spaniards by the local inhabitants (Paso y Troncoso 1905–1906 vol. 7: 26). In a later report, the Spaniards relate that the mine had been previously worked by the Indians living in that region.[14]

Information about the more northerly cluster of deposits (figure 2.8) describes veins of copper and silver in the Ayutla area in central Jalisco.[15] However, they were not being worked because the veins were too "coppery" [por ser cobrizos]. The *Relación de Tenamaxtlán* reports the same problem; Tenamaxtlán was located just to the north of Ayutla. It is likely that these ores were high-silver tetrahedrites or freibergite. The *Relación de Cuzalapa* mentions that there were copper mines approximately three miles from the town.

Mining also took place in the southeastern region of Jalisco, near the modern towns of Tamazula and Tuxpan. Carl Sauer (1948) maintains that certain sixteenth-century documents point to Tamazula as the region where the Tarascans obtained silver, and that Hernán Cortez had claimed that area as well as Amula and Tuxcacuesco because of the silver deposits. Yet that cannot be the only region where Tarascans obtained silver; silver mines were reported in northeastern Michoacán and in the La Huacana area, and certainly they were under Tarascan control at that time. The aim of one of the expeditions reported in the *Lienzo de Jucutacato* was to find mines in the Coalcomán region, which is in this same general zone. Schöndube (1974) argues from the *Relaciones* and other sixteenth-century documents that Tamazula had silver, copper, and gold mines, and that plates or *tejuelos* (ingots) of gold and silver were manufactured there. The *Relaciones* report silver mines in Zapotlán as well.

This brief survey indicates that some of the copper and silver deposits in the Huétamo–La Huacana area that are shown on the *Mapa Metalogénico,* and in the data assembled in figure 2.5, were exploited prior to or at the time of the conquest. Some of the ores may have been chalcopyrite-arsenopyrite associations or their weathered products; others may have been oxides of copper. The only suggestion that the metalworkers may have been producing copper-arsenic alloy metal comes from Sinagua, a region where we know chalcopyrite is associated with arsenopyrite. These data also point to the southern part of the state of Mexico and northern Guerrero as the probable and closest sources of tin for the area. That zone may also have been a source of arsenopyrite and native arsenic, as well as of silver. Silver was apparently also mined at Temascaltepec. To the northwest, in the southwestern region of modern Jalisco, copper and silver deposits that appear in the maps in figures 2.4 and 2.5 were also exploited.

I have now established the metals and alloys that defined the metallurgy of ancient West Mexico and the most likely native metals, ores, and processing regimes employed. The metallurgy was copper-based, although silver and gold objects also constituted a significant facet of the technology (see chapter 6 for a more extended treatment of this topic). Metalsmiths worked with the two bronze alloys (copper-arsenic and copper-tin), with a copper-arsenic-tin alloy, and with alloys of copper and silver. Copper objects were fashioned from native copper but also from metal smelted from copper oxide ores, such as malachite, and copper sulfide ores, most commonly chalcopyrite. Copper-arsenic alloys were usually made by smelting arsenopyrite with chalcopyrite or their weathered products. These two ore minerals are the primary arsenic- and copper-bearing ores in the west; they occur in close association in particular mines, and artifact chemistries strongly suggest that these constituted the parent materials for this alloy. Artifacts containing arsenic in extremely high concentrations, that is, greater than 12%, may have been made using native arsenic. Tin bronze alloy metal was made by adding tin to copper or by co-smelting cassiterite with chalcopyrite or its weathered products. Not only is cassiterite rare in this region, but the range of indium-to-tin ratios in the artifact metal suggests that more than one source was exploited. This would indicate that the West Mexican peoples imported at least some cassiterite from outside the zone.

I examine the archaeological and laboratory evidence concerning how, when, and for what social ends smiths chose to use these metals and alloys, and the South and Central American origins of certain facets of this metallurgy, in the following chapters. I begin by treating the earliest period of this technology.

3

Period 1 of West Mexican Metalworking

A.D. 600 to A.D. 1200/1300

The metallurgical technology that developed in this zone flourished for approximately 900 years. West Mexican metalsmiths incorporated elements introduced from the metallurgies of Central and South America during this long period and elaborated them, inventing totally new ways of handling the material. The timing of these technical events and the historical circumstances surrounding them are difficult to reconstruct. The archaeology of the zone is not as well known as that of other Mesoamerican areas, and analyzed, dated assemblages of metal objects are rare. Nevertheless, I was able to chronologically order major components of the technology by assessing the results of laboratory studies of the RMG collection artifacts (appendix 2)[1] in light of information about those metal objects for which dates are available.

Two technological periods emerged. The first extends from A.D. 600 to A.D. 1200/1300, when metalworkers principally used copper. During this time they fashioned an array of objects by lost-wax casting and by cold work with annealing.[2] The constellation of artifacts makes clear that these artisans were interested in items that visually and aurally expressed their conceptions of the sacred and that reinforced elite status. They disregarded for the most part metal's many utilitarian applications. The property of metal that most intrigued them and became most central to their technical experiments was its ability to sound, an interest manifested in myriad bells of different sizes, shapes, and pitches. Smiths also crafted tools from metal during this time, but in far fewer numbers.

By Period 2, which begins between A.D. 1200 and A.D. 1300, metalworkers in this zone became equally intent on developing another property of metal: its color. Their focus on metallic color is dramatically apparent in the range of copper alloys they used (copper-tin, copper-arsenic, copper-arsenic-tin, copper-silver)[3] for bells, large ornamental tweezers, and other ritual and sumptuary items. In most cases the alloying element, tin, arsenic, or silver, was present in high enough concentrations to transform the color of the artifact metal to differing hues of gold or silver. They also used bronze alloys (copper-tin and copper-arsenic) for tools, incorporating the alloying element in appropriately lower concentrations. Hot working, a fabrication method required to avoid brittleness when working objects made from copper-tin alloys, was added to the technical repertoire. Most objects made from these alloys were variations on types made from copper during Period 1. However, the superior properties of the alloys allowed artisans to explore new design possibilities: they cast larger, more intricate bells; made axes thinner and harder; and crafted wider, thinner, and larger tweezers with symmetrical, tightly wound spirals emerging from each side of each blade. In other words, they used these alloys to develop alternatives to existing and familiar objects rather than to devise uses for metal outside of New World cultural experience, such as for metal armor, for example. Period 2 smiths also enlarged the range of ores and extractive and smelting technologies, as they expanded their repertoire of designs and fabrication techniques.

The Technological Chronology

The method used to order the distinct components of this metallurgy chronologically is unique to this study. It depends heavily on engineering principles, as well as physical measurements commonly used in the field of materials science and engineering. I divided West Mexican metallurgy into two chronological periods based on a striking and systematic relationship apparent between the chemical compositions of the RMG collection artifacts and their formal design (that is, types or functional classes). By "formal design" I mean the arrangement and specific dimensions of an object's structural elements, for example degree of curvature, height, length, width, thickness, and internal volume. Each of the major artifact classes in the RMG corpus—bells, needles, tweezers, rings, axes, and others—contained some specimens made from copper, and others, with different design attributes, made from the bronze alloys: copper-tin, copper-arsenic, and copper-arsenic-tin. For instance, most bronze bells differ systematically from bells made from copper with respect to key parameters such as thickness and length. Sheet metal ornaments are exceptions to this pattern, but they are made nearly exclusively from gold and gold alloys, from silver, and from alloys of silver and copper.

Materials engineering explains these patterns in terms of the very different physical and mechanical properties of copper and bronze. These property differences decisively influence the possibilities for formal design. They also influence the functional capabilities of copper and bronze objects. For example, a copper tweezer of a given length, width, and thickness may fail in use while a bronze tweezer of precisely the same dimensions can be completely functional. Fabrication method is also crucial because it directly affects the properties and mechanical behavior inherent to a given metal or alloy. Particular fabrication methods may be required to achieve and to ensure the functional success of a certain formal design. Such interdependence among formal design, material composition, and manufacturing technique can be experimentally verified.

As I evaluated the formal designs of each artifact type using dimensional and mechanical property data, the systematic associations between formal design attributes and choices of materials and manufacturing techniques became explicable and predictable. I used these predictable associations as I discovered them in analyzing the RMG artifacts to establish the chronology. This meant examining formal design (and artifact chemistry when available) of other Mesoamerican metal artifacts formally identical to those in the RMG study collection. Some of these artifacts have been recovered from dated archaeological contexts, although their chemical compositions have not yet been determined. Other artifacts have been chemically analyzed but they are undatable, their provenience is unknown, or they derive from undatable archaeological contexts. A small proportion of metal objects can be dated and also have been chemically analyzed.

The laboratory studies of the RMG collections, based on the materials engineering principles discussed here, predicted correctly that chemistries of dated artifacts would replicate the RMG pattern (Hosler 1988a). In the RMG assemblage, certain artifact types are virtually always made from particular metals and alloys. Dated artifacts, identical in formal design to objects analyzed in the RMG study, show that they are made from the same metals and alloys and fashioned using the same manufacturing techniques. Such objects then served as key reference points in the chronological scheme. The two periods of the technology, and the fabrication methods and metals and alloys that characterize them, are set forth in table 3.1. I describe Period 1 technology in this chapter and the technology of Period 2 in chapter 5.

Archaeological Evidence for Period 1 Metallurgy

The initial evidence for metalworking in Mesoamerica appears between approximately A.D. 600 and A.D. 800, predominantly from sites along the western coastal plain or with riverine access to the coast (figure 3.1 and table 3.2). No metal objects are known from Mesoamerica before this time apart from a single lower Central Ameri-

Period 1: A.D. 600 to A.D. 1200/1300

Table 3.1. Summary Chronology for the Development of West Mexican Metallurgical Technology

Characteristic	Period 1, A.D. 600–1200/1300	Period 2, A.D. 1200/1300–1521
Metals and Alloys		
Copper	X	X
Gold	X	X
Silver	X	X
Copper-arsenic (low arsenic)	X	X
Copper-arsenic (high arsenic)		X
Copper-arsenic-silver		X
Copper-arsenic-tin		X
Copper-gold		X
Copper-silver	?	X
Copper-silver-gold		X
Copper-tin		X
Artifact Types		
Fishhooks	X	X
Awls (unipointed)	X	X
(bipointed)	X	X
(blade)		X
Bells (smooth walled)	X	X
(wirework)	?	X
Needles (perforated eye)	X	X
(loop eye)	?	X
Rings (round cross section)	X	X
(rectangular cross section)	X	X
Tweezers (beam design)	X	X
(shell design)		X
Axes	?	X
Axe-monies		X
Sheet metal ornaments	?	X

Characteristic	Period 1, A.D. 600–1200/1300	Period 2, A.D. 1200/1300–1521
Methods		
Annealing	X	X
Cold work from an initial cast blank	X	X
Lost wax casting	X	X
Hot work		X

can trade piece found at the Maya site of Altun Ha in Belize. Tomatlán, a site in Jalisco, provides what seems to be the earliest object: a piece of metal sheet dated to A.D. 600, or possibly before (Mountjoy and Torres M. 1985). Two lost-wax cast bells dating to between A.D. 650 and A.D. 750 have been excavated at the inland site of Cerro de Huistle (Hers 1990), also in Jalisco. The technology that subsequently developed in West Mexico predated metalworking in any other area of Mesoamerica by three to four hundred years.

Many investigators (for example, Arsandaux and Rivet 1921; Meighan 1969; Mountjoy 1969; Pendergast 1962b) have maintained that metallurgy was introduced to West Mexico from Central or South America via maritime trade. Not surprisingly, most Period 1 sites are located near the coast. Metal objects appear later at settlements in the higher-elevation interior lake basins, for instance at Cojumatlán and Tizapán on Lago Chapala. As I mentioned, metallurgy seems to have moved inland up the river systems, such as the Lerma-Santiago, the Balsas, and their tributaries. Yet the movement was slow. Archaeologists have excavated no metal objects at Teuchitlán in Jalisco nor at Tingambato in highland Michoacán or other large centers in the lake region during this period.

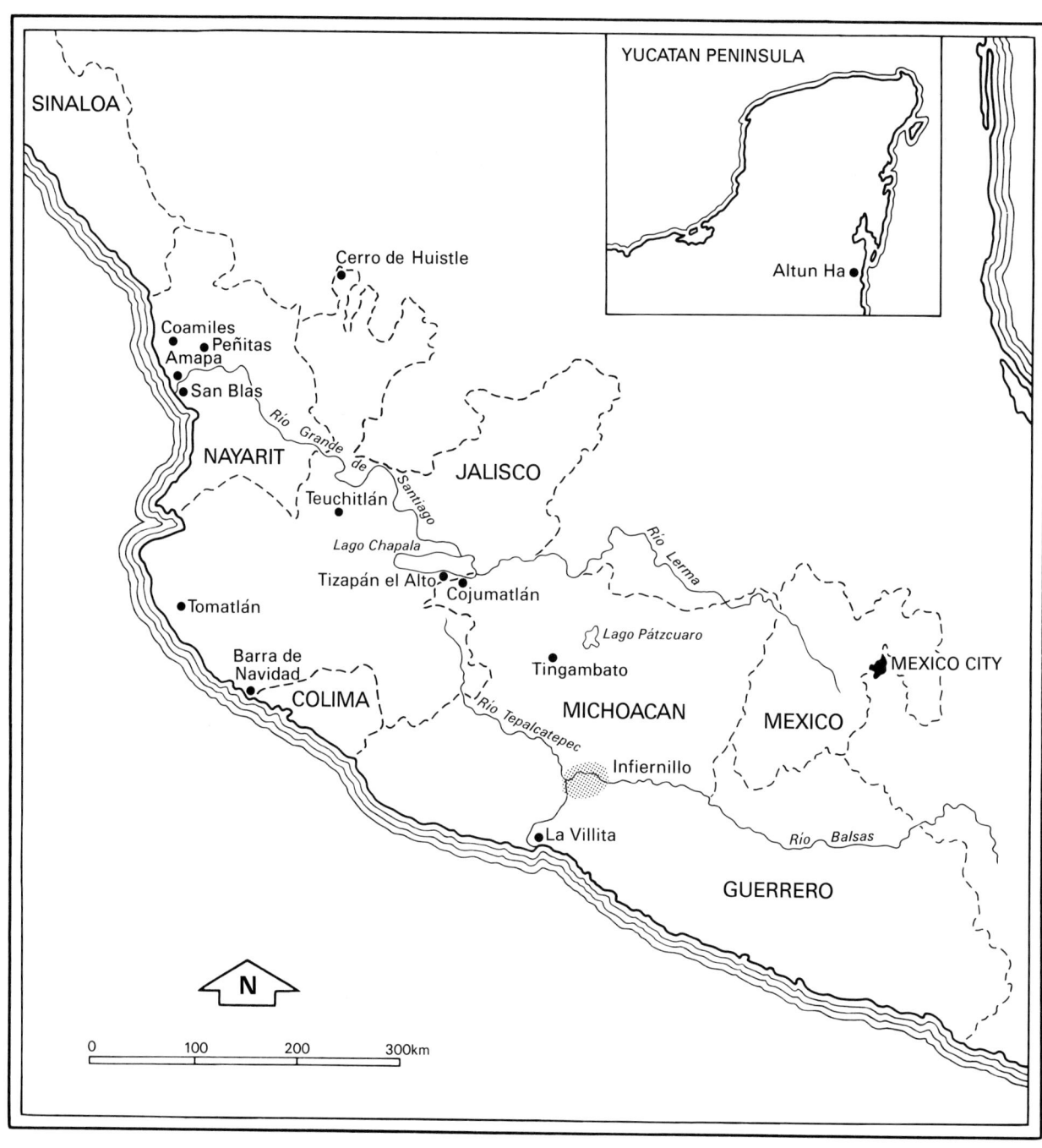

Period 1: A.D. 600 to A.D. 1200/1300

3.1

Period 1 archaeological sites and regions mentioned in text. Those where metal objects have been recovered include Cerro de Huistle, Coamiles, Peñitas, Amapa, Tizapán, Cojumatlán, Tomatlán, Barra de Navidad, the Infiernillo area, and La Villita.

Metal objects remain uncommon in the lake basin and highland areas of the metalworking zone until after about A.D. 1250, when the Tarascan state coalesced in the basin of Lago Pátzcuaro.

As the subsequent discussion shows, apart from the presence of metal objects, Period 1 sites where metal items appear have little in common with one another with respect to size, settlement pattern, and possible cultural affiliations. The Postclassic Period throughout West Mexico is one with many independent cultural complexes; most have their own characteristic pottery, for example (Cabrera C. 1986). Metal's appearance does coincide with the Aztatlán polychrome ceramic tradition at some sites, but these ceramics cover a long time span, they vary regionally, and their origins are debated (see for example Mountjoy 1982; Nicholson 1960; Smith and Heath-Smith 1980). That these two classes of material culture—a particular ceramic style and metal—sometimes co-occur merits further attention; for the present, the interpretation remains obscure.

Metal appears first at Cerro de Huistle (Hers 1989, 1990), Tomatlán (Mountjoy 1982; Mountjoy and Torres M. 1985), Amapa (Meighan 1976), and possibly the Infiernillo sites along the Balsas river (Cabrera C. 1976, 1986; Maldonado C. 1980), although the Infiernillo dates are problematic (figure 3.1). Tomatlán, located along the Tomatlán river drainage, consists of numerous small habitation sites with house foundations and associated patios. Tomatlán lacks public architecture. Mountjoy (1982) estimates that some 1,500 people may

Table 3.2 West Mexican Sites with Dated Metal Assemblages

Site	State
Period 1: A.D. 600–1200/1300	
Cerro de Huistle	Jalisco
Tizapán el Alto	Jalisco
Tomatlán	Jalisco
Cojumatlán	Michoacán
Amapa	Nayarit
Coamiles	Nayarit
Peñitas	Nayarit
Period 1 and Period 2: A.D. 600–1521	
Infiernillo	Guerrero/Michoacán
La Villita	Guerrero/Michoacán
Period 2: A.D. 1200/1300–1521	
Bernard	Guerrero
El Chanal	Colima
Lo Arado	Jalisco
Tuxcacuesco	Jalisco
Apatzingán	Michoacán
Hundacareo	Michoacán
Milpillas	Michoacán
Tzintzuntzan	Michoacán
Urichu	Michoacán
Culiacán	Sinaloa
Guasave	Sinaloa

have lived in the largest settlement that existed at the time when metal objects initially occur (A.D. 600–1000). Aztatlán ceramics appear at the same time. Mountjoy excavated 199 copper-based items from household contexts and from burials. He thinks that metal may have been worked at the site, classifying several objects as metalworkers' tools.

Amapa, where numerous metal objects were excavated, covers one square kilometer, has structures oriented like those of Central Mexico, and was occupied from 300 B.C. to at least A.D. 1300 (Meighan 1976). Metal objects first appeared between A.D. 600 and A.D. 1000, during the period when the buildings reflecting Central Mexican influences were constructed. Aztatlán pottery, some identical to that found at Tomatlán (Mountjoy 1982), also dates to this period. Excavations have yielded 205 metal artifacts as well as evidence for metalworking in the form of a few bits of slag. In addition, archaeologists recovered objects described as "slugs" (Meighan 1976; Pendergast 1962a) that may be cast blanks or items of ceremonial regalia. Metal artifacts were found in burials and test pits.

The only inland site where metal artifacts date this early is the ceremonial center of Cerro de Huistle on the Jalisco-Zacatecas border, occupied between approximately A.D. 100 and A.D. 900 (Hers 1989, 1990). Cerro de Huistle is linked to the Chalchihuites culture in northwest Mexico. At Cerro de Huistle, Hers recovered three small bells that she places before A.D. 850. Two of the three date to between A.D. 650 and A.D. 750, and one of these two may date to slightly prior to A.D. 650.

Metal objects occur slightly later at Peñitas, a center northeast of Amapa, also on the coastal plain. Peñitas was occupied from A.D. 400 to A.D. 1300. Excavations revealed a cobblestone platform mound, burial mounds, and several habitation mounds. The pottery associated temporally with metal objects dates to A.D. 1000–1200 (Meighan 1960)[4] and is related to Central Mexico and to complexes to the north in Sinaloa (Bell 1971). Only a few metal items were excavated, but looters' reports suggest that many others, for example bells, came from the same site. Archaeologists also recovered several ceramic sherds with metal residues; they may have been used as stirrers to skim slag from molten metal (Carriveau 1978), suggesting that metal objects were fashioned at Peñitas. Coamiles, a site 16 kilometers west of Peñitas, produced a needle from the one test pit dug there (Meighan 1960).

Metal also occurs slightly later along the Río Balsas (figure 3.1). An extensive salvage operation by the Mexican government in the Infiernillo region identified numerous sites along both sides of the river. They range from ceremonial centers to isolated house foundations. At least 200 metal objects were recovered, primarily from burials. Some archaeologists estimate that metal dates from the Early Postclassic Period (A.D. 900–1200) through the Spanish invasion in the sixteenth century (Maldonado C. 1980), although other specialists argue that metal appeared in the region as early as A.D. 700 (Chadwick 1971: 680, citing J. Lorenzo). Pottery shows links to sites both in Michoacán and in Jalisco and Nayarit further to the north.

Another salvage project, carried out at the lower Balsas site of La Villita, yielded metal objects and possible evidence for metal production. Archaeologists identified residential as well as ceremonial and public structures there. The metal objects generally date to the Postclassic Period (A.D. 900–1521), but some appear in the early phase (A.D. 900–1200) and continue to the end of the Late Postclassic (Cabrera C. 1976, 1986; Litvak K. 1968). Amorphous fragments of metal adhere to one artifact identified as a crucible, leading Cabrera (1986) to think that metal may have been melted or smelted at La Villita. The cultural affiliations of these Balsas peoples is difficult to determine.

Metal was virtually absent elsewhere in Mesoamerica during the period before about A.D. 1000. Metal artifacts occasionally are recovered at sites on the southeastern periphery (see chapter 4). They are related to the vigorous casting technology of lower Central America and Colombia and may represent imports, or the beginnings of the local southeast Maya metalworking tradition. However, it is surprising that metal has not been recovered at other coastal locations in the west, even those like San Blas (Mountjoy 1970), which lies in close proximity to Amapa.

Inland, at the higher elevations of the lake region of Jalisco, the earliest metal comes from two sites on Lago Chapala, Tizapán el Alto (Meighan and Foote 1968) and Cojumatlán (Lister 1949). At both, metal was associated with Aztatlán ceramics. These centers were occupied between A.D. 800 and A.D. 1100. Excavations were not extensive at either. Tizapán is large; it consists of one square kilometer covered with mounds. Not all mounds were occupied simultaneously. Some were probably rubbish dumps, others burial sites. Still others were pyramids or temple platforms, but show no indication of the cardinal orientation that typifies Central Mexican structures. Tizapán produced 12 metal artifacts, one of which, described as a "slug," may constitute raw material for subsequent elaboration. Cojumatlán has no surface architecture and consists simply of sherd scatters; 13 metal objects were recovered. Archaeologists excavated all metal objects at both sites from burials or cemetery areas. The earliest securely dated metal appears at both sites around A.D. 900, although at Tizapán excavators think that at least one metal item could date to before A.D. 800.

Apart from these sites, few metal artifacts have been recovered in the West Mexican metalworking zone dating to Period 1. Nonetheless, the technical studies outlined in the description of the technological chronology indicate that many of the objects in museum collections—needles, awls, bells, and tweezers made from unalloyed copper—date to this period.

THE METALLURGICAL TECHNOLOGY OF PERIOD 1

The objects recovered at Tomatlán and Cerro de Huistle mark the beginning of metallurgy in this metalworking zone. Copper was the primary metal used for Period 1 objects. Although other native metals and ore minerals were present in the region (see chapter 2), metalworkers did not develop the alloy systems (except possibly a low-arsenic copper-arsenic alloy represented by a few objects found at Tomatlán and at Peñitas) until a few hundred years later.[5] Such innovations resulted, at least in part, from technical know-how gained from the Andean area (see chapter 6).

The laboratory data indicate that during this time West Mexican metalsmiths used native copper but also obtained copper metal by smelting. Chemical analyses (table 3.3) of the three small cast bells excavated at Cerro de Huistle show them to be made from Group IA materials (see chapter 2): either from native copper or from very pure copper oxides and carbonates, such as malachite or azurite.

Table 3.3 Quantitative Chemical Analyses of Bells from Cerro de Huistle

ID No.	Composition (weight percent)								
	Ag	As	Au	Fe	In	Ni	Pb	Sb	Sn
J-1A	0.02	—	—	0.008	—	0.004	—	0.01	—
J-1B	0.07	—	—	0.01	—	0.005	0.005	0.02	—
J-1C	0.07	—	—	0.07	—	0.004	0.007	0.02	—

Microstructural studies were not helpful in determining whether these particular bells were fashioned using native copper or smelted metal, because when heated for casting the metal recrystallizes. Metalworkers sometimes did use native copper, however. Certain cold-worked RMG objects reveal the characteristic microstructure of unannealed copper in its native state (Hosler 1986). Smiths also smelted chalcopyrite ores during this period. Copper objects from Amapa (Meighan 1976) and Tomatlán (Mountjoy and Torres M. 1985) contain tin in trace amounts, a characteristic of chalcopyrite ores. Some Tomatlán artifacts also contain arsenic, indicating that the objects were crafted from very low-arsenic copper-arsenic alloy metal. The presence of arsenic even in low concentrations suggests that Period 1 metalworkers were using some of the smelting regimes described in chapter 2 to produce the alloy. Period 1 artisans very occasionally worked with silver (or a copper-silver alloy) and gold as well; a few silvery artifacts are reported from the Infiernillo burials. Researchers identified gold flakes at Amapa and Tomatlán.

Period 1 metalsmiths shaped metal in two very different ways, casting bells using the lost-wax technique but cold-hammering other metal objects—including needles, tweezers, rings, awls, and axes—from an original cast blank. This metalworking zone was the only region in the Americas where artisans exhibited a sustained commitment to handling metal by these physically and conceptually very different methods: one in which they shaped metal in the liquid state, the other in which they formed it as a solid. Artisans fashioned ritual objects using both methods; tools were formed by hammering. However, by Period 2, metalworkers employed a very wide and unusual range of metals and alloys for both object classes.

Lost-Wax Casting: Bells

Bells were to become a distinctive, imaginative, and consuming interest of the West Mexican metalworkers. The earliest lost-wax cast bell yet recovered appears at Cerro de Huistle (figure 3.2). Lost-wax casting was introduced to West Mexico from lower Central America and Colombia (chapter 4); some Period 1 West Mexican bells are replicas of formal designs that appear in those more southerly regions. At Amapa, metal bells outnumber the next most common metal artifact type by more than two to one. Bells are most frequently found in burials, worn by the deceased in bunches around the ankles, wrists, or neck. One Amapa burial contains 33 bells associated with a single individual. In fact, bells make up 101 of the 205 metal objects excavated at that site (Pendergast 1962a), and three-fourths of these come from burial contexts. Bells have also been recovered in burials from Cojumatlán, the Infiernillo sites, and La Villita. One group was recovered from a looted burial at Peñitas.

Virtually all West Mexican and Mesoamerican bells I have examined are designed in the same way. They are small, usually measuring 1–8 cm in height, although very

3.2

Bell from Cerro de Huistle, from before A.D. 750.

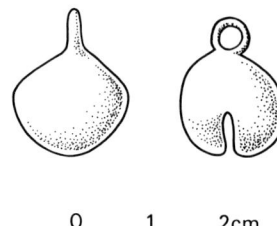

occasionally they are much larger. They range from round to oval to cylindrical in shape. All have a narrow, elongated opening (or slit) at the base and are suspended by a loop at the top. Most contain loose clappers, usually made from pebbles but sometimes from ceramic or metal. The bells sound when the clapper strikes the interior wall of the resonating chamber, or when the bells, worn attached to ankle or wrist bands or sewn onto clothing, strike one another.

All Mesoamerican bells were cast in one piece using the lost-wax technique, at least according to the extensive evidence on hand. Fray Bernardino de Sahagún in the Florentine Codex (Sahagún 1950–1982 book 9: 73) has described the lost-wax casting process as practiced in sixteenth-century Mexico. Sahagún reports that the artisan first ground charcoal to powder, which he mixed with a little potter's clay, kneading it so that it became a cohesive mass. After it had dried for two days in the sun it was sculpted and carved into precisely the shape of the desired metal object, including all of the decorative detail. Then beeswax was mixed with copal, so the mixture would harden; this was strained and rolled out very thin. The carved model was then covered with this beeswax mixture, and powdered charcoal was placed in a thin layer over it. The wax-covered model was then enveloped in a coarse mixture of charcoal and clay, and set out to dry for another two days. During this time a tube or pouring sprue was attached to it. Then the model was placed in a brazier or crucible to melt out the wax and the molten metal was poured through the sprue. After the metal solidified, the charcoal and clay mold was broken away. In the case of hollow objects, like bells, the interior clay core was then dug out using a sharp tool.

The microstructural evidence from an RMG collection bell, identical to types excavated at Amapa and other Period 1 sites, demonstrates certain aspects of the casting

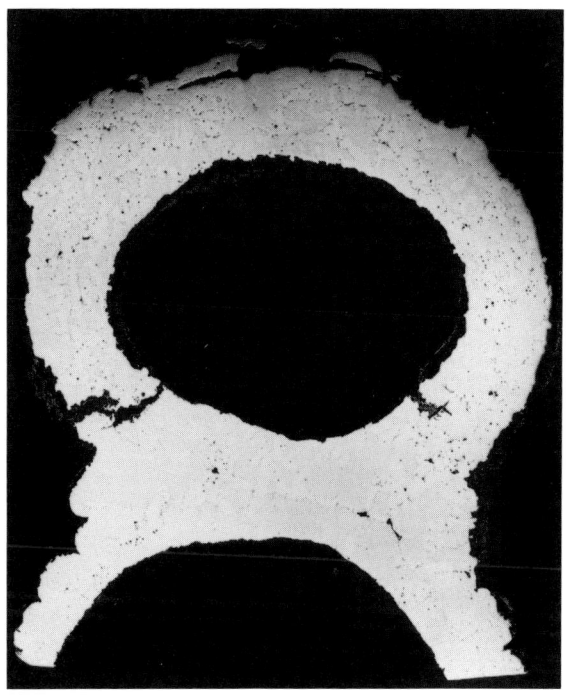

3.3

As-polished section from suspension ring of an RMG Period 1 copper bell (type 11b) (mag.: 12.5).

technique. The bell belongs to Type 11b, illustrated in figure 3.5. Figure 3.3 shows a polished section of the suspension ring and top of the bell resonator. A break appears at the base of the suspension ring. Figure 3.4 illustrates the microstructure typical of the entire section.

The metal exhibits the equiaxed, undeformed grains that characterize a casting. This bell was cast in one piece. The microstructure is continuous, with no join, fissures, porosity, or other indication that the ring and resonator were made individually, then joined. This particular bell contains oxygen in fair concentrations, as the copper oxide eutectic reveals, but the casting procedures were evidently controlled to minimize the development of

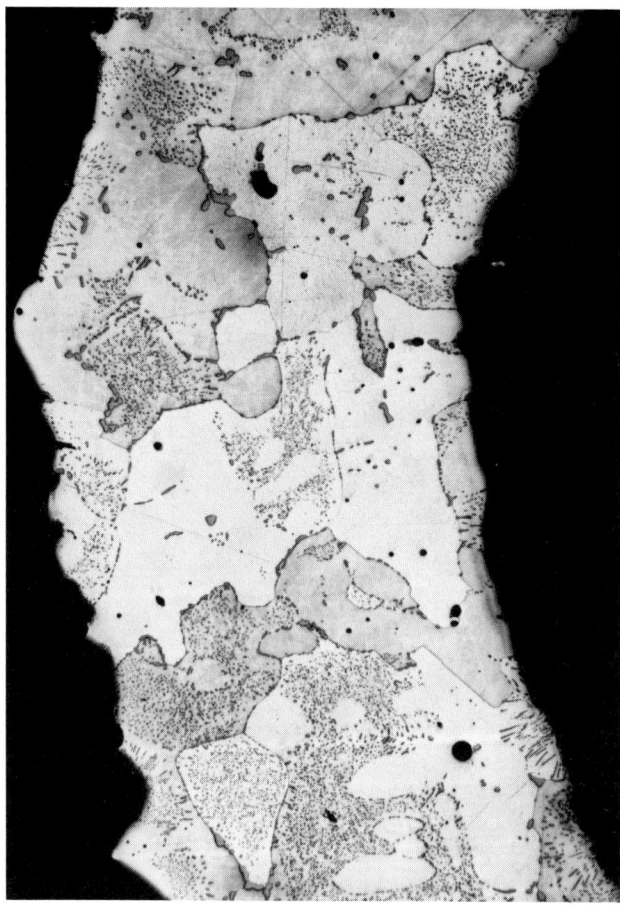

3.4

Longitudinal section from suspension ring of RMG type 11b bell. Note equiaxed grains characteristic of as-cast structure and considerable copper oxide as eutectic. Sample etched in potassium dichromate (mag.: 50).

bubbles and macropores in the solid metal. They can develop as the liquid copper dissolves oxygen. Metallographic studies carried out on 26 bells, examples from the 11 types (formal designs) identified in the RMG collection, show that West Mexican bells were invariably made by casting them in one piece, including those bells with elaborate decorations that initially appear to be appliquéd.

During Period 1, smiths cast at least seven different bell varieties in five formal designs (figure 3.5). I defined the types (formal designs) on the basis of design attributes, including bell height and shape (round, oval, globular/conical, etc.) (Hosler 1986). Subcategories (a, b, c, etc.) reflect the presence of distinctive external decorative characteristics.

Table 3.4 lists bell dimensions, the numbers of bells sampled for analysis, and the overall chemical composition of the RMG collection bells. Those made from copper are datable to Period 1. The table also lists archaeological sites where bells of these types were recovered and the gross chemical compositions of the excavated specimens in cases where analyses were performed. As the table shows, all analyzed archaeological site examples found in Period 1 contexts are made from copper. All RMG examples sampled from these seven designs, except for some of type 1a, are also made from copper. All types but one were sampled for chemical composition; that one type (11a) was recovered at Amapa, and Root's analyses (Meighan 1976: 116–118) showed that these bells, too, were cast from unalloyed copper.

The dimensions of these seven RMG Period 1 bell varieties vary, and the bells are usually plain and relatively thick-walled. All can be cast in copper without difficulty except type 6b, which, when found in the RMG collection, was usually miscast; it was often asymmetrical, for example. Nonetheless, the use of copper exercised real constraints on formal design even in bells where no difficulties in casting were apparent. This becomes clear if we compare the height-to-thickness ratios of bells of equal size from two types: one fashioned during Period 1 and made from copper, the other made from copper-tin bronze during Period 2. The comparative data in table

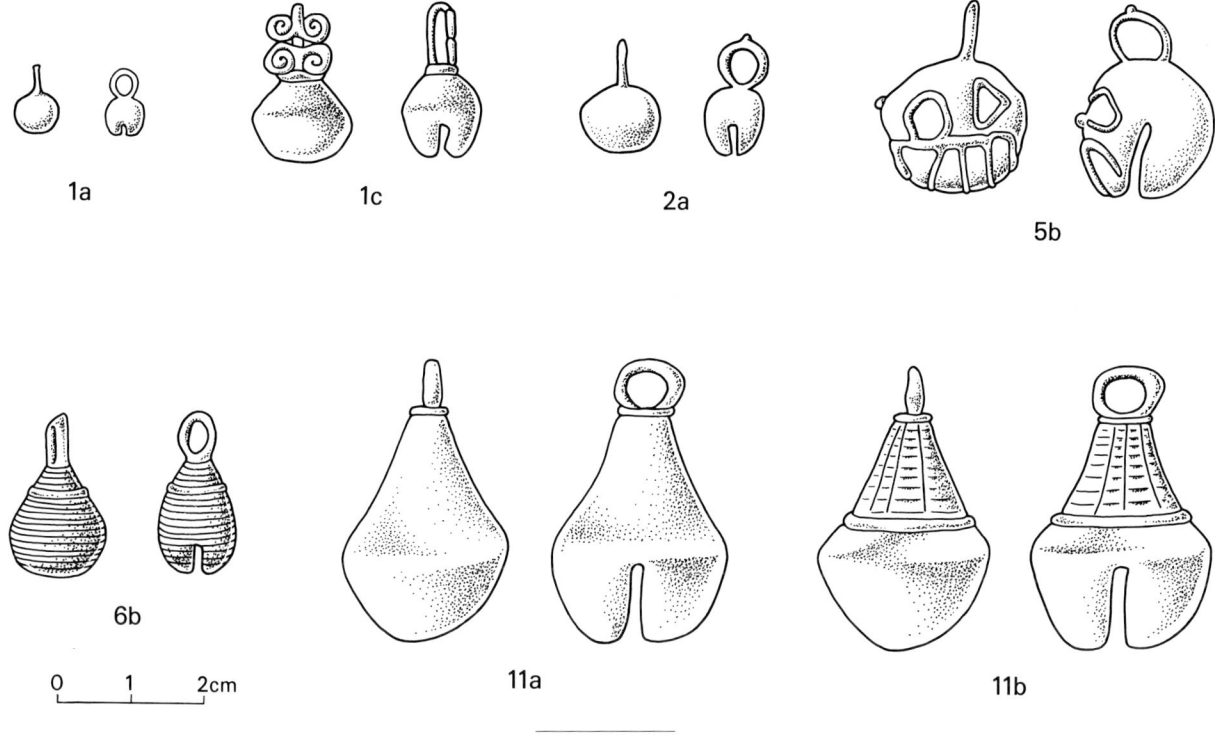

3.5

Period 1 bell types identified in RMG collection and present in datable archaeological contexts.

3.5, particularly the height-to-thickness ratios, demonstrate that smiths could cast bronze bells with far thinner walls than when casting in copper due to the increased strength of the bronze alloy.

Two properties of copper explain the relatively thick-walled, plain designs of Period 1 bells. Copper lacks the mechanical strength of its alloys, so that metalsmiths could not cast the large, thin-walled designs that they later achieved in bronze. Second, copper solidifies at a single temperature. When the liquid metal is poured into a mold, such as the one Sahagún describes, it begins to cool; when it reaches the freezing point of copper the entire melt solidifies. The liquid metal near the mold wall solidifies sooner than the liquid at the center of the mold, but overall solidification time is extremely short. Casting plain, thick-walled bells in copper is therefore technically quite feasible, because in a mold with ample free passages the liquid metal can run in and fill the entire mold cavity before it freezes. However, this same characteristic of copper, indeed of any pure metal, makes the casting of extremely intricate designs, such as the bell in figure 3.6, particularly difficult, because the liquid metal does not have time to enter and fill in the detail before freezing. Narrow or constricted channels in a mold like those used for thin-walled castings become clogged, preventing the flow of liquid through the mold.

Table 3.4 Period 1 RMG Bell Types (Lost-Wax-Cast): Dimensions, Composition, Number Analyzed in RMG Collection, and Archaeological Sites of Appearance

Datable Specimens from Archaeological Sites		Specimens from RMG Collection				
RMG Type	Site	Average Wall Thickness (cm)	Average Height (cm)	Number Copper	Number Analyzed	Number in Collection
1a	Amapa (Cu)	0.05	0.61	5	8	894
	Cerro de Huistle (Cu)					
	Cojumatlán					
	Infiernillo					
	Tomatlán					
1c	Cerro de Huistle (Cu)	0.05	0.90	1	1	2
	Cojumatlán					
2a	Amapa (Cu)	0.06	0.90	9	9	136
5b	Amapa (Cu)	0.09	1.60	11	11	33
6b	Amapa (Cu)	0.07	1.40	9	9	79
11a*	Amapa (Cu)		6.50		0	15
11b	Amapa (Cu)	0.13	3.00	6	6	10

Note: (Cu) indicates specimens analyzed from these sites were made from copper.
* RMG examples not analyzed.

Bells cast during Period 1 vary considerably in pitch; tonal variety may have been a priority from the beginning. Pitch is defined as the quality that makes a sound seem high or low on a musical scale. Pitch is determined by the internal volume of the bell resonator cavity; the larger the bell the lower the pitch. The other design variable in bells such as these is the width of the opening at the base of the bell.[6] In West Mexican bells this dimension need not be considered since it increases systematically with bell size. In spite of the constraints on formal design imposed by using copper, smiths could and did vary the size of bells, hence their pitch.

The pitch of bells (or of any instrument) is produced when force is applied to a vibrating system. Once set into vibration the system produces sound waves, the frequency of which determines pitch. Bells vibrate in a particularly complex fashion because of their shape. When the bell is struck by the clapper it vibrates in many modes. A mode is one component of a sound, and the sound of a bell is produced by many modes of vibration and, therefore, at many frequencies. Pitch in bells of the West Mexican design is determined by the frequency of their lowest mode of vibration. Bell metal is a factor in transmitting vibrations; copper is not an ideal material for bells because

Table 3.5 Differences in Formal Design: Copper and Tin Bronze Bells

Type	Material	Number of Specimens	Average Thickness (cm)	Average Height (cm)	Ratio of Height to Thickness
11b	Cu	6	0.13	3.0	23
8a	Cu-Sn	6	0.05	3.2	64

gas bubbles are generated in the liquid metal during casting. If they do not escape, the bubbles become trapped and remain as pores in the solid metal. Experimental evidence has shown that the relative porosity of a metal influences pitch substantially because the pores dampen vibrations. The greater the porosity, the duller-sounding the bell. However, only a few of the copper bells studied metallographically exhibit porosity, suggesting that West Mexican metalsmiths had solved this problem in most cases.

In Western musical systems, instruments are tuned to produce particular pitches; for example, a given string on a violin produces a clearly defined fundamental pitch. By contrast, in ancient Mesoamerica individual instruments, including these bells, generally were not tuned to particular pitches.[7] Musicologists term such bells "untuned." When these bells were sounded together they produced many pitches simultaneously, and the kind of textured sound that resulted is one of the characteristics of the indigenous musical systems of the Americas. The archaeological evidence suggests that during Period 1 this sort of "textured sound" may have been intended and achieved by casting metal bells of different sizes. Bells varying in size and shape, producing different pitches, accompany individual burials both at Amapa and Cojumatlán. Ethnographic accounts from the Huastec region

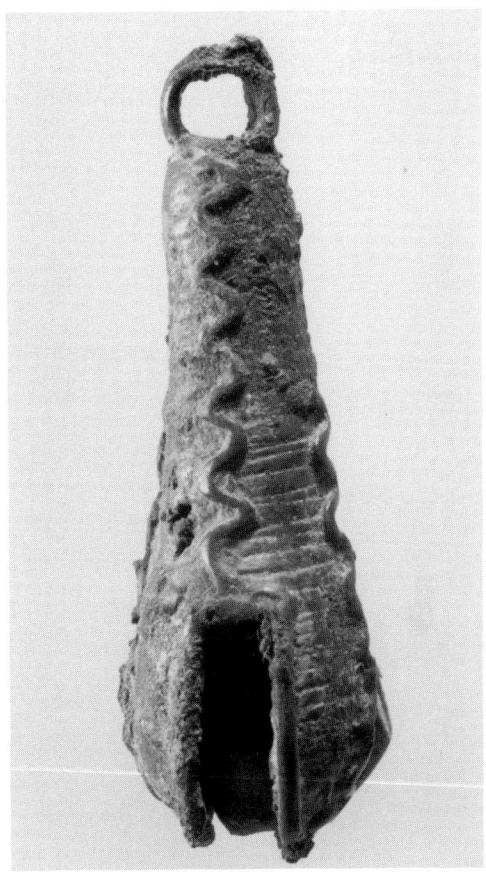

3.6

RMG wirework bell dating to Period 2 and made from a copper-tin bronze alloy. The properties of the bronze alloys are technically necessary to achieve the intricate thin-walled (circa 0.05 cm) casting.

of eastern Mexico relate that dancers in the recent past fastened small bells around their ankles, which produced a high-pitched sound. Large bells, producing lower pitches, were worn around the waist (Hosler and Stresser-Péan 1992) suspended from a belt referred to as a *cinturón-sonaja* (bell-belt) in other areas of Mexico. In the

Huastec region local people collected these ancient bells from archaeological sites.

In spite of the fact that copper is not an optimal material for lost-wax castings, the record shows that during Period 1, bells were crafted in far greater numbers and varieties than any other object class, judging from their relative abundance in excavated assemblages and in collections. They usually appear in burials containing high-status items, suggesting that they marked elevated social position and rank. Bells comprise 1,936 of some 3,200 RMG metal artifacts, and slightly over 1,200 of these belong to Period 1 designs. Although the number of Period 1 bells in the RMG collection cannot be determined precisely because some of these designs continued to be made after A.D. 1200, bells do constitute 59% of the entire collection, and these proportions generally hold in other assemblages.[8]

Table 3.6 Period 1 Cold-Worked Objects: Composition and Number Analyzed in RMG Collection, and Archaeological Sites of Appearance

Object Type	Datable Archaeological Sites	Specimens from RMG Collection			Object Type	Datable Archaeological Sites	Specimens from RMG Collection		
		Number Copper	Total Analyzed	Total in Collection			Number Copper	Total Analyzed	Total in Collection
Rings, round cross section	Amapa (Cu) Infiernillo Tomatlán (Cu, Cu-As)	16	45	499	Awls, unipointed	Amapa Infiernillo Tomatlán	3	4	6
rectangular cross section	Infiernillo Tomatlán (Cu, Cu-As)	6	23	188	bipointed	Amapa (Cu) Tomatlán	6	7	9
Tweezers, beam	Amapa (Cu) Tomatlán	10	10	10	Fishhooks	Amapa (Cu) Infiernillo Tizapán el Alto	2	3	14
Axes	Peñitas (Cu-As) Tomatlán	17	35	38					
Needles, perforated-eye	Amapa (Cu) Coamiles** Infiernillo Tizapán el Alto Tomatlán	13	13	87*					

Note: (Cu) indicates specimens analyzed from these sites were made from copper. (Cu-As) indicates specimens analyzed from these sites were made from low-arsenic copper-arsenic alloy.
*Only 54 of the RMG needles could be identified by eye type.
**Eye type unknown.

Cold Working: Sumptuary and Ritual Objects

During Period 1, West Mexican metalsmiths also fashioned objects by cold-working them from a cast blank. This approach to shaping metal and a circumscribed array of specific artifact types were introduced from the metallurgies of southern Ecuador and Peru, as documented in chapter 4. Although metalworkers did fashion tools, most worked objects were ritual and status items. Table 3.6 lists the principal artifact classes made by cold work during Period 1, their relative numbers in the RMG collection, and how many of the artifacts sampled were made from copper. The table also lists the archaeological sites where such artifacts were excavated and the metal or alloy type when chemical analyses were performed.

Rings. Small open rings worn as earrings, hair ornaments, or hair braid holders were the most common Period 1 metal objects made by cold work. Their diameters range from about 1.2 to 4.0 cm. Open rings were found at Tomatlán in household contexts, but also in burials, near the base of the skull (Mountjoy and Torres M. 1985). Groups of rings also appear in burials at the Infiernillo sites. A few occur at Amapa. At Tomatlán, these rings, either linked together as chains or occurring singly, make up 175 of the 192 objects in the metal assemblage. Tomatlán and Infiernillo burials contain two varieties of rings, some with round cross sections, others with cross sections that are rectangular. In fact, rings comprise 400 of the approximately 450 artifacts in those two assemblages. Maldonado (1980) thinks that some were worn as hair ornaments, because they are found on the cranium of the skeleton. A composite object in the RMG collections consisting of multiple adjacent metal

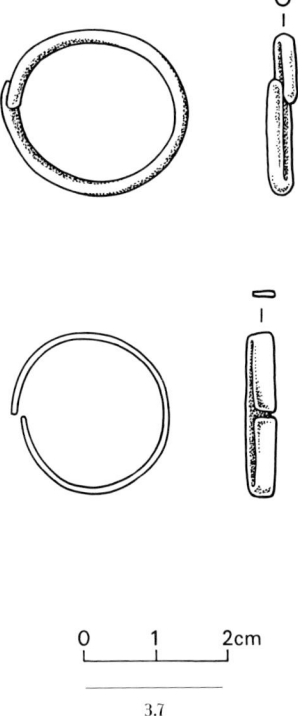

3.7

RMG open rings. Two types have been identified dating to Period 1: one with a round cross section (thickness varies between 0.1 and 0.2 cm), the other with a rectangular cross section forming a band that varies in width from 0.1 to 0.7 cm.

loops encircling a fabric braid may be such an ornament; the object could have served as a kind of hair band.

The *Relación de Michoacán*, a sixteenth-century illustrated manuscript that deals with the Tarascan empire at the time of the Spanish invasion, repeatedly refers to an object, called a *guirnalda* (wreath), made of woven fibers and worn on the head. The *Relación* makes clear that one entire guild of workers devoted itself to making these objects, and frequently depicts individuals wearing them such as the priest illustrated in figure 3.16.

Open rings are also common in the RMG collection, making up 21.4% of the total or some 687 objects (see

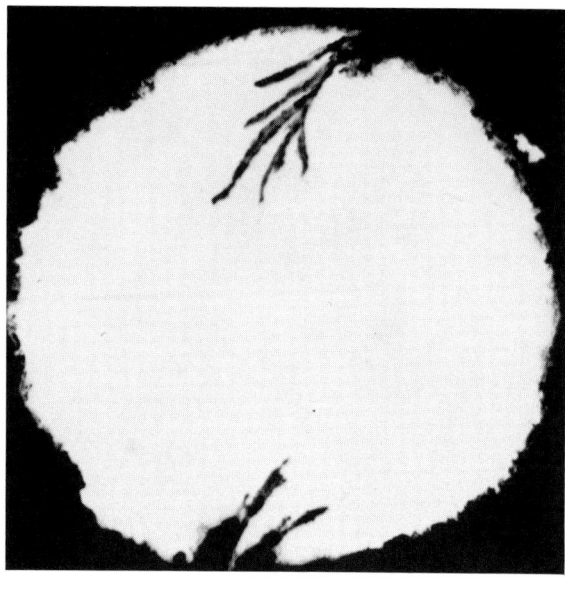

3.8

As-polished cross section of RMG round cross section ring (mag.: 25). Note fissures formed in folding and hammering metal.

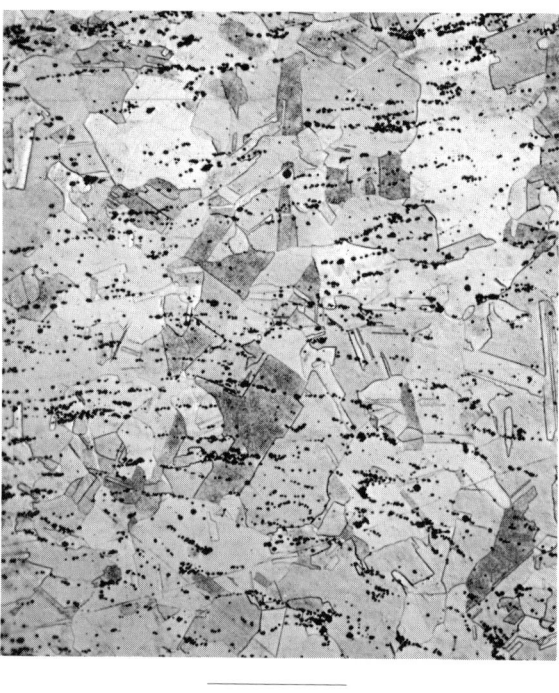

3.9

Microstructure of a longitudinal section through an RMG round cross section ring; shows annealing twins and highly oriented and elongated cuprous oxide inclusions. Sample etched in potassium dichromate (mag.: 200).

appendix 1 and table 3.6). Some proportion of these were made during Period 2, but they clearly were common earlier.

These rings vary in diameter as well as in the shape of their cross sections. Figure 3.7 illustrates the two cross-sectional types. The metallographic evidence from the RMG collection rings shows that those with round cross sections were shaped from an original cast rod that was probably square. The surface fissures visible in the cross section (figure 3.8) were introduced as the square rod was hammered round. That rod was cold-hammered and then annealed through several sequences as the metal was shaped to its desired length and thickness. The photomicrograph in figure 3.9 shows highly oriented, elongated inclusions from the initial cold work, equiaxed grains, and annealing twins. The annealing twins provide evidence that the metal was heated after the final working, and the small grain size indicates that several working/annealing sequences occurred. At some point during the shaping process the elongated and flattened rod was folded along its longitudinal axis to thicken the metal, creating an internal fissure (figure 3.10). The orientation of the inclusions circumscribing the central fissure reflects the movement of the metal that resulted from such directional hammering.

Metalsmiths also fashioned rings with rectangular cross sections, but the fabrication sequence differed. Rectangular cross section rings lack internal fissures, as the photomicrograph in figure 3.11 makes clear. They were

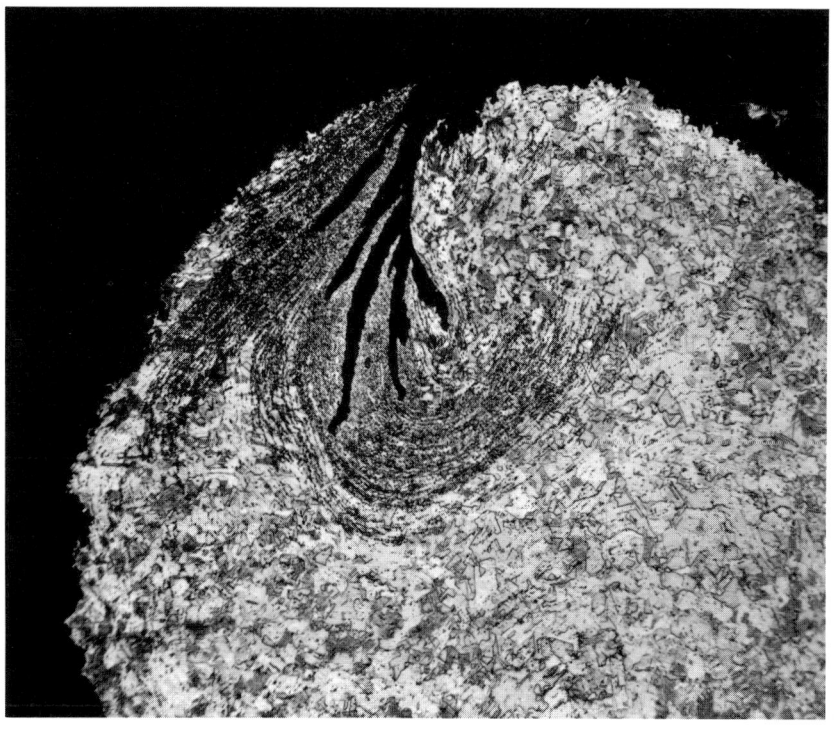

3.10

Microstructure of the ring illustrated in figure 3.8, showing central fissure. Note the equiaxed grains, annealing twins, and highly oriented inclusions, strung out around the fissure in the direction of working of the metal. Sample etched in potassium dichromate (mag.: 50).

made from an original cast blank that was rectangular in section. These objects were simply elongated and bent through cold work and annealing; they were left in the annealed condition. The photomicrograph in figure 3.12 shows large equiaxed grains and annealing twins, reflecting the final anneal.

Nearly all Period 1 rings are made of copper. Some, from Tomatlán, contain arsenic at levels above 0.50% (Mountjoy and Torres M. 1985) and hence are copper-arsenic alloys. The RMG data and these data from Tomatlán suggest that for the same diameters, metalworkers made thicker rings of both designs using copper and thinner rings using copper alloys. Table 3.7 shows dimensions and compositions of Tomatlán and RMG rectangular cross section rings. I include the copper-tin bronze rings for the sake of comparison; copper-tin alloys were not used widely until after A.D. 1200. As the absolute values of dimensions and their ratios show, the bronze alloys permitted thinner bands. The ratio of diameter to thickness for RMG copper rectangular cross section rings

3.11

As-polished cross section through an RMG rectangular cross section ring (mag.: 20). Note absence of fissure.

3.12

Microstructure of cross section through rectangular cross section ring, showing equiaxed grains and annealing twins. Sample etched in potassium dichromate (mag.: 100).

Table 3.7 Dimensions and Materials of Rings

ID No.	Material	Diameter (cm)	Band Width (cm)	Band Thickness (cm)	Ratio of Diameter to Thickness
Rectangular Cross Section Rings from RMG Collection					
36a	Cu	3.0	0.50	0.15	20
36b	Cu	3.0	0.40	0.13	23
635	Cu-Sn (12.44% Sn)	2.6	0.70	0.07	37
620	Cu-Sn (8.50% Sn)	2.5	0.60	0.05	50
Rectangular Cross Section Rings from Tomatlán					
98	not analyzed	2.8	0.30	0.10	28
102	not analyzed	2.9	0.25	0.10	29
103	not analyzed	2.9	0.30	0.10	29
5	Cu-As	2.7	0.40	0.05	54
Selected Tomatlán Round Cross Section Rings					
9	Cu	2.6		0.15	17
151	Cu	3.9		0.13	30
3	Cu-As	2.9		0.05	58
8	Cu-As	3.1		0.05	62

clusters around 17 (Hosler 1986); the ratio is considerably higher for the rings made from alloys. Copper constrains formal design options. The rings from Tomatlán may demonstrate the same pattern, although we do not have analyses for all of them. When made from alloys, these rings were made thin. Similar patterns occur in the Tomatlán rings made with a round cross section, as table 3.7 suggests. The data from Tomatlán are particularly important because they indicate that, even during Period 1, people were already managing a low-arsenic copper-arsenic alloy, and, at least in some instances, the alloy was selectively used for thinner formal designs.

West Mexican smiths also made other objects from copper during Period 1, although they made them in far smaller numbers than either bells or rings. All were cold-worked to shape. They include tweezers, axes, awls, sewing needles, and fishhooks. People used tweezers and axes as tools, but archaeological and historical evidence indicates that both items also served as symbols of sacred and secular power.

Tweezers. Facial and bodily depilation was and is a widespread practice among indigenous Americans; depilatory tools have been made in a variety of configurations and from various materials, including metal, wood, bone, and

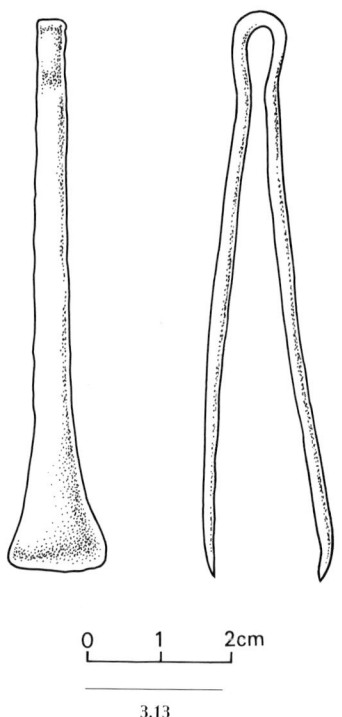

3.13

West Mexican beam tweezer. Front and profile view. In profile view in some specimens, the blade begins to angle in toward the tip at approximately the midpoint (see tweezers illustrated in Meighan 1976, for example).

conch shells. Contemporary highland Andean men sometimes use two separate coins to remove facial hair; at other times they make tweezers from sheet metal or acquire commercial tweezers. During Period 1, West Mexican metalsmiths fashioned two tweezer designs, variants of a design that mechanical engineers describe as a beam.

West Mexican beam tweezers consist of two symmetrical blades joined by a hinge, fashioned from a continuous piece of metal (figure 3.13). The RMG collection contains ten, ranging from 4.5 to 7.8 cm in length. All are made from copper. They are crafted so that the flat sheet metal of the tweezer, both the blade portion and the hinge portion, changes in one plane only—across the length of the tweezer, not across its width. In profile view, the blades have a kinklike indentation at the top of the hinge; the blades splay out and then angle inward, sometimes at the midpoint and at other times close to the tip (figure 3.13). In the second design the blades of beam tweezers are parallel to one another. Apart from overall length, the tweezers also vary in the other dimensions crucial to tweezer performance: hinge width, hinge thickness, and tip opening.

A fragmentary beam design tweezer dating to Period 1 was excavated at Tomatlán, along with two non-Mesoamerican tweezer types made from bent wire. Three tweezers recovered at Amapa are of the beam design; all three are made from copper (table 3.6). The wire tweezers from Tomatlán and one specimen from the Barra de Navidad site in Jalisco (Long and Wire 1966) do not resemble any examples excavated in Mesoamerica during this time or subsequently. All three have analogues in South America.

Metallographic studies showed that all RMG beam tweezers were fashioned by cold-working them to shape. In two cases, the original blank was made from native copper; in the other specimens, the metal was smelted from a copper ore. Some tweezers were left in the cold-worked condition, others were left annealed. A photomicrograph of a longitudinal section from the blade of one of the RMG collection tweezers (figure 3.14) shows a microstructure typical of cold-worked native copper. Grains of extremely irregular shapes and sizes are interspersed throughout the entire section. Overall, the microstructure is elongated from the cold work the metal underwent during shaping, and some extremely elongated annealing twins are visible. These twins are typical of native copper. They are believed to form through

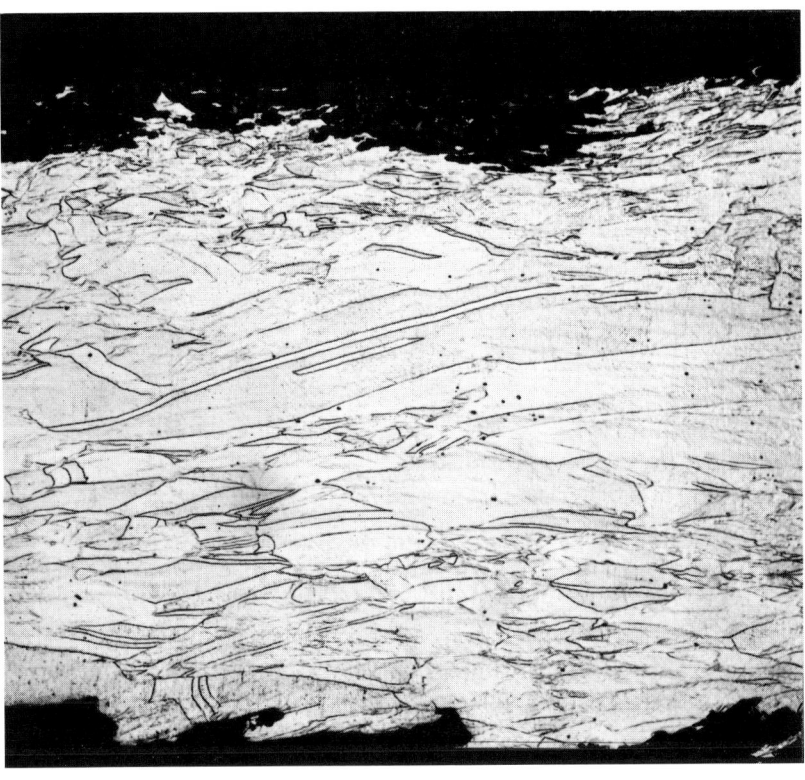

3.14

Longitudinal section through blade of a Period 1 RMG beam tweezer. The structure is typical of worked native copper. Sample etched in potassium dichromate (mag.: 200).

heating (annealing) that results from rock movement occurring at the geological rock fault where the native metal was emplaced.

This particular tweezer was shaped by hammering out the entire form from a lump of native copper, represented by (1) in figure 3.15. The resulting flat form (2) approximated the final length and width of the hinge, including the sharp angle necessary to form the kink, and described the splayed, open blades. Microhardness readings taken on the sampled section indicate that the original metal was reduced at least 50% in thickness during these operations. Next, the metal was placed over a model of wood or some forming device in the shape of the hinge, and bent double along the midplane. (3) At this stage, the hinge was cold-worked to its final shape. The last step (4) aligned the two blade tips; excess metal along the edges was cut or abraded away in regions where final truing up was necessary.

Sixteenth-century documents and other data (see chapter 8) make clear that tweezers were used to remove beard and other facial hairs, but were also a kind of sacred tool worn by priests, other religious functionaries, and perhaps leaders. The Amapa tweezers were recovered from burials, as were tweezers dating to Period 2; the Period 2 tweezers often occur in contexts showing that they were worn suspended around the neck. In fact, one such tweezer type, a design worn by Tarascan priests, became a symbol of office (see figure 3.16 and chapters 5 and 8). West Mexican tweezers are often so large and elaborate as to raise the question of whether they were intended as functional tools. I address that question in the following section.

Determining Tweezer Function. The most direct way to establish the functionality of the West Mexican tweezers would be to subject them to a series of straightforward laboratory tests that measured their mechanical properties and response in use. However, the tweezers cannot withstand mechanical testing because they exhibit extensive surface and intergranular corrosion and the metal has become brittle. Such tests would yield spurious results even if they were carried out without causing brittle frac-

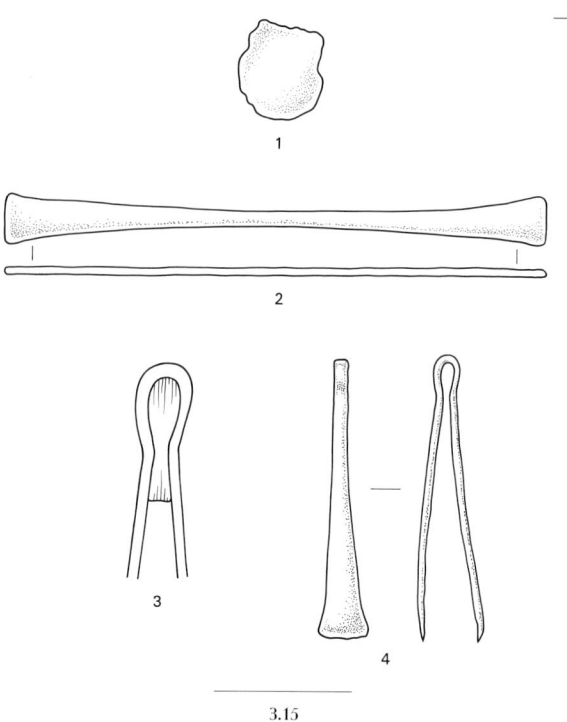

3.15 Reconstruction of the possible fabrication sequence of a beam design tweezer: (1) lump of copper hammered flat, to approximate the tweezer form (2) in its planar configuration; (3) flat form bent over model to complete the hinge detail; (4) final alignment and truing of the blades.

Table 3.8 Beam Design Tweezers: Attributes Critical to Performance

ID No.	Material	Length (cm)	Hinge Thickness (cm)	Hinge Width (cm)	Tip Opening (cm)	Final Fabrication Phase
5	Cu	6.2	0.2	0.50	0.50	annealed
6	Cu	7.8	0.1	0.60	—	annealed
224	Cu	6.9	0.2	0.40	0.05	annealed
225	Cu	7.0	0.1	0.65	—	annealed
2345	Cu	4.5	0.1	0.30	0.38	cold-worked
2346	Cu	7.5	0.1	0.48	1.00	cold-worked
2678	Cu	7.7	0.1	0.40	0.38	cold-worked
2679	Cu	7.0	0.1	0.55	0.18	cold-worked
2686	Cu	5.0	0.2	0.35	0.20	—

3.16

Scene from the Chronicles of Michoacán illustrating "the general administration of justice" (Craine and Reindorp 1970, plate 19). The Tarascan priest wears a large tweezer with a spiral emerging from opposite sides of each blade.

ture. Another means of establishing whether or not these implements were used is to examine their microstructure. Since the action required to pluck beard or other hairs does not generate sufficient force to deform the metallic microstructure, standard metallographic interpretation could not establish evidence of use. A particularly fruitful approach in cases such as these is to examine the functionality of the tweezers through computer simulation methods. Through simulation it was possible to determine whether these particular tweezers, given their chemical compositions, dimensions, and fabrication methods (see table 3.8), possessed the requisite physical and mechanical properties and design attributes to have performed as tweezers.

The method used to evaluate tweezer functionality is known as finite element analysis. This is a technique commonly used in the fields of materials and mechanical engineering to simulate the mechanical behavior of solid structures (Reddy 1984). The method is based on a body of experimentally derived theories dealing with the behavior of materials when subjected to stress. The usability of the tweezers was evaluated by mathematically modeling their designs, then subjecting those designs through simulation to the stresses a tweezer undergoes as it opens and closes and a hair is plucked. The results revealed whether the particular tweezer was functional by indicating whether those stresses exceeded limits the tweezer material (metal or alloy) can support.

Finite element analysis is also useful in studies of archaeological artifacts because critical parameters can be changed (for example, dimensions, composition, manufacturing technique). By altering key tweezer parameters and observing model behavior under new conditions—for example, the use of a 2% tin bronze versus copper, or a hinge thickness of 0.10 versus 0.05 cm—it is possible to distinguish attributes that are mechanically necessary for tweezer function—technical requirements—from others that are superfluous, hence express aspects of the technology that may be determined by other factors.

I used finite element analysis to ascertain whether these particular tweezers were capable of closing around a facial hair and plucking it, given their manufacturing technique, design characteristics, and chemical composition. I also used it to identify the ranges within which these tweezer attributes (dimensions, ratios of dimensions, chemical composition, manufacturing technique) could vary and still allow functional success.

The first step was to construct a mathematical model of a given tweezer, known as a finite element mesh, based on measurements taken from the actual object. The model of the tweezer was then subjected, through simulation, to loadings (stresses) equivalent to those required to close a tweezer of its dimensions (blade length, hinge width, etc.) and material composition, and to pluck a facial hair. As a consequence of the loadings, or stresses, the simulated tweezer model deforms as the real tweezer would during use (figure 3.17). The resulting strains generated in the tweezer are plotted as stress distributions throughout the tweezer model. If these stresses exceed the limits the material can support, the tweezer fails in use.

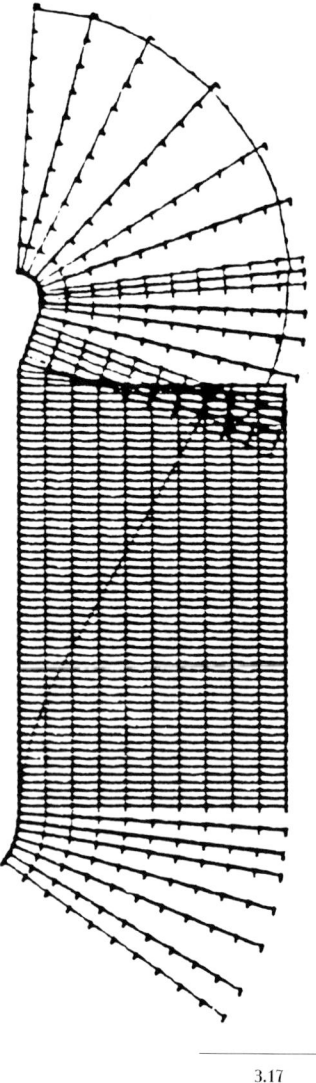

3.17

Finite element model of a beam tweezer. This deformed mesh plot presents a view of the tweezer model as it deforms in use, in response to the force applied to operate the tweezer.

What are the design attributes and mechanical properties of a serviceable tweezer? When a tweezer is closed and a hair is pulled, maximum stress is felt at the tweezer hinge. The magnitude of that stress (and the ultimate success or failure of the tweezer) is a function of the specific values of certain design variables (tweezer dimensions) and one material property variable, the elastic modulus of the metal. Critical tweezer dimensions are blade length, hinge thickness, and the distance between the two tweezer blade tips where contact is made as the hair is plucked. The elastic modulus of the material affects the magnitude of the stress tolerable at the tweezer hinge. Elastic modulus values specify the elastic strain that occurs in a particular metal under a given stress: in this case, as the tweezer is closed. Elastic modulus values are taken from standard tables for copper and its alloys. Stresses that result from simply closing the tweezer are amplified when additional force (load) is applied to pluck the hair. This second set of stresses is a function of tweezer length, the width of the tweezer hinge, and blade thickness.

The equations that generate the model include values for these critical tweezer dimensions and for the elastic modulus of the particular metal or alloy. The solution to the equations for each tweezer, called the "stress excursion," expresses numerically the amount of stress that occurs at the hinge of that implement when tweezing. That value cannot exceed the value for the yield strength of the metal or alloy used to make the tweezer. Yield strength expresses the real limits beyond which failure ensues. If the tweezer exceeds its yield strength, it will undergo plastic deformation during use and will fail. To determine whether a particular tweezer could tweeze, we compare the value for the stress excursion (the stress generated in that tweezer given its dimensions and composition) to the yield strength, which depends on manufacturing technique (whether the tweezer is in the annealed state or has been hardened through cold work). If stress excursion is less than yield strength the tweezer was functional; otherwise the tool was not. Other factors can also enter into the mechanical performance of the tweezer, however. For example, bending the metal in forming the hinge can introduce anisotropic residual stresses in the material, which can increase the elastic range of stresses at the tweezer hinge up to 1.5 times the initial yield strength (see for example table 3.10, artifacts 2345 and 2346).[9]

Another essential property of a metal used to make a tweezer is its "springiness": the range over which the material behaves elastically and does not yield or deform plastically. Springiness is governed by a set of material elastic properties. One is the elasticity of the specific metal or alloy, another is its toughness. Elasticity is the property of matter that allows the tweezer to return to its original shape and dimensions after an applied force (the force required to close the tweezer and pluck a hair) is removed. Toughness is the resistance of the metal to brittle fracture. Toughness is conferred by the inherent characteristics of the metal or alloy but is affected by fabrication technique. Repeated applications of force—opening and closing of the tweezer—can cause brittle fracture through the propagation of short cracks on the tweezer hinge, which, if the metal is not sufficiently tough, will grow until the metal fails and breaks.

How do we determine if these tweezers were sufficiently springy to tweeze? Springiness is measured by the ratio between the yield strength of the metal or alloy and its elastic modulus. This ratio is known as yield strain. The yield strength of many metals, including copper and its alloys, is approximately one-third the hardness of the metal. Here, yield strength was determined by measuring the microhardness of samples removed from 8 of the 10 beam design tweezers in the RMG collection whose con-

Table 3.9 Hardness, Yield Strength, and Yield Strain for Beam Tweezers

ID No.	Hardness (Vickers hardness number, kg/mm^2)	Yield Strength (psi × 10^1)	Yield Strain (kg/mm^2)
5	108	51.1	0.0031
6	91	43.1	0.0026
224	113	53.5	0.0032
225	102	48.3	0.0029
2345	131	62.0	0.0037
2346	124	58.7	0.0035
2678	127	60.1	0.0036
2679	125	59.2	0.0036

dition allowed microhardness determinations (table 3.9). The yield strain values calculated for each tweezer indicate that all were sufficiently springy to have functioned successfully. Ideally, values for a good, springy material vary from 0.01 to 0.005.[10]

The other mechanical properties essential to tweezer function include toughness (in this case, how fast a hinge crack propagates) and fatigue strength (how many times one can tweeze before a crack is initiated). Roughly speaking, these two properties prescribe "endurance limits": the ability of tweezer metal to withstand many cycles of imposed stress and strain as the tweezer is opened and closed before it fails. Both depend on the inherent properties of the metal or alloy and on fabrication methods. The materials used for these tweezers, copper and its alloys, are tough and do not fatigue easily. Yet fabrication methods do influence tweezer success in various ways. Cold working increases the hardness of these tweezer metals or alloys. As table 3.9 shows, the yield strength of a metal increases when the metal is cold-worked. If a tweezer is hot-worked or is left in an annealed state, the metal becomes softer, the yield strength is lowered, and the material usually becomes less springy as well.

Simulation Studies of Beam Tweezers. Finite element models (see figure 3.17) were constructed for individual beam tweezers, and the stresses the real object would undergo if used were determined through computer simulation. Simulation data show that all but one of the tweezers modeled were functional, given their particular design attributes and fabrication techniques. Among these tweezers, dimensions crucial to tweezer function—length, hinge thickness, and hinge width—vary significantly (see table 3.8). Fabrication technique also varies. Four tweezers were left annealed, making the metal rather soft; four others were left cold-worked and are significantly harder.

The simulated performance of each beam tweezer design appears in table 3.10 and can be evaluated by examining the columns for yield strength, maximum stress excursion, and contact force. Yield strength is a constant for a given metal or alloy treated in a particular manner. For example, 207 MPa is the value of yield strength for copper left in the annealed condition, 265 MPa for cold-worked, unalloyed copper, and 377 MPa for a 5% tin bronze left annealed. Contact force is the amount of force required to close a tweezer of given dimensions and to pull a hair. While the amount of force required to pull a hair is essentially a constant,[11] the force required to close a particular tweezer is a function of its design. Stress excursion represents the amount of stress that occurs in a tweezer of given dimensions when it is closed and a hair is pulled.[12]

Table 3.10 shows that stress excursion values for two tweezers exceed yield strength. However, tweezer no. 2345 was operable as a result of residual stresses intro-

Table 3.10 Performance Data: Beam Tweezer Simulation

Type	ID No.	Contact Force (g)	Yield Strength (MPa)	Stress Excursion (MPa)	Years in Service
Parallel blades	224	213	207	80	$> 5 \times 10^5$
Kink hinge	2345	107	265	397	330
Kink hinge	2346	24	265	856	nonfunctional
Kink hinge	2678	110	265	233	1.6×10^5
Kink hinge	2679	124	265	135	$> 5 \times 10^5$

Note: MPa = Megapascals (10^6 Pascals); 1 Pascal = 0.145×10^{-3} psi.
"Years in Service" means years before failure, at 50 cycles/day.

duced in bending the hinge. Tweezer no. 2346 was nonfunctional because the opening at the blade tips was unusually wide; however, this appeared to have occurred after deposition, and these tweezers probably were originally functional.

Parameter Changes. To what extent were the formal design attributes, fabrication techniques, and material composition of these tweezers necessary to their successful function? Which could vary and to what extent could they vary? In other words, what were the choices these smiths were making in formal design and manufacturing process, and how were these choices constrained by the inherent physical and mechanical properties of the material?

All beam tweezers were made from copper. The simulation data show that the design easily could have been executed using alloys and, in fact, that the use of alloys would have allowed greater flexibility in design, for instance in dimensions such as length, width, and thickness. For example, consider the performance of tweezer no. 2679. Tweezer performance was modeled with all dimensional attributes held constant but with chemical composition changed from copper to copper-tin and cop-

Table 3.11 Parameter Change in Beam Tweezer Simulation: Copper Tweezer Made from Bronze Alloys

ID No.	Composition	Fabrication Technique	Yield Strength (MPa)	Stress Excursion (MPa)
2679	Cu	cold-worked	265	135
	5.0% Sn	annealed	377	135
	10.0% Sn	annealed	418	135
	3.5% As	annealed	345	135

per-arsenic bronze. The results are shown in table 3.11. The use of a 5% copper-tin bronze alloy substantially increases the yield strength of the metal, even when left annealed, from 265 MPa, which represents the actual cold-worked copper tweezer, to 377 MPa. This constitutes an increase of about 40%.

Copper, even when cold-worked, is not an optimal material for these tweezers, but it is adequate. The yield strength is far lower than the typical yield strength values for the bronzes, and copper implements of these particular dimensions cannot tolerate equal stresses. When the

Table 3.12 Parameter Change in Beam Tweezer Simulation: Hinge Dimensions of Tweezer No. 2345

	Hinge Width (cm)	Hinge Thickness (cm)	Contact Force (g)	Yield Strength (MPa)	Stress Excursion (MPa)
Original	0.3	0.10	107	265	397
Altered	1.0	0.01	100	265	450

Table 3.13 Beam Tweezer Simulation: Length and Functionality

ID No.	Length (cm)	Hinge Width (cm)	Hinge Thickness (cm)	Tip Opening (cm)	Contact Force (g)	Yield Strength (MPa)	Stress Excursion (MPa)
2345	4.5	0.3	0.1	0.38	107	265	397
2678	7.7	0.4	0.1	0.38	110	265	233

same tweezer is made from bronze alloys, yield strength increases in the annealed condition (table 3.11) and would increase still further if the material were cold-worked, as long as alloy concentration is not so high as to induce brittleness.

Two design variables critical to tweezer function are the width and thickness of the tweezer hinge. Within what ranges could these dimensions have varied without compromising function? I simulated the performance of tweezer no. 2345, made from copper, but with a decrease in hinge thickness and an increase in hinge width. Table 3.12 shows the effect on performance of altering those particular parameters.[13]

This tweezer would have failed had the hinge thickness been decreased and its width increased as specified. Although the case chosen represents an extreme, many similar simulations showed clearly that unless the material maintains a certain minimum thickness and hinge width the tweezer fails in use. At issue is the strength of the material. The simulated performances of copper tweezers nos. 2345 and 2678, which are alike in nearly all dimensions except length, is shown in table 3.13. From their lengths and the values for stress excursion and yield strength, it is evident that when other variables are held nearly constant, beam tweezers' functional capacity decreases as they become shorter. The greater length of tweezer no. 2678 distributes the stresses over a larger area, hence decreasing the stresses developed at the hinge.

Using this method we can determine real limits on formal design (dimensions and ratios of dimensions) for beam tweezers when made from copper. The models show that these tweezers were fully functional but that copper constrains their design as it did in the case of the bells and rings. The real limits lie in ratios of critical dimensions such as length to hinge width and length to thickness. As a result, extremes in one dimension do not necessarily render the object nonfunctional; an extreme value can be compensated for elsewhere in the design.

3.18

RMG axes and other cutting tools.

The beam design tweezers that characterized this initial period of metallurgy in West Mexico thus met two social requirements simultaneously: they were symbols of rank and sacred power—if the Period 2 evidence can be generalized—and also served as fully operational depilatory tools.

Axes. Only a few axes or axelike cutting tools have been found in reasonably secure Period 1 contexts. One has been analyzed chemically. The largest group is a cache from Tomatlán, but we have no chemical analytical data for them.[14] The RMG collection contains 32 axes; 14 of them are made from copper, the others from bronze alloys. The copper axes range in length from 8.0 to 17.4 cm. Since copper was the primary metal used during Period 1, I will assume that at least some of these RMG copper axes were made then and will describe their production technology here. Results from future excavations might possibly alter their temporal placement, but information concerning their fabrication history remains the same. Figure 3.18 illustrates selected RMG axes and cutting tools.

Substantial evidence concerning these tools' use exists in sixteenth-century documents, and I will refer to that literature in the discussion in chapter 5. These sources indicate that axes were indeed used as cutting implements, but that metal axes also marked social rank and sacred and political power.

The metallographic studies of the copper axes provide unambiguous evidence concerning their use. Some were used as cutting tools. Others were not used at all. Six of the 14 specimens show no macroscopic evidence

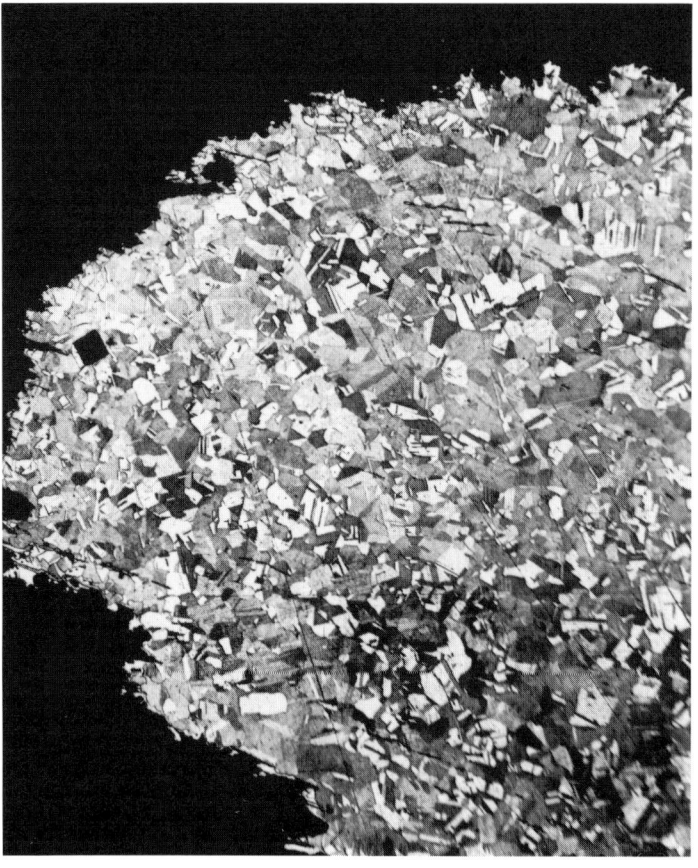

3.19

Longitudinal section through the blade of an RMG copper axe left in the annealed condition. Note equiaxed grains and annealing twins. Sample etched in potassium dichromate (mag.: 200).

for use, and the microstructures of three of these six confirm that observation. Two were left in the annealed condition, one was left cold-worked. The two annealed axes were far too soft to have served as cutting tools. Metallographic studies of a section removed from the blade of one reveals a few elongated inclusions and a fully annealed and recrystallized structure in which none of the grains is deformed (figure 3.19). Hardness tests of this blade section and of a section removed from the butt of the axe gave values ranging from 60 to 80 Vickers Hardness Number (VHN). These values are consistent with standard hardness data for annealed, unalloyed copper.[15] The blade is sufficiently soft even in the one cold-worked example (VHN = 128 maximum) that microstructural and macroscopic evidence would be visible if the axe had been used for cutting. Metallographic studies were carried out on sections removed from the blade and butt end of all axes.

Why were metalsmiths making axelike objects that either were mechanically incapable of use or were never used? In Mesoamerica, the axe form traditionally represented divine authority. The axes described here provide concrete evidence that such "symbolic tools," made from metal, actually existed. These particular examples could represent utilitarian specimens that were recycled because they were unusable for mechanical reasons not apparent in the microstructure (e.g., the presence of an internal fissure or some other casting defect). As flawed tools, they were later annealed and used in ritual or for status ends. These axes also may have been intentionally made as ritual items. They do differ in design from axes showing evidence for use: they are thinner in proportion to length (Hosler 1986). Since copper is not an especially strong material, any usable copper axe needs to maintain a certain minimum thickness. These axes did not do so, either by accident or by design.

Eight of the RMG copper axes (table 3.6) were unquestionably made to be used, and laboratory examination shows use wear. The photomicrograph in figure 3.20, which illustrates a section through the tip of the blade of one of these axes, exhibits somewhat elongated grains with annealing twins in the interior portion of the metal that become highly elongated at the blade edge, where some of the metal exhibits plastic flow and severe distortion. A section farther in from the blade edge exhibits completely equiaxed grains. This microstructure reveals that after the axe was cast roughly to shape it was heavily cold-worked, subsequently annealed, then cold-worked again, a process that left the grains deformed and hardened the metal. At the extreme tip of the blade, the metal folds upward to form the blunted edge that is also visible macroscopically. The hardness data confirm the metallographic data; microhardness values for the blade range between 83 and 128 VHN. The highest values occur at the cutting edge, the lowest values at the center of the sample. These microhardness values demonstrate that this tool possesses the range of properties required: it must be harder at the cutting edge and tough at the center. Toughness is a measure of a metal's resistance to brittle fracture. The axe must be able to absorb impact during use.

All axes in this group were made in the same way, and the microstructural and macroscopic evidence demonstrate unequivocally that they were used. Metal as soft as that used to make these axes invariably shows evidence of deformation if any occurred. Maximum hardness values at the blades measure only 130 VHN. However, even work-hardened copper axes with a cutting edge hardness slightly higher than that—of about 135 VHN—are not hard enough to cut hard woods or to fell trees. Copper is an unlikely material for an axe if the purpose is to cut. It is far more probable that these tools were used for splitting. Wood splitting requires the tool to be sufficiently tough to resist brittle fracture but not especially hard or sharp. Because the metallographic studies showed no deformation of the butt ends of these tools, they probably were hafted in such a way as to prevent it. Sahagún (1950–1982, book 11, plate 371) shows a tool apparently used for log splitting that is hafted like an adze, and some of these copper axes were probably used in this way for splitting logs or branches.

Cold Working: Utilitarian Objects

Needles. Sewing needles comprise one of the most abundant types of Period 1 utilitarian copper objects. Before metallurgy developed, needles had been made

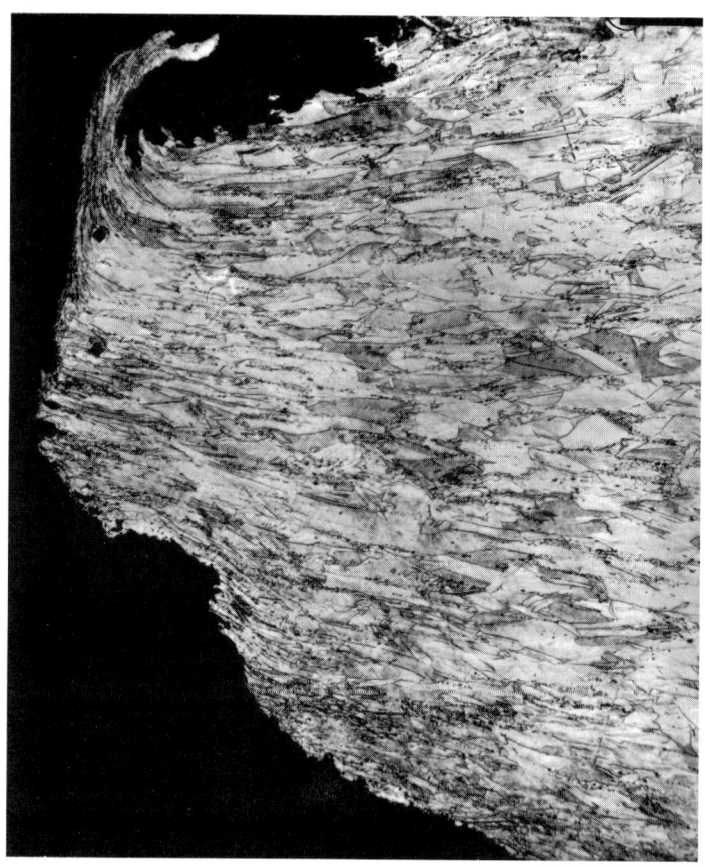

3.20

Longitudinal section through blade tip of a cold-worked copper axe. Grains are highly elongated and compressed from intensive cold work. Sample etched in potassium dichromate followed by ferric chloride (mag.: 50).

from bone (as some continued to be), but, at least in some areas, metal needles began to replace them. At Amapa, for example, apart from bells, metal needles were recovered most often: they made up 32 of a total of 205 objects. Archaeologists reported only four bone needles. Differential preservation may account for such numbers, but the fact that needles frequently were being made from copper is noteworthy. At Tizapán, seven copper needles were excavated and only one from bone.[16] There and at the Infiernillo sites, copper needles occur in burials, suggesting either that metal as such was considered sufficiently important to constitute a burial good, or that the needles represented some activity of the deceased. At Amapa, they were recovered both from cemetery areas and in stratigraphic pits.

Copper's toughness and elasticity make a needle that is far superior to one made from bone. Bone fractures easily; copper is preferable for long needles, curved needles, or needles with extremely thin shafts. The marked variation in length and configuration is a particularly striking characteristic of the Period 1 needles. The copper needles excavated from Amapa, for instance, range from approximately 6 to 19 cm in length and 0.065 to 0.2 cm in thickness and in some cases are curved; an example from Tizapán measures a remarkable 35 cm in length. These differences in formal design suggest a broad range of specialized cloth, netting, or basket production activities. The variety of lengths and thicknesses also occurs in the RMG perforated-eye needles, which range in length from 7 to 22 cm (Hosler 1986).

West Mexican smiths made two distinct needle designs, distinguished by the eye type. One of these was rare in Period 1. The eye of the most common Period 1 type is fashioned in the form of a concave groove, with the eyehole punched or drilled in the center (figure 3.21). These needles have been excavated at Tizapán, Coju-

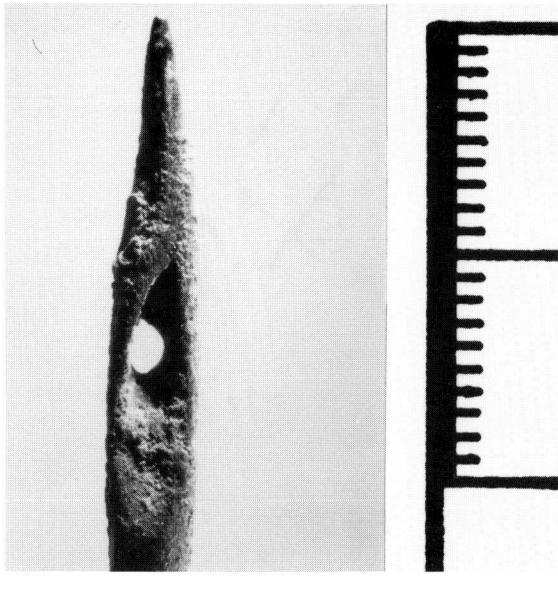

3.21
Period 1 perforated-eye needle: eye.

matlán, Infiernillo, and Amapa. The other type, with a loop eye, may appear at Amapa, although the evidence is equivocal.[17] These two designs differ sufficiently to suggest their use for distinct tasks. The eye end of the perforated-eye needle is narrow and tapers to an extremely sharp point, which could easily pierce fabric as the needle is drawn through. This motion would have been virtually impossible with the loop eye needle because the loop maintains the thickness of the needle shaft. I discuss the loop design in chapter 5, which treats the development of the technology after A.D. 1200; most needles of this particular design were made during that period and from bronze alloys. Regardless of how the loop eye needles were used, the evidence indicates that the aspects of cloth production or weaving that required this needle design were relatively unimportant during Period 1.

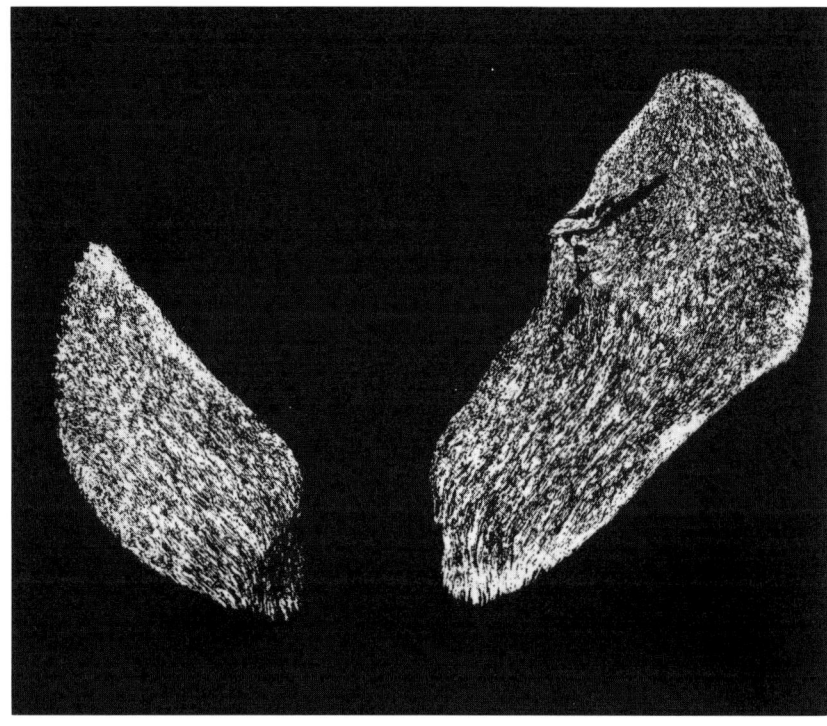

3.22

Section through needle eye; elongated, compressed grains show extreme cold work. Plastic flow and dragging of the metal along both edges of the eyehole indicate that the hole was punched, not drilled. Sample etched in potassium dichromate (mag.: 50).

All RMG needles with a perforated eye were made from copper, as were all needles analyzed from Amapa. The groove, the eyehole, the region above the eye, the top of the groove, and the tapered end of such a needle are visible in figure 3.21. The shaft of the needle is rectangular in cross section at the base of the eye, then becomes gradually rounded toward the tip. The groove for the eye was formed by flattening the eye end somewhat and then cold-working it roughly into a U shape. At the extreme end of the eye, the flattened edges were hammered inward until they met, forming a solid pointed end. The two parallel sides and the eyehole are shown in the section in figure 3.22. The eyehole was punched with a tool like an awl; the photomicrograph showing one side of the eyehole (figure 3.23) reveals the marks of such a tool as it was used to punch through the metal. Microhardness values from the eye section range from 119 to 141 VHN, reflecting the intensive cold work as the needle was shaped. All the needles of this type were fashioned in precisely the same way.

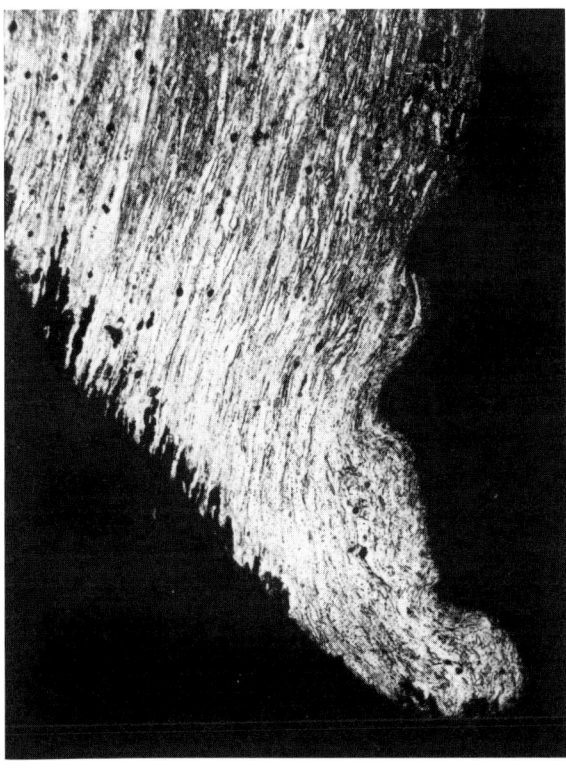

3.23

Cross section through one side of needle eye groove, showing indentions from tool used to punch eyehole. Sample etched in potassium dichromate (mag.: 200).

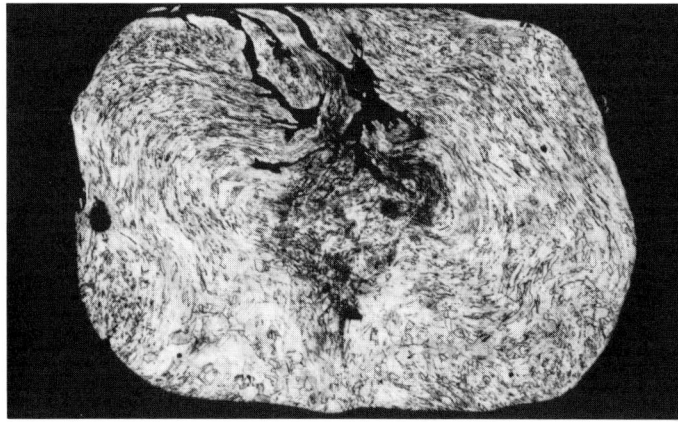

3.24

Cross section through perforated-eye needle shaft just below the eye. Note central fissures, rectangular form, and extremely elongated and compressed grains. Sample etched in potassium dichromate (mag.: 50).

These needles began as long, rectangular pieces of copper that were flattened, then folded over along their longitudinal axis to thicken the metal that would become the needle shaft. The folding resulted in a thick, rodlike form that retained a roughly rectangular cross section but that also created an internal fissure running the length of the rod. That portion of the needle was further cold-worked to flatten it. The orientation of the flow lines shows how the metal moved as it was being worked (figure 3.24). The section went from rectangular to round by hammering the metal along its length while rotating the rod. Figure 3.25, a cross section cut from the midpoint of the shaft, shows the fissure.

The needles made during Period 1 present a mixed picture with respect to control over formal design and material properties. A usable needle must achieve a balance between toughness and strength. It must be sufficiently tough so that the stresses to which it is subjected (in the slight bending that can occur when the needle is drawn through heavy cloth or other material) do not induce fracture. The needle must also be strong enough so that it does not permanently deform when used; thus, some elasticity is a distinct advantage. In addition, the needle must be hard enough and sharp enough to pierce quite resistant material.

All but eight of the seventeen Amapa perforated-eye needles I examined were deformed; that is, the needle shaft was bent. Many were also broken at the eye. Seven

3.25

Cross section through perforated-eye needle shaft at approximate midpoint along the shaft. Sample etched in potassium dichromate (mag.: 100).

of the RMG perforated-eye needles analyzed were deformed and broken at the eye or elsewhere. The damage to these needles does not occur randomly but is a function of the needle's length and thickness. It is clear from table 3.14 (group A) that copper needles thinner than 0.09 cm will deform regardless of length. It is also clear from group B that needles thicker than 0.19 cm tend not to deform. Similarly needles ranging in thickness between 0.09 and 0.19 cm will deform if made longer than 12 cm (group C); if made shorter than 12 cm they tend not to deform (group D). Only a few RMG needles of the other design, with a loop eye, are made from copper; they are also deformed or damaged in all but one case, due to the inherent fragility of the loop when made from copper.

Smiths thus faced a major technical problem if they wanted to make long, thin copper needles: the metal, even when cold-worked, lacks the necessary mechanical strength to sustain the design. Copper is relatively soft, and even when cold-worked is not the optimal material for the design and function of these particular needles. Period 2 smiths did fashion long, thin needles but used the bronze alloys to do so, suggesting that concomitant changes in some facet of cloth production technologies may likely have been taking place.

These metalworkers shaped both of the rodlike forms examined here in a similar way: by casting a blank, flattening and extending the metal, then folding and hammering it around its longitudinal axis. They made round cross section rings and needles in this fashion, and, as I will show, they also made fishhooks and round cross section awls following the same principle. This approach forms an internal fissure running along the longitudinal axis of the object. The fissure does not compromise functionality for the rings or needles because they are not impact tools where toughness is essential. However, function unquestionably is compromised in the small copper tools such as awls.

Awls: Unipointed and Bipointed. At least two awl designs (figure 3.26), one unipointed and the other bipointed, were made during Period 1. These are found at Amapa in Nayarit, at Tomatlán in Jalisco, and in the Infiernillo burials (see table 3.6). The RMG tools range from 6 to 13 cm in length. Within each group, some have

Table 3.14 Copper Needles: Breakage or Deformation as a Function of Thickness and Length

Thickness (cm)	Length (cm)	Condition
Group A		
0.04	9.4	deformed
0.05	8.3	deformed
0.08	6.0	deformed
0.08	9.1	deformed
Group B		
0.2	8.9	intact
0.2	14.7	intact
0.2	22.5	intact
0.3	12.9	intact
Group C		
0.10	11.1	deformed
0.10	13.4	deformed
0.12	13.0	deformed
0.12	14.4	deformed
0.13	14.0	deformed
0.15	12.2	deformed
0.15	12.8	deformed
0.15	14.0	deformed
0.15	14.0	deformed
0.15	14.8	deformed
0.18	13.5	deformed
Group D		
0.09	8.2	intact
0.10	6.0	intact
0.10	8.0	intact
0.10	8.8	intact
0.10	10.8	intact
0.10	11.3	intact
0.12	10.2	intact
0.12	11.5	intact
0.15	8.4	intact
0.15	10.5	intact

round cross sections and others are rectangular; manufacturing regimes for the two cross-sectional types differ slightly.

Metallographic studies show that artisans fashioned the round cross section awls as they fashioned needles and round cross section rings: by hammering a cast blank roughly to shape, then folding the metal and hammering it around its longitudinal axis. After cold-working the resulting rod into an awl- or punchlike shape, they left

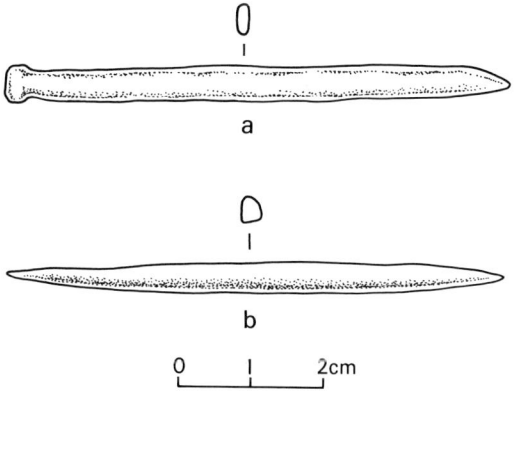

Period 1 awl designs: unipointed (a) and bipointed (b).

3.27

Unipointed awl: microstructure of cross section through the shaft, illustrating portion of internal fissure, equiaxed grains, and annealing twins. Sample etched in potassium dichromate (mag.: 50).

(129 to 146 VHN) reflect this. A longitudinal section from the tip of a second copper awl (figure 3.29) also shows elongated grains some of which are perpendicular to the longitudinal axis, reflecting deformation of the metal that could only result from use.

Fishhooks. Fishhooks are unusual in the metalworking zone; only one has been recovered from each of the sites listed in table 3.6, and they are not abundant in the RMG collection. The basic design is shown in figure 3.30. The central fissure that characterizes these objects is visible in the photomicrograph in figure 3.31.

3.29

Longitudinal section of awl near blade tip. Orientation of grains at the surface orthogonal to the longitudinal axis is the result of deformation from use. Sample etched in potassium dichromate (mag.: 100).

the metal annealed except at the working ends where it was worked further to increase hardness. Figure 3.27, a photomicrograph of a cross section through the shaft of a small awl, shows the internal fissures as well as the flow lines in the metal, which indicate the directional flow from the initial cold work. The grains surrounding the fissure exhibit annealing twins. The working tip of the object was cold-worked to harden it after the final annealing operation, as the extremely elongated and deformed grains indicate (figure 3.28). Microhardness readings at the tip

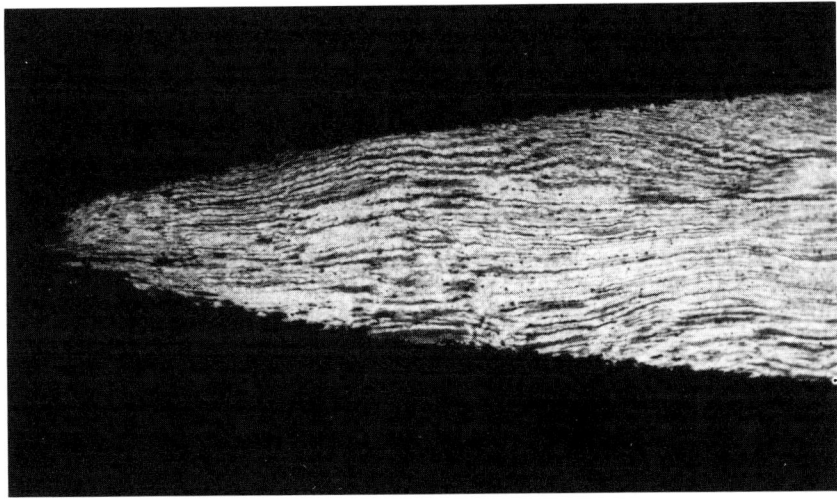

3.28

Longitudinal section of working tip of an awl. The thin, highly compressed grains have been severely distorted by cold work in hardening the tip (mag.: 50).

Metallographic studies showed fishhooks to be made in the same way as the round cross section open rings, needles, and awls. First a cast blank was hammered roughly to shape, and the metal was folded and hammered around its longitudinal axis. The fishhook was bent to its final form through sequences of hammering and annealing. Following the final anneal the fishhook was worked slightly to true it up.

Summary and Observations

Period 1 extends from approximately A.D. 600 to A.D. 1200/1300. During this time artisans worked primarily with copper, both the native metal and metal smelted from copper ores. Evidence from Tomatlán and Peñitas suggests that metalworkers also were beginning to experiment with a low-arsenic copper-arsenic alloy. At Amapa, Tomatlán, Peñitas, La Villita, and Cojumatlán, metal may have been worked *in situ*: Mountjoy describes metalworkers' tools; slag appears at Amapa; a crucible is reported from La Villita; and so on. Period 1 artisans across a large region seemed to be making the same kinds of objects, from similar materials, and in the same way.

One of the most unusual features of Period 1 metallurgy is that these smiths employed two such distinct approaches to shaping the material: lost-wax casting, and cold-working from an original form cast in an open mold. Artisans most often made bells, casting them using the lost-wax method. Bell designs were quite varied even in Period 1, when at least seven distinct types appear. A second significant class of metal objects (with respect to their sheer numbers and their sumptuary context) consisted of small open rings, cold-worked to shape and worn

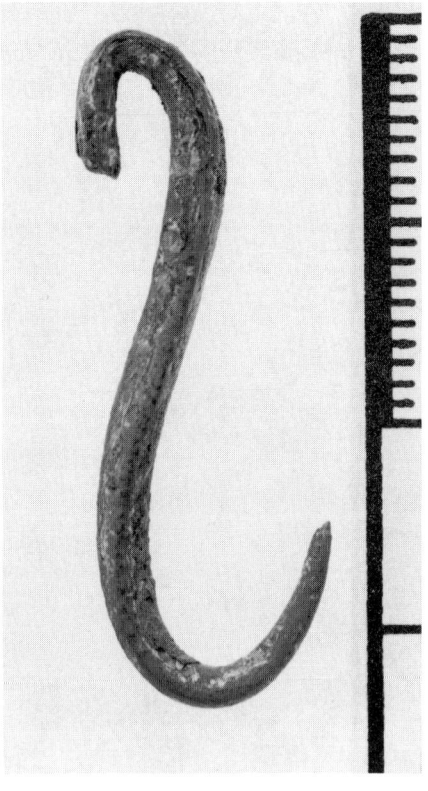

3.30

Fishhook, RMG collections.

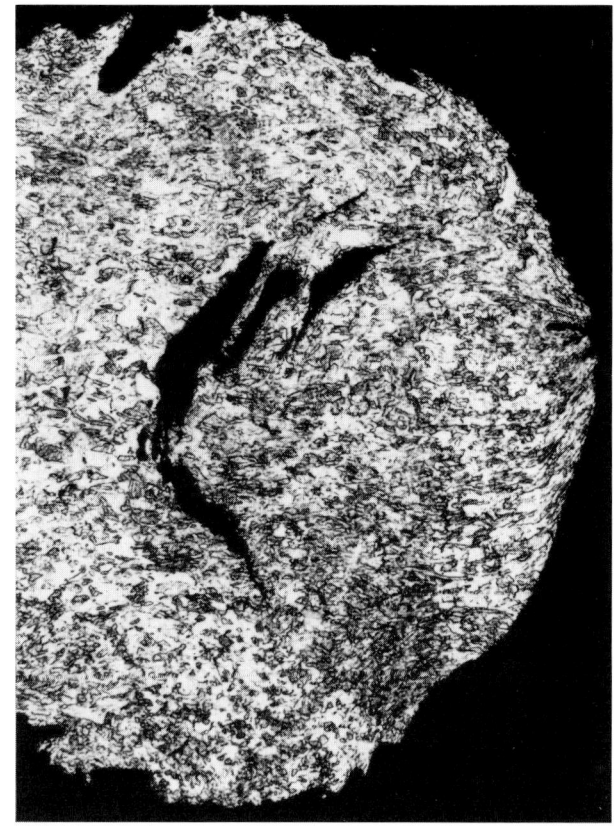

3.31

Cross section of fishhook, showing central fissure. Sample etched in potassium dichromate (mag.: 50).

as hair and ear ornaments. These two artifact types constitute a majority of all objects made from metal during Period 1. Bells were principally ritual items; the rings marked social status. Tools were far less common, and even some of these (for example, tweezers and axes) sometimes served symbolic rather than utilitarian ends.

All Period 1 metal objects have counterparts in other materials. Bells had been made from fired clay; the RMG collection contains ceramic bells, and others have been excavated. Rattles made from seed pods, deer hooves, and other materials are also widely known ethnographically and are illustrated on archaeological pottery and in murals (see chapter 8) throughout the Americas. Needles, fishhooks, and awls made in bone, and tweezers and rings made from shell and wood, also appear. Stone axes are commonplace. The superior physical and mechanical properties of metal would have optimized the design and function of any of these objects. Nonetheless, when metallurgy developed, metal replaced only one of these to any significant degree, namely bells. The range of pitches that metal allowed and metal's superior resonant properties became the focus of Period 1 technology. These people explored and developed this quality (within the constraints imposed by copper) in an entirely new material.

The significance of these technical choices becomes clear when we consider that the physical and mechanical properties of metal could have offered great advantages for many other classes of sumptuary objects, as well as for tools and implements. Smiths did fashion some metal implements, but never in large numbers. They largely ignored tools and implements other peoples made from metal—knives, spear points, and other weapons, and certain agricultural tools like hoes and digging stick points.

As I show in chapter 4, many Period 1 objects are nearly identical in formal design, fabrication techniques, and materials to corresponding objects found in the casting tradition of lower Central America and Colombia, or the hammering/working tradition of the central Andes. The evidence is unambiguous that technical know-how, and sometimes artifacts, were imported from those two regions and sparked the development of metallurgy in West Mexico. Nonetheless, West Mexican artisans did not incorporate the full range of Central and South American objects and techniques into their own metallurgical repertoire, although they clearly did possess the technical expertise and the region the mineral resources to have done so. The central emphasis of Period 1 metallurgy, on bells, certainly was not borrowed from either of the two metallurgies to the south. The components of those two technologies that West Mexican smiths chose to integrate and emphasize—some of which were choices among technical alternatives—reveal their own world views, attitudes, and concerns. I discuss this issue in the concluding chapter.

4

Origins of Period 1 West Mexican Metallurgy

Scholars have suspected for many years that West Mexican metallurgy originated in the metallurgies of Central and/or South America (Arsandaux and Rivet 1921; Meighan 1969; Mountjoy 1969; Pendergast 1962b). In the central Andes, artisans were crafting metal objects by 1500 B.C., although metalworking did not come into its own until the Early Horizon (700 B.C. to 200 B.C.). Colombian metallurgy began to take shape around 200 B.C. or before; the metallurgy of lower Central America several hundred years later. In the West Mexican metalworking zone the technology was flourishing by A.D. 800, although metal objects have been recovered dating to several hundred years prior to that. The late appearance of metallurgy in West Mexico suggests that it was introduced from outside. Also, some metal artifacts from Andean South America and lower Central America and Colombia are identical in formal design to later West Mexican types. Such design similarities can signal historical connections, but they represent only one measure of them. To thoroughly explore this issue we need to investigate all aspects of the technology: the kinds of objects made and their design characteristics, the manufacturing techniques, the metals and alloys used to make them, and, where possible, their meaning in specific social contexts. This evidence, taken as a whole, unambiguously indicates that Period 1 West Mexican metallurgy is so similar to that of certain South and Central American metalworking zones that some elements of West Mexican metallurgy clearly were introduced from those areas. The metallurgies of two regions played primary roles: lower Central America and Colombia, and the southern part of the modern nation of Ecuador.

The people in these two areas handled metal in very different ways. In southern Ecuador and in the central Andes, artisans formed metal as a solid material by working it either hot or cold. They generally used metal sheet, even for three-dimensional figures (figure 4.1), but also for breastplates, masks, diadems, pendants, and metallic

4.1

Jaguar from the north coast of Peru made from 12 individual pieces of metal sheet. Photograph courtesy of the Virginia Museum of Fine Arts.

4.2

Figure pendant from Colombia: Tairona culture. Photograph courtesy of the Museo del Oro, Bogotá, Colombia.

Table 4.1 Chronological Development of Ecuadorian Metallurgy: First Appearance of Object Types, Metals, and Alloys

Relative Chronology	Type	Metals and Alloys
Integration (A.D. 800–A.D. 1530)	axe-monies axes	copper-tin (southern highlands only; Inka import)
Regional Developmental (500 B.C.–A.D. 800)	awls beads bells needles nose rings open rings pendants plaques tweezers	copper-arsenic copper-gold copper-silver
Formative (1500 B.C.–500 B.C.)	nose rings sheet metal	copper gold silver

feathers—objects exhibiting flat, highly reflective, and brilliantly colored golden, silvery, and sometimes bimetallic surfaces.

These Andean smiths did cast certain forms; for example, symbols of state such as star-shaped maces, decorative *tupus*, and massive solid nose rings. They sometimes used open molds and sometimes piece molds, and occasionally they used the lost-wax method. Bronze tools were also common; some, for example axes and hoes, were cast in molds and subsequently cold-worked to achieve their final form and to increase hardness. Metalworkers made smaller implements as well (needles, tweezers, awls, axes, and fishhooks) by hammering them to shape from an initial cast blank.

Lower Central American and Colombian metallurgy centered on casting, particularly by the lost-wax method and other techniques. *Tumbaga*, the copper-gold alloy, was extremely common, and metalsmiths used it to cast elegant and intricate earrings, pendants, and nose rings worn by elites (figure 4.2). Other, larger objects were employed in ritual. Artisans fashioned relatively few tools and implements from metal.

Each region lent certain elements to the metallurgy that emerged in the West Mexican metalworking zone. I will consider those that derive from southern Ecuador

Table 4.2 Chronological and Geographical Placement of Ecuadorian Sites

	Coast			Highlands	
Relative Chronology	Central	Southwest	Interior	Northern	Southern
Integration (A.D. 800–A.D. 1530)	Salango (Manteño)	Ayalán Cerro Alto (both Milagro-Quevedo) Loma de los Cangrejitos (Manteño)	La Compañía Peñón del Río (Milagro-Quevedo)	Ingapirca (Inka)	
Regional Developmental (500 B.C.–A.D. 800)	Salango (Bahía)	El Azúcar Cerro Alto La Libertad OGSE-Ma-172 Palmar (all Guangala)	San Lorenzo (Jambelí)		
Formative (1500 B.C.–500 B.C.)	Salango (Chorrera)			La Florida	Pirincay

first. We now have a reasonable chronology for Ecuadorian metallurgy from laboratory investigations of dated objects and a comprehensive study of the metal collections in the Museo Antropológico del Banco Central in Guayaquil (MAG). I compare that information to the laboratory and other data from West Mexican artifacts to evaluate Ecuadorian contributions to the technology. The section "Setting, Sites, and Objects" provides background and context for the comparative evidence presented here. In that section I discuss the development of metallurgy at particular Ecuadorian sites, and present chemical analytic data for objects from them (see also tables 4.1 and 4.2).

Ecuador and West Mexico

Remarkable similarities exist between Period 1 West Mexican metallurgy and that of ancient Ecuador prior to and during Period 1. They appear in many facets of these technologies: in the kinds of objects made from metal, in their particular design characteristics, and in the fabrication techniques and raw materials used in both areas to realize those designs. The most common Period 1 West Mexican metal objects were bells, open rings, perforated-eye needles, beam design tweezers, axes, awls, and fishhooks made from copper and occasionally from low-

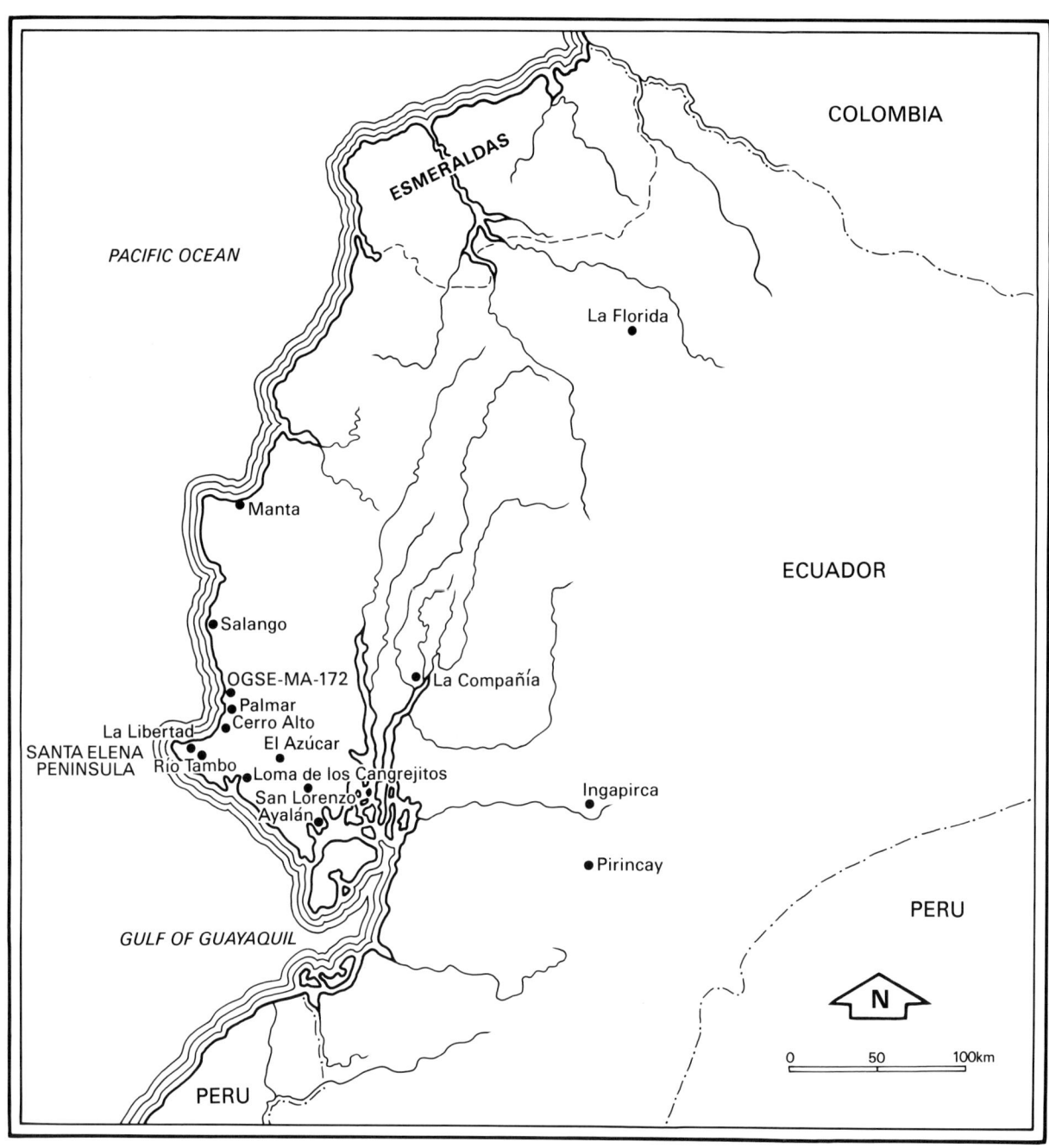

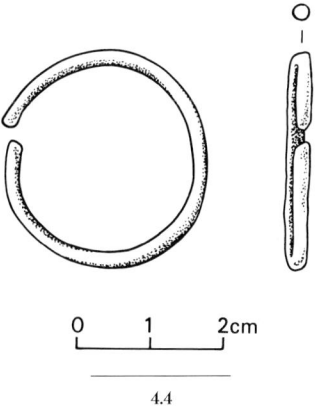

4.4 Open, round cross section ring from MAG collections, Guayaquil, Ecuador. Compare to figure 3.7.

arsenic copper-arsenic alloys. In Ecuador prior to and during this period, metal was used for these same artifact classes made from these same materials; they were fashioned in the same way in all cases but the bells, which in Ecuador were cold-worked to shape. Ecuadorian smiths did, of course, craft many other objects (nose rings and other ritual items, for instance) that never made their way to West Mexico. Nonetheless, the striking congruence in design, materials, and fabrication methods in the particular constellation of objects I have mentioned makes a strong case for historical connections. Tables 4.1 and 4.2 present the chronology for the development of Ecuadorian metallurgy, and list the sites where metal objects appear that I discuss here. Figure 4.3 locates those sites.

The metal objects found most frequently in Ecuador are solid, open, round cross section rings made from copper, and from copper-arsenic and copper-silver alloys.

4.3 Ecuador: coastal, inland, and highland archaeological sites where metal artifacts have been recovered. Manta and Salango were active centers in the Pacific maritime network.

Laboratory studies show that these artifacts are identical in composition, manufacturing techniques, and design to later West Mexican rings. They also are found in analogous archaeological contexts. Figure 4.4 illustrates one of these rings; as table 4.8 shows, the MAG study corpus contains 954 others. Diameters vary between approximately 1.0 and 5.0 cm, a range similar to that of rings from West Mexico. In Ecuador the earliest examples of these artifacts date to between 500 B.C. and A.D. 500 at San Lorenzo (Ubelaker 1983). Rings recovered at El Azúcar (Masucci 1992), a site on the Santa Elena Peninsula, date to between A.D. 150 and A.D. 300. One is made from copper, the other from a copper-silver-gold alloy (see table 4.4). Ubelaker recovered rings from Ayalán dating to A.D. 750 (Ubelaker 1981). Metallographic studies show that Ecuadorian rings were fashioned in precisely the same way as those from West Mexico (see figure 3.9): by hammering a strip of metal around its longitudinal axis to achieve a cylindrical, rodlike form. The fissure this produces is visible in the photomicrograph in figure 4.5. The etched section in figure 4.6 shows equiaxed grains and annealing twins as well as the elongated inclusions reflecting the cold work involved in hammering. As the rod was bent into a ring it underwent several sequences of hammering and annealing but was left in the annealed condition.

These rings are usually found in interments both in Ecuador and in West Mexico. The specific depositional context, which is the same in both cases, suggests that they were worn: as hair ornaments, nose rings, or earrings. At Ayalán, archaeologists discovered them in urn burials; they were found on the cranium and in the nose area of male individuals, and in the ear area of females. In West Mexican burials, at Tomatlán and the Infiernillo sites, archaeologists also recovered rings near the cranium and the ear area. Burials recently excavated by Helen Pollard at Urichu (Michoacán) also contain rings in

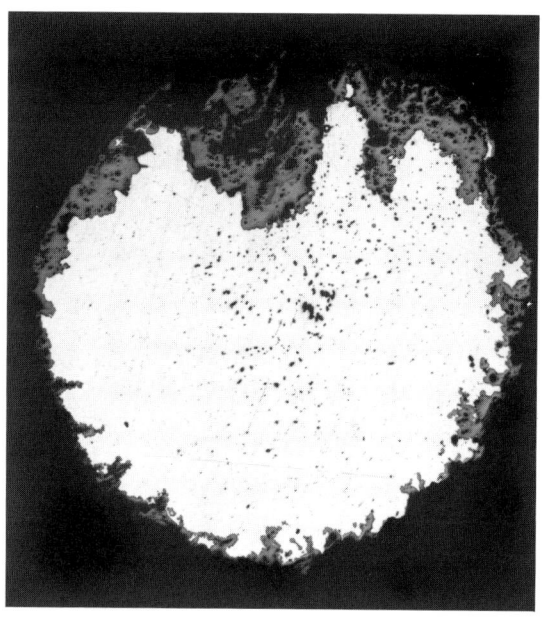

4.5

As-polished cross section of round cross section ring. Fissure formed by folding metal along longitudinal axis has been partly obliterated by internal corrosion. The cross section in figure 4.10 illustrates a similar and uncorroded fissure (mag.: 30).

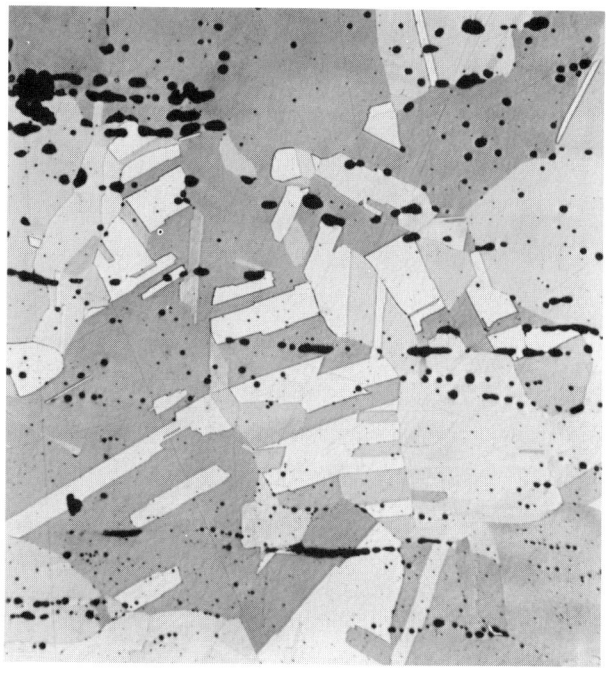

4.6

Longitudinal section through round cross section ring showing equiaxed grains with annealing twins and elongated inclusions. Sample etched in potassium dichromate (mag.: 100).

groups of two to three near the ear.[1] The earliest securely dated Ecuadorian example of this artifact class comes from El Azúcar and dates to between A.D. 150 and A.D. 300; its West Mexican counterparts, from Tomatlán and Infiernillo, date to after A.D. 700.

Needles are another Ecuadorian artifact type with precise, later counterparts in West Mexico. They are made with a perforated eye. The example shown in figure 4.7 is made from unalloyed copper. In Ecuador these needles first appear between A.D. 100 and A.D. 500, at sites on the south coast (tables 4.1 and 4.2). One, from El Azúcar, is made from a relatively pure copper (see table 4.4); another, from Cerro Alto, also on the Santa Elena Peninsula,

is made from a copper-arsenic alloy containing 3.24% arsenic (see table 4.5). This same design is also reported from Palmar, and after A.D. 500 it becomes extremely common to the north, in Manabí (Holm 1963). Perforated-eye needles also occur sporadically in coastal and highland Peru and in certain regions of the southern Andean highlands. In West Mexico the first specimens appear at Tomatlán, in the Infiernillo burials, and at Amapa; they date to about A.D. 800, or some 400 years later than the earliest Ecuadorian examples. All West Mexican perforated-eye needles datable to Period 1 so far are made from copper.

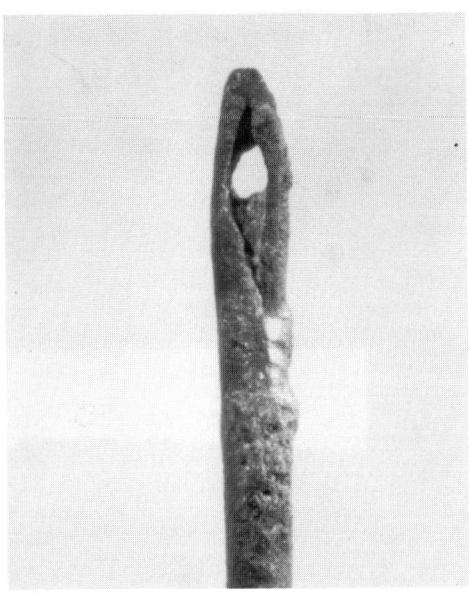

4.7

Ecuadorian perforated-eye needle. Compare to figure 3.21.

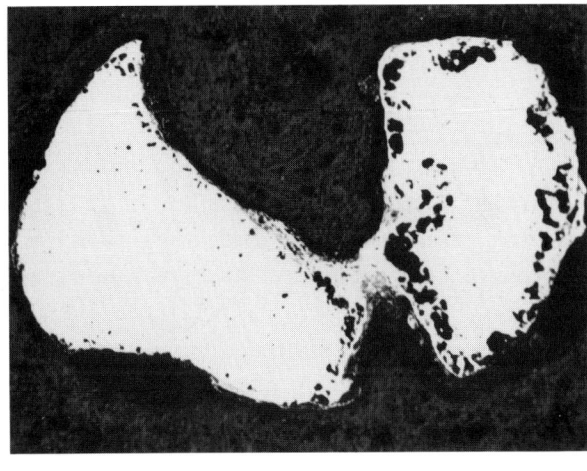

4.8

Ecuadorian perforated-eye needle. As-polished cross section through the eye. Eye is filled with corrosion product (mag.: 50).

Metallographic studies show that the Ecuadorian needles, like the rings, were made in exactly the same way as the later West Mexican designs (see figures 3.22 to 3.24). The solid cylindrical shaft of each was formed by hammering a rectangular strip of metal around its longitudinal axis, leaving a central fissure. The groove for the eye was made by first flattening one end of the shaft, then hammering up the opposite sides of the flattened area. The photomicrograph in figure 4.8 shows the two sides of the groove just below the eyehole. The eye was made by punching or drilling a hole in the groove.

Depilatory tweezers are also common to both regions and are nearly identical in design. The Ecuadorian design, like the early West Mexican model, is the beam type (figure 4.9). These tweezers first appear at Cerro Alto on the Santa Elena Peninsula (see tables 4.1 and 4.2)

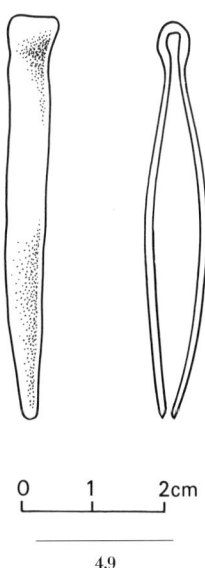

4.9

Front and profile view of beam tweezers from MAG collection, Guayaquil, Ecuador. Compare to figure 3.13.

and at La Libertad (Bushnell 1951); at both sites they date to before A.D. 800. Caches sometimes occur in burials.[2] The specimens from La Libertad and Cerro Alto are made from copper.

Ecuadorian and West Mexican beam tweezers are similar in overall dimensions, although Ecuadorian tweezer dimensions and their ratios (length to thickness) differ sufficiently from those of West Mexico that the two groups can be readily distinguished. Fabrication techniques were identical. Both were hammered to shape from a single piece of metal. Metalworkers fashioned the hinge by bending and hammering the metal over some kind of former, perhaps made of wood. In both cases the tweezers were left either in the cold-worked condition or annealed. The Ecuadorian tweezers, like the West Mexican ones, were functional depilatory tools. Their dimensions and dimensional ratios conform to those that computer simulation studies (Hosler 1986) show were necessary for tweezer function. The West Mexican beam tweezers were made from copper; they replicate their earlier counterparts, the standard Ecuadorian design, in all respects. In Ecuador this design persisted until the Spanish invasion.

Some Ecuadorian tweezers have holes for suspension at the top of the hinge and apparently were worn around the neck. Tweezers from the coast of Peru, with suspension cords still attached, are found in the collections of the American Museum of Natural History (AMNH) in New York. Their design differs from the Ecuadorian type, but the AMNH artifacts with their attached suspension cords do indicate that in the Andean region tweezers were intended to be worn, as they were in West Mexico. Facial depilation using tweezers is still common in the highlands of Peru and Ecuador.

The two Period 1 West Mexican awl designs (found at the Infiernillo sites and at Tomatlán) have Ecuadorian

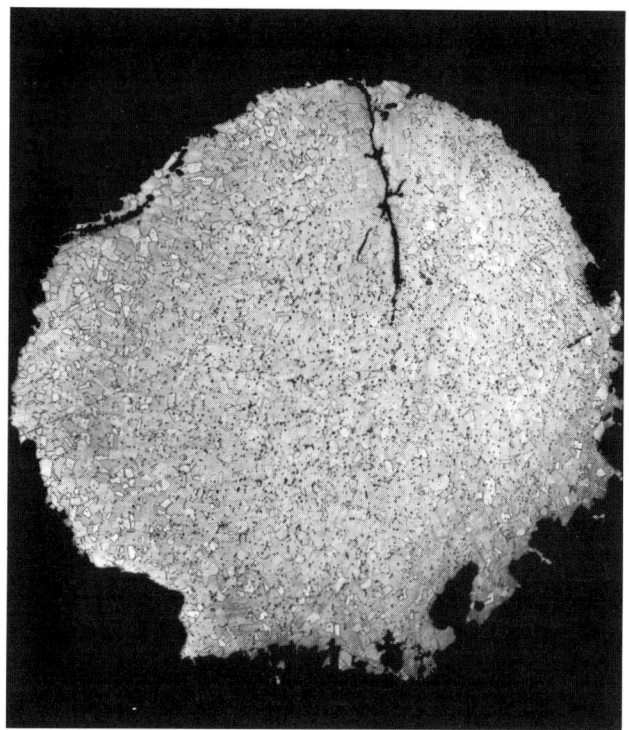

4.10

Cross section of awl from Ecuador, showing central fissure. Sample etched in potassium dichromate (mag.: 17).

analogues at the sites of Río Tambo and La Libertad (Bushnell 1951). They date to the Guangala period.[3] The Ecuadorian examples were hammered around their longitudinal axis, leaving a central internal fissure. Like the awls from West Mexico (see figure 3.29), they were left in either the cold-worked or annealed condition. Figure 4.10 shows the cross section from an awl fragment excavated at the coastal site of El Azúcar. This awl was left annealed but shows signs of deformation at the blade tip. The central fissure is clearly visible in figure 4.10. These tools are virtually nonexistent in the MAG collections, but they do appear in the Guayas basin at Loma de los

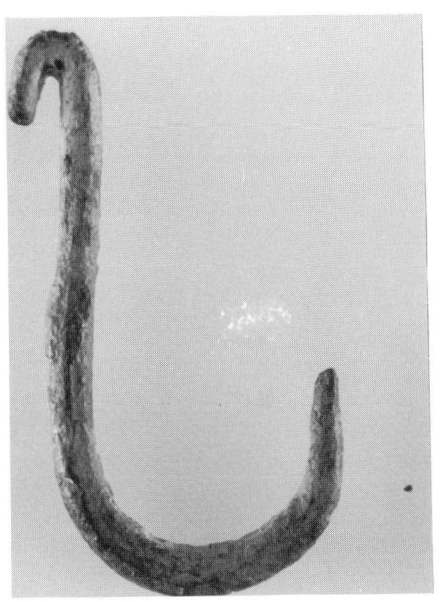  

4.11

Fishhook from MAG collections, Guayaquil, Ecuador.

4.12

Cross section of fishhook from Ecuador. Note central fissure. Sample etched in potassium dichromate (mag.: 31).

Cangrejitos (see table 4.7), and at La Compañía. At both sites, they are made from a low-arsenic copper-arsenic alloy.

Fishhooks are also common in Ecuador and West Mexico. They first appear in Ecuador between 500 B.C. and A.D. 800, the Guangala period, at the OGSE-Ma-172 site in the village of Valdivia,[4] and at La Libertad on the Santa Elena Peninsula (Bushnell 1951). Fishhooks from both sites are made from copper. Like beam tweezers, fishhooks are both formally similar to the West Mexican implements (see figures 3.30 and 4.11) and were made in the same way: by folding a strip of metal along its longitudinal axis and hammering it round, leaving a central, internal fissure (see figures 3.31 and 4.12). Final shaping was achieved through sequences of hammering and annealing. Ecuadorian fishhooks are made from copper; some in the MAG collection are made from a very low-arsenic copper-arsenic alloy.

One of the striking technical links between the metallurgy of this Andean area and that of West Mexico lies in their approach to fashioning a round, rodlike form. These forms served as the starting point in elaborating needles, fishhooks, rings, and round cross section awls. Metalworkers hammered and folded metal stock along a longitudinal axis to thicken it, creating a long internal fissure. The fissure is visible in cross sections removed from all four of these artifact types, whether made in Ecuador or in West Mexico. These procedures reflect a long-standing central Andean practice of shaping metal by deforming it as a solid. There is nothing inherent in this design that requires this fabrication sequence; rods

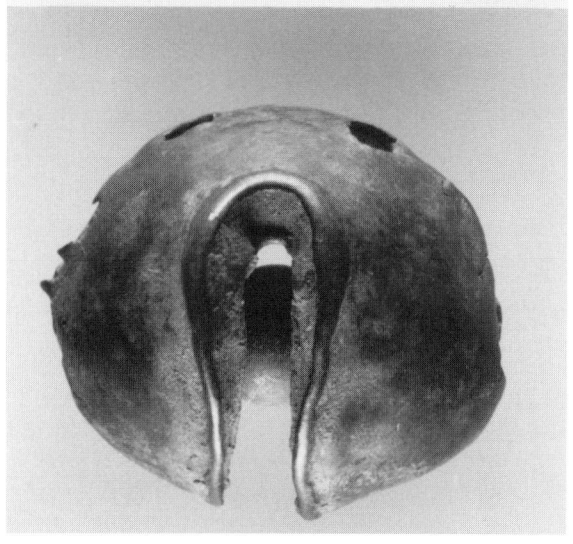

4.13

Bell from MAG collections, Guayaquil, Ecuador.

could have been cast close to their final shape and thickness, requiring only minor truing up to achieve the final form.

Bells were the objects West Mexican smiths made most often. Ecuadorian metalsmiths also made large numbers of bells, as the information indicates in the section "Setting, Sites, and Objects." Nonetheless, Ecuadorian bells are technically unrelated to the West Mexican variety; they are fashioned by cold hammering rather than lost-wax casting and are suspended from two holes at the top. MAG specimens generally are made from a low-arsenic copper-arsenic alloy, although a few bells are made from alloys of copper and silver (see table 4.9). Figure 4.13 illustrates a typical Ecuadorian bell. The extremely

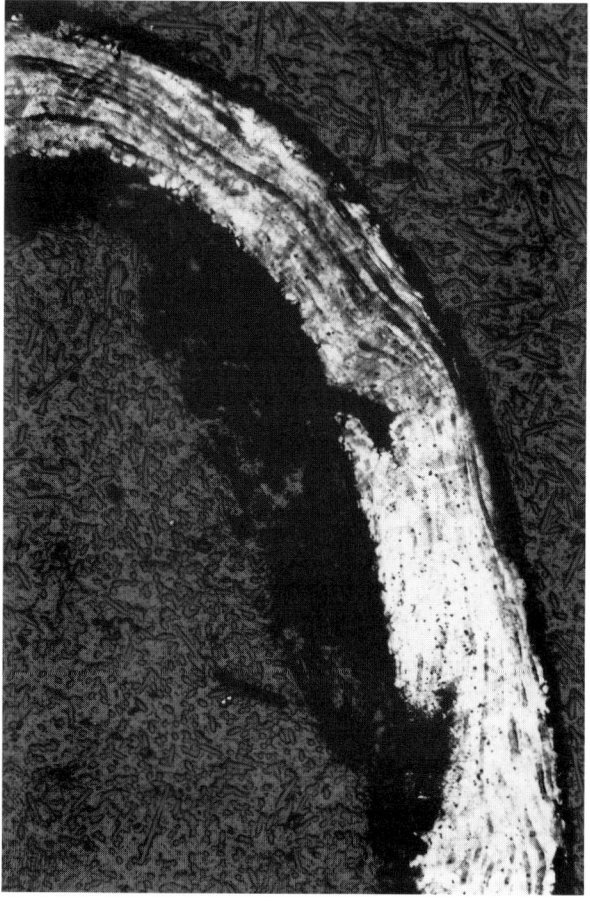

4.14

Longitudinal section from bell wall showing narrow and highly compressed flow lines from intensive cold work. Sample etched in potassium dichromate and ferric chloride (mag.: 50).

elongated grains in the longitudinal section from the bell wall shown in figure 4.14 reflect the intensive cold work.

Bells in ancient Ecuador and Peru, as in West Mexico, were worn around the ankles and wrists in ritual dances. A drawing by Poma de Ayala depicts Andean musicians and dancers wearing bells (1936: 322); how-

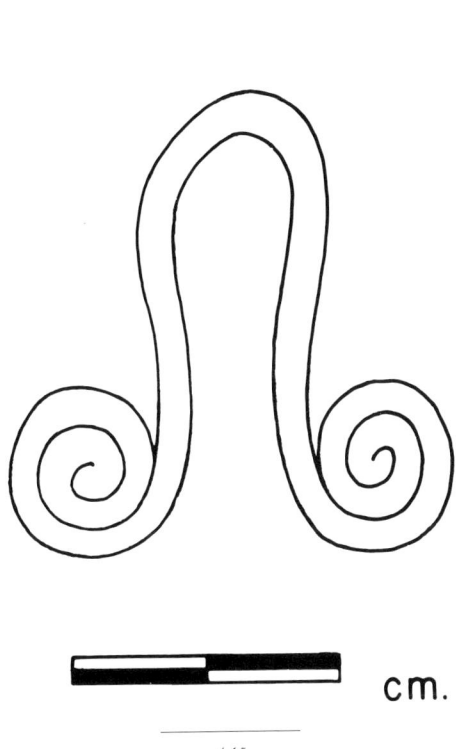

4.15

Spiral ornament from Tomatlán, Jalisco, Mexico.

4.16

Spiral ornament from MAG collections, Guayaquil, Ecuador.

ever, the dancers are from the eastern lowlands and may have used the dried husks of fruit, with a pebble clapper, rather than metal bells.[5] Elites wore bells in the Andean area as well as in West Mexico; a 1598 document cited by Salomon (1986) describes Ecuadorian highland nobles wearing headdresses of hammered silver, and suspended from each headdress were 36 little pendant bells. Elite individuals buried at the Moche site of Sipán in north coastal Peru wore elaborate composite bell forms at the waist (Alva and Donnan 1993).

A few artifact types appear only occasionally in West Mexico but are common in Ecuador. Two spiral orna-ments and the unusual wire tweezers excavated at Tomatlán figure among them. The spiral ornaments are extremely common in Ecuador, especially in the La Tolita area to the north in the province of Esmeraldas. Others were excavated to the south, at Ayalán, from an urn burial dating to A.D. 730; analyses indicate that at least one was made from copper and the others from copper with a gilt surface (Ubelaker 1981) Figure 4.15 illustrates the Tomatlán form and figure 4.16 its counterpart from coastal Ecuador. Maldonado (1980) reports a similar ornament from a burial at Infiernillo in an Early Postclassic Period context. The wire tweezers from Tomatlán (figure 4.17) do not resemble depilatory tools. Numerous similar ex-

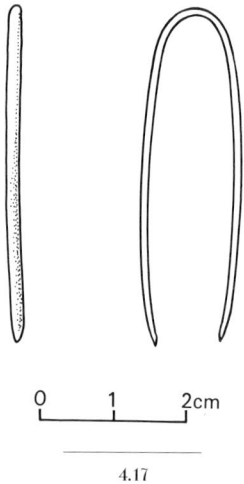

4.17

Wirelike tweezers from Tomatlán, Jalisco, Mexico.

amples have been reported from the La Tolita area. Bergsøe (1937) suggested that they were used for metalworking. Neither this tweezer type nor the spiral ornaments ever were common in West Mexico, and, as suggested in chapter 3, it is entirely possible that they were made in Ecuador and imported.

COLOMBIA AND LOWER CENTRAL AMERICA

The volume of archaeological and laboratory-analytical data for Colombian and lower Central American metallurgy does not yet compare to that for Ecuador. Nonetheless, the evidence we do have shows unambiguously that the later metallurgy of West Mexico is related to it. The lower Central American and Colombian tradition is very different from that of Ecuador, since smiths formed metal primarily by casting it to shape (Bray 1978, 1981, 1985), especially using the lost-wax method.

Colombian metalsmiths used a more restricted range of metals and alloys than their counterparts in Ecuador

Table 4.3 Number of Artifacts by Functional Type: Museo del Oro, Bogotá, Colombia

Type	Number in Collection
Bells	5,000
Gongs	100
Fishhooks	500
Needles	20
Other tools (awls, axes, etc.)	200
Ritual and status objects (nose rings, figurines, plaques, pins, effigy pendants)	27,320
Total Elite/sumptuary	32,420
Total tools	720
Total	33,140

or West Mexico. Gold and copper-gold *tumbaga* alloys were their primary materials. They only occasionally used unalloyed copper. The Museo del Oro in Bogotá contains the largest collection of metal artifacts in the country; 59% percent are made from *tumbaga*, 30% from gold, 10% from copper, and 1% from platinum.[6] The museum's vast holdings (table 4.3)[7] reflect the extraordinary numbers of ritual and ornamental objects fashioned prior to the European invasion.

The low proportion of metal tools is related to that region's mineral resources, which do not include the large deposits of arsenic-bearing copper ores found in southern Ecuador and northern Peru that are useful for such items. Small hand tools sometimes were made from *tumbaga*, however. Reichel-Dolmatoff (1988) argues that the very large repertoire of Colombian ritual and status objects— figurines, zoomorphic and anthropomorphic figures, and effigy pendants—communicated a complex of ideas associated with shamanism, unified by the common theme of transformation. Metal bells and other musical instruments, played during rituals in which shamans ingest hallucinogenic drugs, form an integral part of this complex. Reichel-Dolmatoff (1981, 1988) also explicitly links the color of gold to the prehispanic sun deity, arguing that the sun was and is viewed in indigenous Colombian societies as a powerful generative male being. Some indigenous groups periodically place prehispanic gold and *tumbaga* objects outside in the sun so that the objects can regain their generative powers.

The first lost-wax cast artifacts made in the Americas are found in this region and date to around A.D. 100 (Duque G. 1964), predating the appearance of the technique in West Mexico by at least 500 years. Between the third and the tenth century, lost-wax casting became widespread in the Calima region and in the Muisca highlands (see figure 6.1) and was used for bells and other objects (Plazas and Falchetti 1985). In the Muisca highlands, the earliest dates for lost-wax casting fall between A.D. 200 and A.D. 500.[8] In central Panama, lost-wax castings are found in contexts that date to before A.D. 500 (Bray 1981).

Colombian, Panamanian, and Costa Rican bells are identical to the earliest West Mexican types in fabrication methods and overall design. They are lost-wax cast, possess a suspension ring, and contain a clapper. One of the Cerro de Huistle bells (a type 1c specimen) exhibits a typically lower Central American design (figure 4.18) and provides a small but compelling piece of evidence for contact between the two regions. Two other bell types from Colombia and Panama are identical to Period 1 West Mexican designs (see figure 3.5): type 1a, found in the Muisca highlands of Colombia,[9] and type 11a, found in the same region and in the Sinú area. Specimens similar to type 1c (the bell illustrated in figure 4.18) also appear in both Colombian regions.[10] Two of these (types 1a and 11a) also are known from Panama. The only dates I know of for the Panamanian examples fall between A.D. 700 and A.D. 900 (Hearne 1992: 19; also see plate 41). The exterior walls of these lower Central American bells are smooth, like the walls of the West Mexican varieties. Figure 4.19 illustrates a Panamanian bell (a) similar to West Mexican type 1a, and a Colombian bell (b) like West Mexican type 11a.

Bell forms identical to West Mexican types 1a and 11a have also been recovered in the Diquís delta region of Costa Rica, but these bells probably date to after A.D. 1200.

As these similarities in design characteristics and fabrication methods indicate, the component of West Mexican metallurgy devoted to the lost-wax casting of bells originated in lower Central America and Colombia. Nonetheless, bells never played the primary role in the metallurgies of those regions that they later did in the metalworking zone.

The Introduction of the Technology to West Mexico

How were elements from these two distinct regional metallurgies introduced to West Mexico? The strongest connections are with Ecuador, where artifact types, fabrication methods, and materials are nearly identical for a variety of object classes. In all cases the earliest dates are in Ecuador. Links also are close with the lost-wax casting technologies of lower Central America and Colombia, but fewer Period 1 West Mexican artifact types have precise Colombian or lower Central American counterparts. The evidence marshaled here demonstrates that an entire technical complex, one component of Ecuador's multifaceted metallurgical technology, was introduced to West Mexico. It was comprised of small portable objects fashioned by cold work—sewing needles, depilatory tweezers, fishhooks, and awls—as well as open rings and other objects.

Most scholars accept the idea that some Ecuadorian traits (including metallurgy) were introduced to West Mexico by the seagoing peoples of coastal Ecuador. I will consider the issue in light of the new laboratory data presented here, and at the same time address the question of how lost-wax casting, which was not typical of Ecuador, may have been introduced to the West Mexican region.

The crucial evidence for a maritime introduction of metallurgy is the discontinuous geographical distribution of the particular technical complex I have identified here: specifically of identical fabrication methods used to craft objects possessing the same design attributes and made using the same materials. During the period in question, roughly A.D. 600 to A.D. 1200–1300, West Mexican and

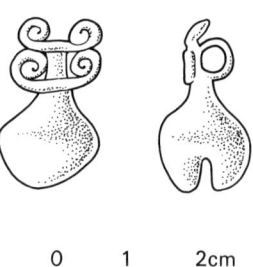

4.18

Cerro de Huistle bell (type 1c); similar bells are found in Colombia and Panama.

Ecuadorian smiths fashioned beam design tweezers, open rings, perforated-eye needles, fishhooks, awls, and other artifact classes by cold-working them to shape from an original cast blank; fabrication methods, design, and materials were similar if not identical. They also shared the same method of fashioning a rodlike form, which was to

4.19

Panamanian bell (a) similar to West Mexican type 1a, and Colombian bell (b) similar to West Mexican type 11a. Objects on display at the American Museum of Natural History, New York.

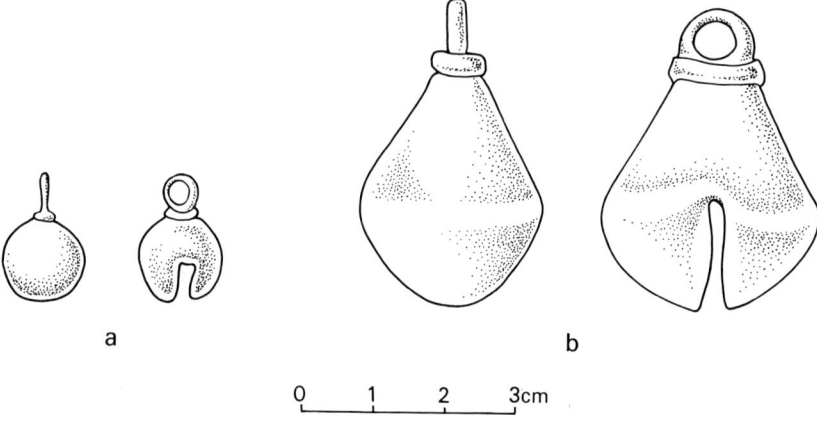

hammer it around, leaving an internal longitudinal fissure. The earliest West Mexican examples of these artifact classes appear some 500 years later than their Ecuadorian analogues. Period 1 West Mexican smiths worked primarily with copper but occasionally also used low-arsenic copper-arsenic alloys. As I document in the subsequent section, Ecuadorian smiths used copper only occasionally, but low-arsenic copper-arsenic alloys extensively. What makes the evidence for a maritime introduction compelling is that this complex of materials, techniques, and designs is virtually unknown during this period in the intervening areas of Colombia, lower Central America, and southern Mesoamerica.[11]

During this same period, metalsmiths in West Mexico and in Colombia and lower Central America were making bells by lost-wax casting. The technique and the particular bell designs predate their West Mexican counterparts by several hundred years. No evidence presently supports diffusion overland, since several of the West Mexican bell designs with the earliest dates have lower Central American and Colombian prototypes but are absent in southern Mesoamerica (Guatemala, Chiapas, Oaxaca, Tabasco), on the Yucatán Peninsula, or on the southern periphery (Honduras, El Salvador).[12] At least initially, lost-wax casting must have been introduced to West Mexico via a Pacific coast maritime route. Later, knowledge of the method moved north, but overland, from lower Central America (see Bray 1977; Pendergast 1962b). Between A.D. 900 and A.D. 1100 at the Bay Islands in Honduras (Strong, Kidder, and Paul 1938) and by A.D. 1150 at Lamanai, Belize, a distinctively southern Mesoamerican casting style took shape, characterized by certain bell designs and other small lost-wax cast objects found at these sites.[13]

The introduction of metallurgy via a maritime route is especially plausible given the seagoing orientation and the sophisticated navigational technologies of coastal Ecuadorian polities. Seagoing merchants operating off the coast of Ecuador, the so-called League of Merchants (Jijón y Caamaño 1940–1945; Marcos 1978; Norton 1986), united the coastal centers before the Spanish invasion, and metal objects were very likely items of trade. Ethnohistoric sources reveal that these sailors navigated balsa wood rafts with sails from the central Ecuadorian coast around Salango and Manta (see figure 4.3) to at least as far south as northern Peru. They also sailed north to Esmeraldas at the modern border of Ecuador and Colombia (E. Estrada 1955; J. Estrada 1988). The inventory of exotic materials brought from Ecuador into Peru makes clear that the maritime activity was extensive. They include emeralds from Colombia, *Spondylus* shell, and *chaquira* or tiny beads elaborated from *Spondylus*. All have been found in elite tombs in the Lambayeque valley of Peru (Shimada 1985). This maritime trade network had considerable chronological depth by the Middle Horizon (circa A.D. 600–1000) in the central Andean area. How early this network may have been operating off the coast of West Mexico is unclear.

These large balsa wood sailing rafts were equipped with movable centerboards and rigged sails. Specialists who have considered raft design and the pattern of Pacific Ocean currents estimate that these rafts were fully capable of making the voyage north to Mexico (Edwards 1969) either nonstop on the open sea or by coastal expeditions. All materials required to construct such oceangoing rafts were available along the Ecuadorian coast. Balsa, a self-sowing tree that grows in humid regions of the tropical Pacific coast of Ecuador, made up the primary construction material. However, bamboo, hemp for cordage and lashings, and cotton for sails were also plentiful along the coast. Figure 4.20 reproduces a sixteenth-century drawing of one of these balsa wood rafts.

Sixteenth-century Ecuadorian sailing raft drawn by Jerónimo Benzoni (Benzoni 1857: 55).

The archaeological record and documentary sources strongly suggest that the single most important item exchanged by these merchants was the seashell *Spondylus* (Murra 1975; Paulsen 1977), which is found in the warm waters of the Pacific Ocean in discontinuous pockets from the Gulf of Guayaquil to the Gulf of California. *Spondylus* was widely used in highland Ecuador for ritual purposes from 3000 B.C. on (Paulsen 1974). Its use became common in Peru, both in the highlands and along the coast, beginning in the Early Horizon (circa 700 B.C. to 200 B.C.; Burger 1984; Marcos and Norton 1981). Because this mollusk cannot survive in the cold, upwelling waters off the Peruvian coast it was acquired from Ecuador. The demand for *Spondylus* increased dramatically around A.D. 900 and reached its apogee among the large expansionist states of the central Andes: the kingdom of Chimor and the Inka empire (Cordy-Collins 1990). Marcos (1978) has suggested that an interest in acquiring *Spondylus* shell may have been the factor that brought Ecuadorian seafarers to the coasts of West Mexico; this is one of the creature's natural habitats. It is not coincidental that most West Mexican sites where metalworking is first in evidence—Tomatlán in Jalisco, Amapa in Nayarit, and the Infiernillo sites along the Río Balsas—lie along the coastal plain or have riverine access to it.

What evidence do we have that metal objects were among the goods that were moved north from Peru and Ecuador in exchange either for the shell or other goods? Bartolomé Ruiz de Estrada, Francisco Pizarro's chief pilot, captured a seagoing balsa wood raft as it sailed northward along the Ecuadorian coast (Oviedo y Valdés 1945; Sámano-Xerez 1937). Ruiz describes a raft with 20 men aboard filled with "trade" goods that included "muchas piezas de plata, diademas, coronas, cintos, tenazuelas y cascabeles . . . todo esto traían para rescatar unas conchas de pescado" [many silver objects, tiaras, crowns, bands, tweezers, and bells, all of this they brought to exchange for some shells] (Sámano-Xerez 1937: 65–66). An account from Zacatula (see figure 6.1), near the mouth of the Río Balsas, may record the end point of such voyages. The document relates that local peoples' grandfathers had traded with mariners from the south who brought rich cargoes in canoes (West 1961).

These merchants may have been acquiring *Spondylus* along the coast of West Mexico, but the data are not available to prove it. As I mentioned earlier, we also do not know when the maritime exchange system might first have begun to link these coastal zones. The Capacha ceramic complex in Colima (see chapter 1) dates to 1500 B.C. Kelly (1980) argues that the Capacha ceramic material is more closely related to ceramic complexes in northern South America than to anything in Mesoamerica at that time. We do know that *Spondylus* grows off the coast of Sinaloa and that it was ritually important at the great city of Teotihuacán (Starbuck 1975). It also figured among tribute items to the Aztecs (A.D. 1450 to A.D. 1521). *Spondylus* provides only one means of explaining the Ecuadorian voyages that precipitated the introduction of Ecuadorian metallurgical techniques to West Mexico. Other goods may have been important as well. Joseph Mountjoy (1969) has suggested that peyote, a more perishable item, may also have interested these South American peoples. That suggestion is also difficult to document, but it must be seriously considered in light of the widespread use in South America of hallucinogens and other mood-altering substances. Traders were most probably acquiring a variety of goods, and *Spondylus* may or may not have figured among them.

Evidence for the physical presence of Andean peoples in West Mexico does appear in other realms. Ornithologists (Crossin 1967; Haemig 1979) point to a curious pattern in the distribution of the painted jay, *Cyanocorax dickeyi*, in the Americas. In West Mexico this bird occupies an extremely narrow range, encompassing some 190 by 30 air kilometers that crosscut a mountainous region including portions of the West Mexican states of Nayarit, Sinaloa, and Durango. The painted jay's taxonomically closest relative is the white-tailed jay, *Cyanocorax mystacalis*, known only from coastal regions of Ecuador and Peru. These two jays are so closely related that one ornithologist proposed they be considered representatives of the same species, an idea rejected by other taxonomists because the two jays are separated by a distance of 4,000 kilometers. However, their habitats differ. The South American jays occupy the tropical coastal plain. The West Mexican bird lives in a mountain habitat above about 1,200 meters. The very narrow range of the West Mexican jays and the fact that their habitat is so different from that of their South American relatives indicate that they are an exotic species in West Mexico. On this basis Haemig (1979) argues that the West Mexican jay was introduced to its highland environment by humans. Birds adapted to one habitat but introduced to another generally reproduce poorly, which would explain the West Mexican jay's unusually narrow range. Perhaps the Ecuadorian maritime expeditions were accompanied by these raucous, exotic white-tailed birds, which were prized as much for their ability as mimics as for their resplendent tail feathers.

How did these jays end up in a mountainous habitat? I can offer two possible explanations, and both or neither may be correct. The region inhabited by the painted jay lies in a zone of mineralization, where geological maps show deposits both of copper and tin. Perhaps the birds escaped as Ecuadorian metalworkers accompanied their West Mexican exchange partners in a search for ore minerals. The other possibility, suggested by Mountjoy, is that the Ecuadorian seafaring voyagers were seeking peyote. Peyote is extremely important to the indigenous peoples of this particular area. It was and is a central element in the religious life of the Huichol people, who undertook long treks from their home territory to the northeast to harvest it.

The maritime exchange system can explain the tweezers, needles, rings, awls, and other Ecuadorian-style objects that appear in West Mexico. The lost-wax cast bells are more perplexing, because lost-wax casting was rare in Ecuador. Nonetheless, the evidence to date indicates that the technique and the prototype bells were probably also introduced via a maritime route. The first known appearance of metallurgy in West Mexico, at Tomatlán, contains elements typical both of Ecuador and of lower Central America and Colombia: a lost-wax cast bell, a probable needle fragment, two tweezers, a small awl-like item, a fishhook, and two metal strips. One of the Cerro de Huistle bells may date to before A.D. 650.[14] However, even these objects and those from Amapa are not imports from Ecuador nor from lower Central America and Colombia; they were copied locally from Ecuadorian, Colombian, and lower Central American prototypes. Lost-wax cast bells made in lower Central America and/or Colombia must have been acquired by Ecuadorian sailors as they stopped off on the voyage north, then replicated by West Mexican artisans.

Several pieces of evidence substantiate this reconstruction. The wire tweezers and spiral ornaments found at Tomatlán are identical to those made in the La Tolita-Tumaco area, the coastal region on the Ecuadorian-Colombian border. The presence at Tomatlán of such objects strongly suggests that Ecuadorian sailors were in contact with these peoples, whose neighbors excelled at lost-wax casting. Emeralds were traded south to Peru from that same area. Costa Rica may have been another stopping-off point for these voyages. Michael Smith and Cynthia Heath-Smith (1980) point out that the ceramics at Amapa, one group of which pertains to the so-called Mixteca Puebla tradition, are very closely related to ceramics on the Gulf of Nicoya in Costa Rica, a logical port in a Pacific coast maritime route. Ceramic assemblages in both areas date to between A.D. 700 and A.D. 800.

None of the data reviewed here—archaeological, ethnohistorical, laboratory-analytical—indicate that either stock metal or artifacts were introduced on a large scale to West Mexico. West Mexican bells, needles, rings, and tweezers differ enough in certain formal design elements that in most cases (rings may constitute exceptions) they can be visually distinguished from their Ecuadorian counterparts. In addition, the presence and concentration levels of certain major, minor, and trace elements varies, and generally distinguishes artifacts from the two regions. Copper-arsenic alloys were common in both areas, for example, but arsenic concentration ranges for a particular artifact class differ, as does the mean arsenic concentration for that class. These differences in the pattern of arsenic alloying also have been demonstrated for Ecuadorian and West Mexican axe-monies (Hosler, Lechtman, and Holm 1990).

The presence in both areas of similar object designs, fabrication methods, and generally similar but not identical chemistries suggests that a few lower Central American, Colombian, and Ecuadorian artifacts were traded into West Mexico and that their designs were copied and subsequently manufactured locally. I have identified a few artifacts that may have been imported. For the most part, however, it was technical know-how rather than objects that was introduced. That knowledge was of ore types, smelting technologies, fabrication methods, and the kinds of objects one could make from metal. Initially, of course, these were the same that were made from metal in Ecuador, Colombia, and lower Central America.

The Evidence From Ecuador

Setting, Sites, and Objects. Archaeologists term the territory occupied by the modern nation of Ecuador as the Intermediate Area, because it is one of the nations that lies between the great ancient civilizations of Mesoamerica and the central Andes. Ecuador is dominated by the parallel mountain chains of the Andes, with peaks reaching to 7,000 meters and a series of high intermontane valleys. A sedimentary coastal plain ranging in width from 50 to 200 kilometers extends to the Pacific Ocean. Semiarid conditions prevail on the southern and central coastal plain; rainfall increases moving north. Tropical rain forest vegetation characterizes the eastern slopes of the Ecuadorian Andes along their entire length. Agricultural productivity on the coast varies with rainfall, although maize, beans, and other New World crops can be grown in most areas. A broad spectrum of crops flourish in the highland basins, including maize, quinoa, and potatoes. Metallic minerals are virtually nonexistent on the coastal plain. The primary metallic ores are located in the highlands and on the mountain slopes to the west and east. Deposits of native metals and ore minerals of gold, silver, and copper, and of arsenic-bearing copper ores such as tennantite and enargite, occur in this zone (Goossens 1972a, 1972b). However, mineral deposits are harder to identify and physically less accessible than on the more arid *puna* of northern Peru, since the Ecuadorian highlands are covered by dense vegetation. We do not yet know whether Ecuadorian smiths obtained their raw materials from their own highland ore deposits or possibly from ore sources in the adjacent northern Peruvian Andes (Hosler, Lechtman, and Holm 1990).

During the latter part of the Integration Period (A.D. 800 to A.D. 1530; see table 4.1) much of the coastal and highland territory was organized into independent chiefdoms. Quito (and the highlands to the south) had been incorporated into the Inka empire. However, the imperial system had no demonstrable effect on the coastal polities. Some coastal centers participated in the loose-knit League of Merchants (Jijón y Caamaño 1940–1945) that managed the coastwise trading voyages discussed earlier. East-west exchange in exotic goods such as *Spondylus* shell, turquoise, and obsidian occurred as early as the Formative Period (1500 B.C. to 500 B.C.) both between regions—the coast, the highlands, and the eastern slopes of the *cordillera*—and intraregionally (Bruhns 1989; Bruhns, Burton, and Miller 1990; Buys and Domínguez 1988; Collier and Murra 1943; Doyon 1988; Villalba O. 1988).

Metal objects have been recovered from many sites along the coast, in the inland valleys of the coastal plain, and in the highland zones. Few derive from datable contexts. Large collections of casual finds and objects that have been purchased are housed in the Museo Antropológico in Guayaquil (MAG) and its counterpart in

Quito. Dated metal artifacts also have been recovered from these same coastal, inland, and highland zones. Although the sample of dated objects is small, the information it offers allows an outline of the development of metallurgy in this region, providing a framework for the comparative evidence presented in the first section (see also table 4.1, which lists the archaeological sites where such artifacts have been recovered, and their culture periods, and figure 4.3 which locates the sites).

The primary information on Ecuadorian metallurgy derives from my laboratory studies of artifacts excavated from five Ecuadorian archaeological sites on the south coast—Salango, El Azúcar, Loma de los Cangrejitos, OGSE-Ma-172, and Cerro Alto—and of the MAG collections. Information about metal artifacts excavated at the Integration Period centers of Ingapirca, Peñón del Río, and La Compañía, sometimes including artifact chemical compositions, has amplified the core study. Table 4.2 identifies sites with their corresponding cultural affiliations.

The Formative Period: 1500 B.C. to 500 B.C. Sixty-three metal objects and fragments have been excavated at Salango, ranging from the Chorrera (1500 B.C. to 500 B.C.) to the Manteño (A.D. 800 to A.D. 1530) culture periods. Artifacts dating to the first two periods, Chorrera and Bahía (500 B.C. to A.D. 800), were too mineralized for chemical analysis but were examined using low-power microscopic techniques. Neither Salango nor other coastal sites shows any significant evidence of metal processing. Metal probably was first worked in the highlands near the ore deposits, and traders then moved the objects (or possibly raw materials in ingot form) to the coast. The coastal zone of Ecuador is far better known archaeologically than other areas, however, and this early coastal evidence for metallurgy probably reflects that reality.

The Salango data show that the first objects fashioned by Ecuadorian metalsmiths were ornaments made using copper, silver, and gold. Metalsmiths shaped these items by hammering them from the native material or from an original cast blank. At Salango, one gold nose ring, a piece of silver foil, and a few pieces of copper sheet date to the Chorrera occupation. Thus the first objects Ecuadorian smiths crafted were status markers, displaying the golden, silvery, and coppery colors of these metals. These objects signal the emergence of a technological style oriented around metal sheet that persisted for at least two millennia. Artisans in the adjacent Peruvian highlands were working sheet metal also, during the contemporaneous Chavín horizon.

The Chorrera occupation at Salango thus offers the earliest evidence so far for metalworking in Ecuador. We know relatively little about Chorrera social forms, village size, or settlement patterns, but we do know that some Chorrera period sites were quite large (Marcos 1986). Salango was relatively small during this time, but by the subsequent Bahía culture period Salango had become one of the major ceremonial centers in the region. Archaeologists believe that Salango and the historically known Manteño polity of Salangone, a member of the League of Merchants (Norton 1986), were one and the same.

The Regional Developmental Period: 500 B.C. to A.D. 800. By the beginning of the following period, regional chiefdoms had emerged in Ecuador, and many of those in the coastal zone were involved in maritime exchange. The distinctive metallurgy of southern Ecuador crystallized at this same time. Metalsmiths added tools to their repertoire, crafting tweezers, needles, fishhooks, and awls from copper. They also began to experiment with the two alloys, copper-arsenic and copper-silver, that became fundamental to the metallurgy of this region. These

Table 4.4 Quantitative Chemical Analyses of Artifacts from El Azúcar

		Composition (weight percent)											
Type	ID No.	Ag	As	Au	Bi	Co	Fe	Ni	Pb	Pt	Sb	Sn	Zn
Awl	3638	0.004	—	na	—	—	na	—	—	na	—	—	—
Needle	3635	0.006	—	na	—	—	na	—	—	na	0.01	—	0.001
Open ring	3640	0.04	—	0.26	—	—	na	0.02	—	—	—	0.01	0.007
Open ring	3642	1.00	—	0.93	—	—	na	0.01	—	0.27	—	—	0.003
Star	3639	0.03	—	na	—	—	na	—	—	na	—	0.02	0.002
Star	3641	0.04	—	na	—	—	na	—	—	na	—	—	0.003
Tweezer	3636	0.007	0.08	na	—	—	na	—	—	na	—	—	—

Note: Analyses carried out by atomic absorption spectrometry. A dash indicates element not detected; "na" indicates element not analyzed quantitatively because not detected in qualitative analysis.

two alloys came into use at approximately the same time in northern Peru. Apart from tools, artisans continued to fashion sheet metal objects. Some were made from gilt copper. A gilt copper-gold sheet metal headband from a Salango Bahía period (circa 300 B.C.) burial provides a spectacular early example of this tradition.

In the southern highlands, one of the earliest metal objects appears at Pirincay (Bruhns 1989), a hilltop center that regulated traffic between highlands and coast. The artifact is a small, flat, cold-worked nose ring, reported as gilt copper. Another contemporaneous item from that site is a very small, hammered, rectangular copper bar (Bruhns 1989; Bruhns, Burton, and Miller 1990) containing about 5% gold by weight.[15] Both date to between 100 B.C. and A.D. 100.[16] La Florida, near the city of Quito, has yielded one of the only other datable (A.D. 340) highland metal assemblages from this period (Doyon 1988), and it also demonstrates the artisans' early interest in sheet metal. The artifacts from La Florida—beads, plaques, and rings—were recovered from burials.

Macroscopic examination shows them to be made from gold, copper, gilt copper, possibly gilt copper-silver alloys, and alloys of copper-gold or *tumbaga* (Doyon 1988).[17]

Tools dating to the Regional Developmental Period come from the coastal sites: Salango, El Azúcar, and Cerro Alto. Metallographic studies reveal that they were cold-worked to shape or were cast in a mold, then worked cold to achieve their final form. El Azúcar, a small Guangala period habitation site, has yielded a varied assemblage of tools and other artifacts. A series of radiocarbon determinations date the levels where metal artifacts were recovered to between A.D. 100 and A.D. 400 (Masucci 1992). Fourteen metal objects were excavated from midden material, a hearth, and a stamped, earthen house floor. The tools consist of two needles, a tweezer, and an awl. The other artifacts, open rings, tiny bells, beads, and small star-shaped ornaments, are sumptuary items. As table 4.4 shows, the smiths used a relatively pure copper for the tools as well as for the star ornaments and one ring. They used other metals and their alloys for some of the status objects: a copper-silver-gold alloy for one of the rings, and a copper-gold alloy for the bells and the beads. (The latter was determined by macroscopic examination.) All objects were cold-worked to shape except the stars, which were cast from copper. One of the most distinctive characteristics of this El Azúcar assemblage is the relative purity of the copper: the metal contains neither arsenic, antimony, nor lead, elements that, in the subsequent period, are present in nearly all artifacts. Masucci (1992) argues that this and other Guangala period sites on the southwest Ecuadorian coast participated in long-distance exchange networks that provided exotic materials manufactured in the highlands, including obsidian, gold, and copper objects. The El Azúcar area, while characterized by small sites with shallow middens, is relatively rich in these nonlocal items.

As I mentioned earlier, none of the artifacts from Salango dating to this period could be analyzed chemically because they were nearly totally mineralized. Most are ornaments; the assemblage contains several nose rings probably made from *tumbaga,* pieces of gold foil, beads apparently made from unalloyed copper, as well as fragments of sheet metal that resemble corroded copper-gold alloy material. Only one tool was recovered there: an object with a narrow shank terminating in a small curved blade. At San Lorenzo, a contemporaneous Jambelí culture period cemetery in the Guayas basin, the only metal items excavated were open rings, described as earrings (Ubelaker 1983). However, at Cerro Alto on the Santa Elena Peninsula, and at the OGSE-Ma-172 site in the coastal village of Valdivia, people were also using metal for implements and tools. The Cerro Alto and OGSE-Ma-172 assemblages include a needle, two fishhooks, an awl, and several tweezers. These implements were cold-worked to shape; the needle, containing 3.24% arsenic, is made from the earliest copper-arsenic alloy yet identified in Ecuador (table 4.5). A few of the copper artifacts from these sites, like those from El Azúcar, are also made from relatively pure metal.

The Integration Period: A.D. 800 to A.D. 1530. During the Integration Period some of the regional chiefdoms along the coast brought large expanses of territory under their control and set in place a vast agricultural infrastructure for building and maintaining thousands of hectares of raised fields, or *camellones* (Holm 1978). The larger of these polities may have supported populations as high as 30,000 individuals. Archaeologists have referred to some of these as "merchant societies" (Salomon 1986), whose political economies were grounded in the long-distance movement of goods. Many of these goods were exotic, supplied by specialist merchants who traveled on land (Salomon 1978, 1986), primarily in an east-west trajectory, or by the seagoing League of Merchants. The sailors dominated traffic in exotic goods along extensive stretches of the Pacific littoral for over half a millennium. As the regional chiefdoms expanded, the production and use of metal increased as well: cemeteries and elite caches in the highlands and on the coast characteristically contain hundreds, sometimes thousands of prestige items, often of copper and its alloys.

At Salango, tools made from metal became increasingly common; among the approximately 50 metal objects that date to this period, half are tools. Table 4.6 presents the quantitative chemical data for the 12 artifacts that could be analyzed. Three are made from copper-silver alloys: two awls and a ring. The awls contain only 4.5% and 9.6% silver respectively and were probably unintentional alloys, but the ring, with 45% silver, would have looked like pure silver after shaping by hammering and annealing. The metalsmiths used low-arsenic copper-arsenic alloys for a needle and a tweezer. Other Salango artifacts from this period were made from unalloyed cop-

Table 4.5 Quantitative Chemical Analyses of Artifacts from Cerro Alto and OGSE-Ma-172 (Santa Elena Peninsula)

Type	ID No.	Ag	As	Au	Bi	Co	Fe	Ni	Pb	Pt	Sb	Sn	Zn
Regional Developmental Period													
Awl	3650	0.05	—	na	—	—	na	—	—	na	—	—	0.001
Fishhook	3645	0.03	—	0.13	—	—	na	—	—	na	—	—	0.002
Fishhook	3651	0.01	—	na	—	—	na	—	—	na	—	—	0.001
Needle	3644	0.09	3.24	na	—	—	na	—	—	na	—	—	0.002
Tweezer	3649	0.05	0.21	na	0.03	—	na	0.05	0.15	na	0.06	—	—
Integration Period													
Needle	3652	0.02	1.37	na	0.02	—	na	0.04	0.18	na	0.05	—	0.001
Star	3647	0.02	—	na	—	—	na	0.02	—	na	—	—	0.26

Note: Analyses carried out by atomic absorption spectrometry. A dash indicates element not detected; "na" indicates element not analyzed quantitatively because not detected in qualitative analysis.

per. The copper-arsenic alloys contain arsenic in concentrations between approximately 0.5% and 1.0%, high enough to have altered the working properties of the metal. Even when arsenic is present in these low concentrations the alloy can be made somewhat harder by cold working. Formal design can be modified, too. Yet, as the MAG data presented subsequently will show, these ancient smiths rarely used the properties of this particular alloy to alter or optimize object design.

At Ayalán, the large Integration Period cemetery southwest of Guayaquil, archaeologists excavated a large corpus of metal objects from urn and extended burials (Ubelaker 1981): rings, nose rings, pendants, tweezers, and axe-monies. Axe-monies are thin, axe-shaped pieces of sheet metal (Hosler 1986; Hosler, Lechtman, and Holm 1990), often stacked and bound in packets. They are almost always made of a low-arsenic copper-arsenic alloy. The earliest axe-monies in this region appear at Ayalán. Axe-monies are also common in West Mexico after A.D. 1200 and are discussed in chapters 5 and 6. Radiocarbon determinations show that the earliest date for a burial with associated metal at Ayalán is A.D. 750. The cemetery continued in use until shortly before the Spanish invasion.

Loma de los Cangrejitos, a ceremonial complex with five small pyramids and an elite cemetery, was occupied from A.D. 900 to the historical period. The 125 metal artifacts excavated there all derive from burials, and include bells, tweezers, needles, axes, axe-monies, knives, and awls. Implements and sumptuary objects are present in almost equal proportions. Three phases are represented at Loma (Marcos 1981); most metal artifacts come from the earliest of these (A.D. 900 to A.D. 1150).[18] Bells are more numerous than any other objects and make up 35%

Table 4.6 Quantitative Chemical Analyses of Artifacts from Salango

		Composition (weight percent)											
Type	ID No.	Ag	As	Au	Bi	Co	Fe	Ni	Pb	Pt	Sb	Sn	Zn
Awl	3656	9.6	0.09	0.12	—	—	na	0.02	0.12	na	0.03	—	0.01
Awl	3662	4.5	0.06	0.07	—	—	na	0.01	0.23	na	0.02	—	0.001
Bell	3658	0.3	—	0.01	0.08	—	na	—	0.16	na	—	—	—
Fishhook	3654	0.03	—	na	—	—	na	0.02	0.05	na	0.03	—	—
Needle	3659	0.06	0.78	na	—	—	na	0.03	0.19	na	0.03	—	—
Needle	3660	0.08	0.38	na	—	na	na	na	0.44	na	0.03	na	0.001
Needle	3661	0.22	0.2	0.004	—	—	na	0.03	0.09	na	—	—	—
Needle	3665	0.85	—	0.02	—	—	na	0.02	0.08	na	0.03	—	0.003
Open ring	3655	45.5	—	0.39	0.11	—	na	—	0.41	na	—	—	0.02
Open ring	3663	0.01	0.70	na	0.06	—	na	—	1.60	na	0.1	—	0.004
Open ring	3666	0.03	—	na	—	—	na	0.02	0.41	na	0.05	—	0.005
Tweezer	3664	0.06	0.46	na	—	—	na	0.03	0.39	na	0.03	—	—

Note: Analyses carried out by atomic absorption spectrometry. A dash indicates element not detected; "na" indicates element not analyzed quantitatively because not detected in qualitative analysis.

of the assemblage; qualitative analyses show that all bells are made from copper or a very low-arsenic copper-arsenic alloy. These Ecuadorian bells are cold-worked to shape, unlike their counterparts from Mexico which are lost-wax cast. All the utilitarian objects analyzed—tweezers, needles, awls, knives, and axes—are also made from a low-arsenic copper-arsenic alloy (table 4.7), which served as a kind of stock material for the metal objects found at this site. The awls and axes contain arsenic in concentrations between 0.5% and 2.0%. Axes and tweezers together comprise 33% of the corpus. The axe-monies recovered at Loma de los Cangrejitos, like all Ecuadorian axe-monies, are made from copper-arsenic metal (Hosler, Lechtman, and Holm 1990). The absence of objects made from gold, silver, and their alloys is unusual and may reflect some aspect of the excavation strategy.

Metal artifacts never were as abundant at the coastal centers as they became in the inland area. We have no evidence thus far that coastal artisans crafted metal objects on any significant scale, although they probably did so occasionally. At Salango, for example, a ceramic mold was recovered that could have been used for casting metal tools. Most items were likely imported to these sites as finished objects. The distributional evidence indicates that by the late Integration Period (A.D. 800 to A.D. 1530), the inland Milagro-Quevedo area was an important zone for metal production, and it is highly probable that earlier this region also constituted a primary production locus. Two large excavated assemblages of metal objects, both dating after A.D. 900, come from sites located on the inland plain: one from La Compañía, and the other from

Table 4.7 Quantitative Chemical Analyses of Artifacts from Loma de los Cangrejitos

Type	ID No.	Composition (weight percent)											
		Ag	As	Au	Bi	Co	Fe	Ni	Pb	Pt	Sb	Sn	Zn
Awl	3496	0.03	1.73	0.008	0.03	na	—	0.39	0.01	na	0.04	—	—
Awl	3497	0.026	1.39	—	0.03	na	—	0.03	0.26	na	0.04	—	—
Awl	3498	0.033	1.22	0.01	0.07	na	0.026	0.27	0.27	na	0.04	—	—
Awl	3533	0.039	0.86	na	na	na	0.006	na	0.02	na	na	—	0.009
Awl	3534	0.027	0.91	0.005	0.05	na	—	0.03	0.16	na	0.04	—	—
Axe	3493	0.058	0.55	—	—	na	—	0.02	0.09	na	0.03	—	0.002
Axe	3499	0.051	0.63	0.002	—	na	0.008	0.03	0.17	na	0.02	—	0.001
Axe	3500	0.062	0.61	—	—	na	—	0.03	0.17	na	0.03	—	—
Baston	3536	0.054	0.68	—	—	na	—	0.03	0.1	na	0.03	—	—
Knife	3535	0.027	1.21	—	0.06	na	0.007	0.03	0.22	na	0.03	—	0.001
Needle	3521	0.032	1.66	0.01	0.05	na	—	0.03	0.36	na	0.03	—	—
Pendant	3494	0.048	2.90	0.007	0.63	na	—	0.02	0.01	na	0.11	—	0.001
Pendant	3495	0.047	3.12	0.006	0.67	na	—	0.02	—	na	0.13	—	—
Tweezer	3523	0.054	0.47	0.31	0.40	na	0.001	0.001	0.83	na	0.01	—	0.001
Tweezer	3524	0.046	0.67	—	0.08	na	—	0.04	0.87	na	0.09	—	—
Tweezer	3525	0.025	0.72	0.02	—	na	—	0.01	—	na	0.03	—	—
Tweezer	3527	0.054	0.12	na	na	na	0.003	na	0.66	na	na	—	—

Note: Analyses carried out by atomic absorption spectrometry. A dash indicates element not detected; "na" indicates element not analyzed quantitatively because not detected in qualitative analysis.

Peñón del Río. Numerous looted sites in this inland zone have also yielded large numbers of metal objects.[19]

Peñón del Río, where 182 metal objects were excavated, was a secondary center that functioned as a redistribution locale for agricultural products and was also involved in commercial interchange (Buys and Muse 1987; Sutliff 1992). It has yielded the largest excavated assemblage of Milagro metal artifacts from domestic contexts, and Sutliff's evidence indicates that the objects were manufactured at the site. People at Peñón del Río used metal primarily for items that marked social status: rings and nose rings comprise 30% of the assemblage. Other objects include tweezers, needles, bells, fishhooks, and awls. Although chemical analyses have not been performed on this material, Sutliff's macroscopic observations (1992) and her metallographic studies indicate that four of the ornaments appear to be gilded and 20 seem to be copper-silver alloys. The remaining objects appear to be made primarily from copper.

Excavation of several mounds at the late Integration center of La Compañía has yielded large numbers of metal artifacts from elite urn and chimney burials. The corpus includes hundreds of bells, nose rings, open rings, sheet metal bangles, and tools, including fishhooks, needles, tweezers, and others. One hundred objects were analyzed from this site from different functional classes (Scott 1988; Battelle Institute n.d.). Artisans used copper-silver alloys to fashion rings, pendants, and other sumptuary objects. Copper or a low-arsenic copper-arsenic alloy was reserved for the utilitarian items.

In summary, the evidence from these excavated assemblages indicates that the earliest metalworkers of this Andean region crafted status objects from unalloyed copper, silver, and gold. Gilding techniques were also developed very early. Through time, the proportion of tools to ornaments made from metal seems to have increased. By A.D. 300, metalworkers were using copper-silver and copper-arsenic alloys, and some of the objects they fashioned were tools. In southern Ecuador (and probably also in northern Peru) these very low-arsenic copper-arsenic alloys (0.5–2.0% arsenic) eventually came to constitute a kind of stock material; smiths systematically employed them for all common artifact classes: awls, tweezers, bells, solid rings, axe-monies, and others.

The generalized use of copper-arsenic alloys also is visible in highland assemblages. Among the 59 objects analyzed from Ingapirca, all but three contain arsenic in concentrations above 0.05%. Objects include rings, bells, a pendant, and miscellaneous ornaments (Escalera U. and Barriuso P. 1978). Often, artifacts made from this alloy contain arsenic in such low concentrations that it either did not affect or only marginally affected the working properties of the metal, and even then only if the metal had been substantially reduced in thickness.

The Museo Antropológico, Guayaquil (MAG), Assemblage. These findings regarding the fabrication techniques and metals and alloys that defined the copper-based portion of this technology were amplified in investigations of the large metal collections of the Museo Antropológico in Guayaquil (MAG). Data from the MAG also provide systematic information concerning how metalsmiths managed alloy-property relations and the extent to which they utilized the properties of the metals and alloys to optimize object design and function. The portion of the museum collection made available for study contained approximately 4,395 recognizable objects and numerous unidentifiable fragments. The total number of metal artifacts reached 7,873 in a subsequent inventory of all museum holdings. Table 4.8 presents the inventory results, listing the most common artifact types in order of relative abundance. The extent to which the percentages shown represent frequencies of occurrence in the prehistoric era varies with artifact class. Implements and tools, such as needles and awls, are underrepresented; they are common in archaeological sites but rare in this collection (less than 10%). The numbers of sumptuary objects with respect to one another probably more closely reflect the ancient situation. Some 50 to 60 MAG objects did not figure in this analysis because they were made from gold, silver, and their alloys and were unique types.

Macroscopic examinations reveal that silver-copper was a primary alloy system in ancient Ecuador; a large proportion of the open rings, hollow rings, and sheet metal ornaments are made from it.[20] The other obvious characteristic is that a very high proportion (60%) of the heavy, cast and worked nose rings are made from gilt copper (see Scott 1986). Their color indicates that some *tumbaga* alloys are found among them, although, apart from a few small, hammered *tumbaga* nose rings from the La Tolita region on the far north coast of Ecuador

Table 4.8 Metal Collection of the Museo Antropológico del Banco Central, Guayaquil (MAG)

Type	Study Corpus		Number of Artifacts	
	Number Available	Number Sampled	Total in Collection	Percent of Entire Collection
Rings (solid)	954	27	2335	29.7
Axe-monies	1993	24	2167	27.5
Bells	883	25	1835	23.3
Nose rings	119	13	394	5.0
Sheet metal ornaments	150	4	377	4.8
Tweezers	224	21	304	3.9
Axes	40	34	174	2.2
Needles	2	2	147	1.9
Rings (hollow)	18	0	120	1.5
Fishhooks	10	3	20	0.3
Star-shaped artifacts	2	1	2	< 0.1
Total	4395	154	circa 7873	

(Bergsøe 1937), no published analyses exist of Ecuadorian objects that verify this observation.

One hundred fifty-four objects were sampled for analyses of chemical composition from the artifact classes listed in table 4.8. The results of those analyses appear in table 4.9. Arsenic is present in all but three artifacts, in concentrations ranging from 0.02% to 4.61%. This arsenical copper, which grades into a low-arsenic copper arsenic alloy, represents the same stock material (containing arsenic, antimony, and lead) used to make the objects analyzed from Loma de los Cangrejitos and from other sites. These 154 objects comprise several compositional subgroups constructed on the basis of the presence and relative concentration of certain elements other than arsenic. The most important of these is silver. Some contain silver in low concentrations (less than 0.40%). In others, the silver content is so high (7.33–42.9%) that the artifacts constitute copper-silver alloys. One object (a tweezer) is an alloy of copper and gold with gold present at 2.98%. There are no alloys of copper-tin bronze.

When arsenic and silver alloy with copper they affect the working properties of the metal. The copper-silver alloy becomes strong and extremely tough, allowing metalsmiths to fashion thin, flexible, silvery-looking objects from sheet metal. The strength and toughness of this alloy increases dramatically after adding approximately 7–8% silver to copper. This is also the approximate lower limit at which silver enrichment occurs at the surfaces of worked and annealed copper-silver alloys. In the copper-arsenic system, the metal becomes progressively harder as arsenic concentration increases; these effects are noticeable at about 0.4–0.5% arsenic.

Table 4.9 Quantitative Analyses of Artifacts in the Museo Antropológico, Guayaquil (MAG)

		Composition (weight percent)										
Type	ID No.	Ag	As	Au	Bi	Co	Fe	Ni	Pb	Sb	Sn	Zn
Axe	3444	0.04	1.06	na	0.07	na	na	0.02	0.38	0.01	na	na
	3445	0.04	3.75	na	0.03	na	na	0.03	0.02	0.03	na	na
	3446	0.04	3.77	na	0.03	na	na	0.03	0.02	0.05	na	na
	3447	0.04	3.81	na	0.03	na	na	0.05	0.01	0.02	0.01	na
	3481	0.04	2.36	na	0.01	na	na	0.03	0.02	0.03	na	na
	3482	0.03	1.53	na	0.06	na	na	0.02	0.09	0.03	na	na
	3483	0.03	1.38	na	0.11	na	na	0.02	0.21	0.02	na	na
	3611	0.04	0.39	na	0.11	—	na	0.1	0.03	0.04	—	—
	3612	0.07	0.92	na	—	—	na	0.04	0.3	0.03	—	—
	3613	0.07	0.04	na	—	—	na	0.03	0.24	0.04	—	—
	3614	0.07	0.21	na	—	—	na	0.03	0.12	—	—	—
	3615	0.08	0.8	na	—	—	na	0.03	0.26	0.03	—	—
	3616	0.06	0.56	na	0.007	—	na	0.03	0.39	0.03	0.69	0.001
	3617	0.02	1.5	na	0.008	0.01	na	0.07	0.09	0.05	—	—
	3618	0.6	0.07	—	—	—	na	0.02	0.04	0.02	—	—
	3619	0.06	0.55	na	—	—	na	0.03	0.15	0.03	—	0.001
	3620	3.64	0.16	0.08	0.02	—	na	0.03	0.11	0.03	—	0.002
	3750	0.08	0.02	na	na	na	0.009	0.03	0.49	0.02	0.03	na
	3751	0.07	0.09	na	na	na	0.002	0.03	0.21	0.02	—	na
	3752	0.04	0.07	na	na	na	—	0.01	2.44	0.01	0.01	na
	3753	0.07	—	na	na	na	0.004	0.03	0.04	0.02	0.003	na
	3754	0.14	0.03	na	na	na	—	0.02	0.62	0.01	0.02	na
	3755	0.19	0.01	na	na	na	—	0.02	2.4	0.01	0.01	na
	3756	0.04	1.22	na	na	na	0.011	0.03	0.14	0.02	0.007	na
	3757	0.21	0.02	na	na	na	0.002	0.03	0.08	0.01	0.011	na
	3758	0.06	0.41	na	na	na	0.001	0.01	2.48	0.02	0.01	na
	3759	0.4	0.05	na	na	na	0.002	0.03	0.11	0.01	0.006	na
	3760	0.06	1.36	na	na	na	—	0.04	0.05	0.01	0.02	na
	3761	0.03	0.56	na	na	na	—	0.04	0.1	0.05	0.01	na
	3762	0.07	1.28	na	na	na	—	0.04	0.25	0.02	0.006	na

continued

Table 4.9 (continued)

		Composition (weight percent)										
Type	ID No.	Ag	As	Au	Bi	Co	Fe	Ni	Pb	Sb	Sn	Zn
	3763	0.09	1.99	na	na	na	0.01	0.04	0.002	0.01	0.004	na
	3764	0.06	0.21	na	na	na	—	0.04	0.91	0.01	—	na
	3765	0.05	0.56	na	na	na	—	0.05	0.25	0.01	0.02	na
	3766	0.05	0.6	na	na	na	0.007	0.03	0.1	0.1	0.01	na
	3767	25.9	0.03	na	na	na	0.01	0.01	0.14	—	0.01	na
	3768	0.09	0.49	na	na	na	0.01	0.03	0.26	—	0.01	na
	3769	0.05	0.4	na	na	na	—	0.02	0.46	—	0.02	na
	3770	0.03	0.59	na	na	na	—	0.03	0.1	0.01	0.02	na
	3771	0.04	0.39	na	na	na	0.01	0.04	0.21	0.02	0.008	na
	3772	33.34	1.25	na	na	na	0.03	0.02	0.13	0.005	0.02	na
	3773	0.09	0.8	na	na	na	—	0.03	0.006	0.01	—	na
Axe-money	3282	0.03	2.43	na	na	na	na	0.14	0.36	na	na	na
	3283	0.03	1.1	na	na	na	na	0.19	0.39	na	0.01	na
	3310	0.02	1.81	na	0.05	na	na	0.02	0.09	0.13	na	na
	3427	0.04	0.71	na	0.06	na	na	0.04	0.38	0.01	na	na
	3428	0.04	2.34	na	0.06	na	na	0.03	0.09	0.02	na	na
	3429	0.03	1.82	na	0.06	na	na	0.03	0.36	0.02	na	na
	3430	0.03	1.18	na	0.03	na	na	0.03	0.03	0.01	na	na
	3431	0.02	3.14	na	0.05	na	na	0.04	0.05	na	na	na
	3432	0.05	1.31	na	0.04	na	na	0.02	0.18	na	na	na
	3621	0.07	0.12	—	0.13	na	0.01	0.02	1.16	0.02	—	0.001
	3624	0.07	0.83	0.01	0.05	na	0.002	0.02	0.004	0.01	—	0.001
	3625	0.24	0.82	0.02	0.05	na	0.005	0.03	0.05	0.02	—	—
	3626	0.05	1.92	0.01	0.05	na	0.001	0.03	0.01	0.03	—	0.003
	3627	0.03	0.72	—	0.24	na	0.003	0.02	0.08	0.04	0.001	0.002
	3687	0.1	1.28	0.53	0.03	na	0.01	0.03	0.001	0.01	—	0.001
	3688	0.04	1.13	0.02	na	na	0.005	0.05	0.08	0.03	na	—
	3717	0.07	2.61	na	na	na	0.007	0.01	3.85	0.02	0.01	na
	3718	0.13	0.62	na	na	na	0.002	0.03	0.29	0.02	0.001	na
	3719	24.58	0.03	na	na	na	0.004	—	0.01	0.01	—	na

continued

Table 4.9 (continued)

Type	ID No.	Composition (weight percent)										
		Ag	As	Au	Bi	Co	Fe	Ni	Pb	Sb	Sn	Zn
	3720	0.06	1.85	na	na	na	—	0.01	0.02	0.01	—	na
	3721	0.07	0.2	na	na	na	0.001	0.03	0.05	0.02	—	na
	3722	0.03	0.12	na	na	na	0.001	0.02	0.02	0.01	—	na
	3723	0.04	0.59	na	na	na	0.005	0.04	0.52	0.02	—	na
	3724	0.08	1.25	na	na	na	0.004	0.03	0.42	0.03	—	na
	3725	0.05	1.67	na	na	na	0.003	0.04	0.05	0.02	0.003	na
	3726	0.09	0.49	na	na	na	0.011	0.04	0.25	0.03	—	na
	3727	21.36	1.02	na	na	na	0.006	0.02	0.13	0.03	—	na
	3728	0.04	0.81	na	na	na	0.01	0.04	0.29	0.03	—	na
	3729	0.06	0.67	na	na	na	0.003	0.02	3.34	0.02	0.008	na
	3730	0.04	0.5	na	na	na	0.002	0.02	0.44	0.02	0.001	na
	3731	0.49	1.19	na	na	na	0.003	0.03	0.02	0.02	0.006	na
	3732	0.06	0.28	na	na	na	0.001	0.01	2.53	0.02	0.02	na
	3733	0.07	0.55	na	na	na	0.011	0.03	0.23	0.02	0.015	na
Bowl	3631	30.86	0.57	0.14	0.07	na	0.005	—	0.36	0.01	0.001	0.001
	3632	28.2	0.85	0.37	0.05	na	0.02	0.007	0.06	0.01	0.002	—
	3633	30.04	0.75	0.14	0.24	na	0.02	0.02	0.76	0.02	—	—
Feathers	3442	0.05	4.61	na	0.15	na	na	0.03	na	0.01	na	na
	3443	0.04	5.2	na	0.05	na	na	na	0.04	na	na	na
Fishhook	3628	0.17	0.13	0.02	—	na	—	0.03	0.12	0.02	—	—
	3629	0.12	0.18	—	—	na	0.005	0.04	0.09	0.03	—	—
	3630	0.07	0.07	0.06	—	na	—	0.05	0.05	0.07	—	—
Hide	3449	na	1.4	na	na	na	na	na	na	na	na	na
	3508	0.04	0.77	na	na	na	na	0.04	0.04	0.01	na	0.001
	3509	0.03	1.06	na	—	na	na	na	0.08	na	na	0.001
Ingot	3502	0.03	0.36	na	0.006	na	na	0.05	0.02	0.003	na	0.001
	3503	0.04	0.3	na	0.008	na	na	0.05	0.02	0.01	na	0.001
	3504	0.16	0.33	na	0.017	na	na	0.02	0.73	—	na	0.002
	3505	0.16	0.5	na	0.031	na	na	0.02	0.48	—	na	0.002
Insignia	3450	na	1.1	na	na	na	na	na	na	na	na	na

continued

Table 4.9 (continued)

Type	ID No.	Composition (weight percent)										
		Ag	As	Au	Bi	Co	Fe	Ni	Pb	Sb	Sn	Zn
Needle	3609	1.65	0.4	0.03	—	na	0.02	0.03	0.09	0.02	—	0.001
	3610	0.03	0.74	—	0.04	na	—	0.05	0.12	0.04	—	—
Nose ring	3670	0.03	0.61	na	0.01	—	na	0.02	0.37	0.02	na	0.001
	3672	0.03	0.4	na	0.04	—	na	0.04	0.06	0.04	—	—
	3674	28.3	0.12	0.55	0.15	—	na	0.02	0.02	0.05	—	—
	3677	0.06	—	0.79	—	—	na	—	—	—	—	—
	3678	0.07	2.97	na	0.11	—	na	0.03	—	0.04	—	—
	3681	42.9	0.12	0.35	0.14	—	na	—	0.14	—	—	0.01
	3689	21.5	0.33	0.37	0.04	—	na	—	0.06	—	—	0.008
Open ring	3575	0.05	1.18	—	—	na	0.002	0.03	0.42	0.03	—	—
	3576	0.22	0.27	—	—	na	—	0.03	0.19	0.03	—	—
	3577	0.03	0.22	—	—	na	0.03	0.04	1.29	—	—	—
	3578	0.04	0.6	—	—	na	—	0.02	0.17	0.03	—	—
	3580	0.18	0.16	0.08	—	na	0.02	0.04	0.16	0.03	—	—
	3581	0.23	0.17	—	—	na	0.003	0.03	0.21	0.02	—	—
	3583	0.04	0.04	—	—	na	—	0.04	0.16	0.04	—	—
	3586	0.26	0.23	0.7	—	na	—	0.04	0.19	0.04	—	—
	3587	0.08	0.74	—	—	na	—	0.03	0.5	0.03	—	0.001
	3588	0.1	0.21	0.06	—	na	—	0.03	0.52	—	—	—
	3683	7.94	0.23	0.1	0.05	na	0.003	0.01	0.09	0.03	—	—
	3684	0.06	0.41	—	—	na	—	0.05	0.37	0.03	—	—
	3686	33.64	0.43	0.38	0.1	na	0.005	0.02	0.12	0.01	—	—
	3734	17.73	1.02	na	na	na	0.03	0.01	0.08	0.01	0.01	na
	3735	0.07	1.19	na	na	na	0.02	0.03	0.47	0.02	0.01	na
	3736	0.06	0.17	na	na	na	—	0.03	0.57	0.04	—	na
	3737	0.12	0.32	na	na	na	—	0.02	0.52	0.03	—	na
	3738	0.05	0.189	na	na	na	—	0.03	0.08	0.03	—	na
	3739	7.33	0.75	na	na	na	—	0.02	0.15	0.02	—	na
	3740	40.18	0.06	na	na	na	0.008	0.02	0.03	0.03	—	na
	3741	27.37	0.06	na	na	na	0.02	0.01	0.03	0.02	—	na

continued

Table 4.9 (continued)

| Type | ID No. | Composition (weight percent) ||||||||||
		Ag	As	Au	Bi	Co	Fe	Ni	Pb	Sb	Sn	Zn
	3742	25.82	0.7	na	na	na	0.007	—	0.05	0.01	0.004	na
	3743	0.07	1.04	na	na	na	0.006	0.03	0.51	0.02	0.006	na
	3745	0.05	0.34	na	na	na	0.02	0.05	0.29	0.01	0.04	na
	3746	24.09	0.15	na	na	na	0.01	0.02	0.1	—	0.03	na
	3747	17.76	0.17	na	na	na	0.008	0.01	0.06	0.006	0.01	na
	3748	0.04	0.88	na	na	na	—	0.03	0.23	0.02	0.01	na
Star	3634	0.03	—	—	—	na	0.02	0.01	—	0.02	—	0.001
Tweezer	3589	0.04	0.49	—	—	na	0.01	0.08	0.05	0.04	—	—
	3590	0.06	0.02	—	—	na	0.02	0.02	0.03	0.02	—	—
	3591	0.05	0.98	—	—	na	—	0.04	0.22	0.04	—	—
	3592	0.05	0.36	—	—	na	—	0.04	0.13	0.03	—	—
	3593	0.07	0.74	—	—	na	—	0.04	0.1	0.03	—	—
	3594	0.58	0.11	0.03	0.06	na	—	0.03	0.02	0.02	—	0.005
	3595	0.04	0.68	—	—	na	0.03	0.06	0.23	0.06	—	—
	3596	0.02	0.31	0.02	—	na	—	0.05	0.77	0.03	—	0.001
	3597	0.03	0.74	—	0.03	na	—	0.03	0.09	0.03	—	—
	3598	0.03	0.38	—	—	na	—	0.04	0.06	0.03	—	0.002
	3599	0.14	0.14	—	—	na	—	0.04	0.14	0.03	—	0.002
	3600	0.05	0.79	—	0.04	na	0.001	0.03	0.08	0.04	—	0.001
	3601	0.12	1.26	—	0.12	na	—	0.04	0.05	0.03	—	—
	3602	0.04	0.21	—	—	na	—	0.04	0.06	0.02	—	0.001
	3603	0.04	2.03	—	—	na	0.02	0.03	0.03	0.04	—	—
	3605	0.08	0.47	—	—	na	—	0.05	0.08	0.03	—	—
	3606	0.12	2.16	0.01	0.18	na	—	0.02	0.03	0.02	—	—
	3607	0.17	0.03	2.98	—	na	0.007	0.03	0.01	0.05	—	0.001
	3608	0.48	0.79	—	—	na	0.02	0.04	0.11	0.02	—	—
	3685	0.03	0.35	—	—	na	—	0.08	0.02	0.03	—	—

Note: Analyses carried out by atomic absorption spectrometry. A dash indicates element not detected; "na" indicates element not analyzed quantitatively because not detected in qualitative analysis.

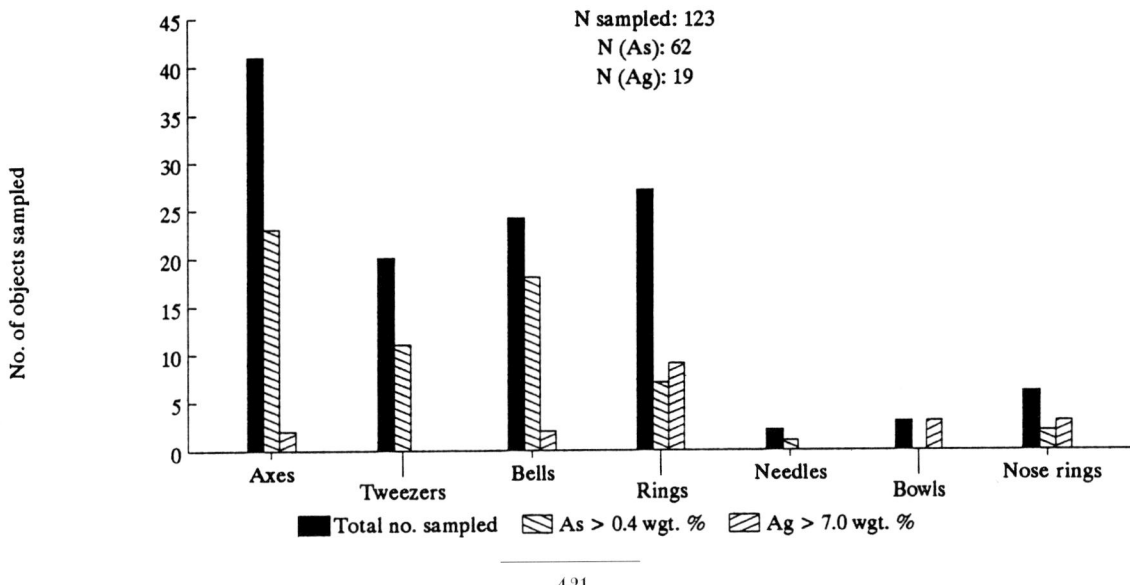

4.21

MAG: Number of artifacts from distinct functional types made from copper-arsenic and copper-silver alloys.

Figure 4.21 and table 4.10 identify the kinds and numbers of artifacts of each functional type containing arsenic at levels above 0.40% or silver at levels above about 7%. The table indicates that Ecuadorian smiths were taking advantage of the color of the copper-silver alloys for some sumptuary objects—bells, rings, bowls, and nose rings—and utilizing the copper-arsenic alloys for implements like axes and tweezers, but also for some rings and bells.

Tweezers and axes contain arsenic in high enough concentrations to alter the working properties of the metal. Arsenic present at these levels (0.40–3.81%) allows the metalsmith to reduce thickness and increase hardness. The other two artifact types made using copper-arsenic alloys, bells and rings, could have been made thinner than when made from copper alone. Nonetheless, despite the opportunities this alloy offers to optimize design, the evidence presented in table 4.11 indicates that in most cases it was not systematically used for that purpose. A good measure of design differences as a function of alloy type is a ratio: of length to thickness for axes and tweezers, of height to thickness for bells, and of diameter to thickness for rings. As table 4.11 shows, the design of tweezers containing arsenic is unaffected by alloy concentration below or above 0.40%. Axe design is affected somewhat; the thickness of axes containing arsenic in higher concentrations is reduced by approximately 20%. Bell thickness is reduced by nearly one-half. The thickness of rings remains the same. Thus, while smiths were certainly optimizing some alloy properties, such as hardness, particularly in the tweezers and axes that were cold-worked to shape, they did not take full advantage of these alloys to improve design—to make thinner, finer axes, rings, and tweezer blades.

Table 4.10 Range in Concentration of Arsenic and Silver Artifacts Containing As > 0.4 weight % and Ag > 7.0 weight %

Type	Range in Alloy Element Concentration (weight percent)		
	Low	High	Mean
Axes			
As	0.41	3.81	1.38
Ag	25.90	33.34	29.62
Tweezers			
As	0.47	2.16	1.01
Ag	—	—	—
Bells			
As	0.49	2.61	1.08
Ag	21.36	24.58	22.97
Rings			
As	0.41	1.19	0.86
Ag	7.33	40.18	22.43
Needles			
As	—	0.74	0.74
Ag	—	—	—
Bowls			
As	—	—	—
Ag	28.20	30.86	29.70
Nose rings			
As	0.61	2.97	1.79
Ag	21.50	42.90	30.90

The ore minerals that likely constituted parent materials for the artifacts from the MAG corpus and from the excavated assemblages can be reconstructed on the basis of artifact chemical composition. The artifact chemistries suggest that the stock copper-arsenic metal was either routinely smelted from arsenical copper ores, such as enargite or tennantite, or that the alloy was produced by co-smelting arsenopyrite with a copper ore. A very brief survey of the geological literature indicates that in Ecuador, arsenic appears in arsenic-bearing copper ores such as enargite, but ores of arsenic, like realgar and arsenopyrite, are also found (Goossens 1972a, 1972b). Although the ore geology of Ecuador has not been studied as systematically or as thoroughly as that of Mexico, the evidence we have suggests that ore deposits in the southern highlands of Ecuador are similar to those in the northern highlands of Peru. There, arsenic-bearing copper ores such as enargite and the tennantite-tetrahedrite series predominate. Whether the stock copper-arsenic material was imported in the form of ingots from northern Peru (Shimada 1985) or was smelted from available ores in the Ecuadorian highlands remains to be determined (Hosler, Lechtman, and Holm 1990).

The metallurgy of the southern region of Ecuador is similar to the technology of adjacent northern Peru with respect to the range of metals, alloys and fabrication techniques employed, although it lacks the inventiveness, sophistication, and scope of central Andean metallurgy. Peruvian metalsmiths used copper-silver and copper-arsenic alloys extensively, fashioning sheet metal sumptuary items from the former and tools and implements from the latter. Artifact classes differ, however. Apart from needles, whose design is identical in the two regions, in northern Peru the distinctively Ecuadorian tweezer and bell designs are rare, and rings are uncommon. The typical Ecuadorian axe-money form is absent. Metalworkers in northern Peru used the copper-arsenic alloy to fashion *naipes,* thin sheet metal objects related to axe-monies, as well as digging stick points, hoes, and other agricultural tools. They used copper-silver alloys for funerary masks, effigy vessels, and related elite items. Copper-gold and copper-silver-

Table 4.11 MAG Collection Copper-Arsenic Alloys: Arsenic Concentration and Artifact Design Characteristics

Type	Number Analyzed	As Level (wgt. %)	Mean As Concentration (wgt. %)	Mean Ratio
Tweezers	11	> 0.4	1.01	42.0 l:th
	9	< 0.4	0.21	42.0
Axes	23	> 0.4	1.38	47.5 l:th
	15	< 0.4	0.13	37.0
Bells	18	> 0.4	1.08	29.0 h:th
	4	< 0.4	0.18	15.2
Rings	7	> 0.4	0.86	10.1 d:th
	11	< 0.4	0.21	9.8

Note: l = length; th = thickness; h = height; d = diameter.

gold alloys were common, as were extraordinary gilding and silvering techniques.

The Character of Ecuadorian Metallurgy. These people fashioned an array of metal objects that make clear they were interested in metal for two purposes: for ritual and status objects that were worn—such as nose rings, rings, and bells—and for axe-monies, items that served as a standard of value or medium of exchange. Technical study of Ecuadorian axe-monies (Hosler, Lechtman, and Holm 1990) has shown that they were consistently made from a low-arsenic copper-arsenic alloy. In addition to their intrinsic social and symbolic worth deriving from their axelike shape, Ecuadorian axe-monies like their West Mexican counterparts may have served as a repository for copper-arsenic metal (Hosler 1986), from which the large majority of Ecuadorian copper-based metal objects were made.

Metal was also utilized for its resonant properties, although far less so than in West Mexico. Metallic sounds were produced by small bells ranging in height from about 0.4 to 3.5 cm, with most measuring between 1.0 and 2.0 cm. They were hammered to shape, and hammering clearly mitigated any significant variation in form. Most are round, and all exhibit smooth exterior surfaces. Since resonator size and shape vary far less than in the West Mexican designs, the range of pitches produced by these bells is correspondingly more limited. Pitch is also compromised as a result of the manufacturing process because these bells were shaped by hammering, and the consequent compression of the metal grains acts to dampen vibrations.

One of the striking characteristics of Ecuadorian metallurgy is the limited range of metals and alloys employed, at least in the copper-based component I have examined here. The range of artifact types is also restricted. Metalworkers focused primarily on three artifact classes—rings, axe-monies, and bells (see table 4.8)—and they usually made them from a stock low-arsenic copper-arsenic alloy. Yet the data also indicate, surprisingly, that Ecuadorian metalsmiths were not systematically using these copper-arsenic alloys to improve artifact design.

One of the common artifacts made from arsenic bronze, axe-monies, served as a medium of exchange.

Others—bells and rings—were display items. Bells were also ritual objects. All three are small and easily portable. All lack distinct iconographic features. Overall, their homogeneity in materials and design suggests that they may have been fashioned for exchange, supporting Salomon's argument that these were merchant societies. Certainly this facet of the technology was not the locus of major aesthetic or technical investments.

The Reinterpretation

West Mexican metalsmiths did not replicate the full array of objects and techniques that characterized the metallurgies of lower Central America and Colombia or of Ecuador, but incorporated only selected elements from each of those technologies. We can identify which elements derived from these two geographic areas, but there is a great deal we still do not know: how the choices of West Mexican smiths may have been constrained with respect to materials and techniques, and what events, encounters, and cultural forms may have encouraged or discouraged the adoption of particular aspects of these external metallurgical traditions. We do not know whether only certain classes of artifacts were exported from the south or which those might have been.

Lost-wax casting became the most important element of these technologies adopted by West Mexican artisans. The West Mexican repertoire of bells shows how extraordinarily interested they were in this technique and in the possibilities it allowed for casting these objects. As I have noted, Ecuadorian and lower Central American and Colombian artisans also made bells, but in far fewer numbers. Tables 4.3 and 4.8 show that bells constitute 15% of Colombia's Museo del Oro collections and slightly more than 23% of the objects in the Museo Antropológico de Guayaquil. With respect to sheer numbers, the most significant artifact classes in lower Central America and Colombia are cast ornaments: staff heads, nose ornaments, ear ornaments, lip plugs, and pendants. In Ecuador the most numerous objects, at least in the museum collections, are axe-monies, rings, and bells, all three of which are present in nearly equal proportions.

Prior to the introduction of metal, Mesoamerican peoples made bells and rattles from nonmetallic materials: seed pods, gourds, and fired clay. Rattles or *maracas* made of ceramic and probably of wood are illustrated in the Bonampak murals and were common in the Maya area between A.D. 700 and A.D. 900 (Healy 1988). In fact, ceramic musical instruments distinguish Late Classic Period sites on the Pacific coast of Guatemala (Shook 1965). One elite burial at El Paraíso, Guatemala (figure 6.1), contained a group of such pottery instruments including nine bells and several drums. As previously mentioned, numerous prehispanic ceramic bells also exist in the collections of the Regional Museum of Guadalajara in Mexico. However, pottery lacks metal's resonant qualities, and Period 1 West Mexican smiths chose to use the new material for objects that optimized properties that could not be replicated using other materials; the most important of these was sound.

West Mexican artisans had probably seen Ecuadorian bells, since bells figured among the objects on the sailing raft described by Bartolomé Ruiz. Yet Mesoamerican bells are lost-wax cast. This method of fashioning bells is particularly appropriate for peoples interested in visual and tonal variety, because such variety is a function of differing bell sizes and shapes. It is far simpler to alter bell size (and hence pitch) by casting bells to shape them than to plastically work the solid material. Junius Bird (1979) has suggested that the availability of certain raw materials might explain why New World metalworkers used the technique extensively in some areas and rarely elsewhere:

the distribution of lost-wax casting in the New World apparently coincides with the distribution of *Meliponidie*, the stingless bee whose wax was used to fashion models for the castings. These bees are adapted to the moist tropical lowlands and cannot survive in more arid zones, such as the coast of Peru. They are present in only certain regions of coastal Ecuador, but are common from Colombia north to Mexico. This may partially explain the distribution of the technique, but it is also true that a variety of resinous materials can serve the same purpose.

West Mexican metalsmiths chose the Central American and Colombian method of making bells, but they did not employ the materials used to cast those bells: commonly gold and copper-gold *tumbaga* alloys. The RMG collection contains only a single gold bell in a corpus of 3,200 metal objects. Metalworkers did use gold extensively, particularly in the Tarascan zone in Period 2, but principally for sheet metal ornaments and ritual items for which color and reflectivity were primary concerns.

A variety of lower Central American and Colombian lost-wax cast objects are conspicuously absent from the West Mexican repertoire. If we take Colombia, for example, these tended to be the objects that in that region carried great symbolic weight: nose rings, vessels, figurines, lip ornaments, pectorals, earrings, and votive objects. Although bells comprise a relatively minor component of that technology, West Mexican smiths cast at least three bell designs that are direct copies of lower Central American and Colombian types and, exploring the possibilities of lost-wax casting, independently invented numerous new bell designs that are unique to the metalworking zone.

West Mexican artisans also incorporated methods, materials, and artifact types that derived from the utilitarian domain of the very different metallurgy of southern Ecuador. They fashioned needles, tweezers, awls, and fishhooks from copper and occasionally from low-arsenic copper-arsenic alloys, employing the same materials and fabrication techniques used in Ecuador but with minor changes in design. However, West Mexican peoples did incorporate at least one Ecuadorian status item, open rings, and made them in very large numbers. These rings were apparently worn and used in West Mexico in the same way as in Ecuador. If my interpretation is correct, some of these objects served as hair band holders, indicating that people may also have adopted what was essentially a foreign item of apparel. In addition, Period 1 smiths apparently began to use a low-arsenic copper-arsenic alloy for rings, the same stock material employed so pervasively in Ecuador for these items as well as for tweezers, needles, and other objects.

Nonetheless, even during these initial stages of West Mexican metallurgy the smelting regimes used to produce copper-arsenic alloys may have differed from those employed in Ecuador. Mountjoy and Torres's (1985) chemical analyses of materials from Tomatlán show tin present in very low concentrations. Disseminated tin is a common constituent of chalcopyrite, so that chalcopyrite was probably the copper ore co-smelted with an arsenic-containing mineral. Chalcopyrite and arsenopyrite typically associate in Mexico (see chapter 2), and metalsmiths were probably using what was most readily accessible. When chalcopyrite associated with arsenopyrite is roasted and smelted, a copper-arsenic alloy is the inevitable product. Five of William Root's 81 analyses of copper objects from Amapa also contain arsenic in low concentration (Meighan 1976: 116–118), and these alloys were probably produced in this same way.

It was the nonelite spheres both of Ecuadorian and of lower Central American and Colombian metallurgy that provided a scaffold for subsequent elaborations in West Mexico. West Mexican smiths incorporated those objects—rings, tweezers, and needles—that lacked distinct and local iconography. In Ecuador these were almost mass-produced and did not convey the most important cultural messages. Ecuadorian sacred and ritual items did not become part of the West Mexican repertoire. They include, for example, exceptionally thin, bimetallic masks and pendants, often with typically Andean raised designs, and gilded or plated with silver. Heavy, solid-cast gilt-copper nose rings, hollow nose rings made from thin copper-silver sheet, earrings made from copper-silver sheet terminating in intricate tightly wound spirals, and other sumptuary items made from gold-platinum, copper-silver, and copper-silver-gold alloys likewise did not enter the West Mexican tradition.

The ritual components of lower Central American, Colombian, and Ecuadorian metallurgy were probably never accessible to West Mexican artisans. The technical know-how required to reproduce such virtuoso achievements was likely restricted to those specialist metalsmiths who crafted such objects for elites, although substantiating data are unavailable. Some of these items belonged to the most sacred domains of experience: they were ritual and cult items, objects that coalesced the most powerful cultural symbols. These objects may not have been defined as items for exchange. However, tweezers, rings, bells, and (as I show in subsequent chapters) axe-monies were so defined, and thus were produced in large quantities and from commonplace materials.

Symbols of elite culture may be exported from the regions in which they develop under certain circumstances. Mary Helms (1979) makes a convincing case that Panamanian chiefs promoted their own power and prestige by using golden objects imported from Colombia. However, the religious ideologies in these two areas were fundamentally the same. The situation in the case of Ecuadorian seafaring peoples and the small-scale Period 1 societies of West Mexico is quite different. The most technically complex, ritual objects were not imported from Colombia or Ecuador, or, if they were, they were not reproduced in West Mexico. The metallurgical evidence we have to date indicates that what West Mexican societies received from the south was primarily knowledge about fabrication methods, materials, and design.

The technical alternatives West Mexican artisans chose to elaborate during Period 1, regardless of the range of materials, techniques, and artifact types they may have seen, reflect and reveal their own perceptions of this new material. These people were only marginally interested in metal for its utilitarian possibilities. Instead, they focused on those properties that served to demarcate ritual and elite realms of cultural activity. But the meaning and form of these physical and material symbols were not appropriated from the elite components of the two southerly source technologies; they arose directly from the West Mexican cultural experience.

5

THE FLORESCENCE OF WEST MEXICAN METALLURGY

A.D. 1200/1300 to the Spanish Invasion

During Period 2, the technical expertise and repertoire of the West Mexican metalsmiths expanded greatly. Metalworkers in Michoacán, Jalisco, Colima, northwest Guerrero, and the southern parts of the state of Mexico began to experiment with a variety of copper alloys. These included copper-tin and copper-arsenic bronze, alloys of copper-silver and copper-gold, and ternary alloys of copper-arsenic-tin, copper-silver-gold, and others. The enhanced physical and mechanical properties of these new materials allowed the artisans to elaborate, refine, and sometimes redesign the same object types that they had previously crafted in copper. In developing these alloys metalworkers mined and processed a host of new ore minerals, including tin, arsenic, and silver, and invented new smelting regimes to win these metals from their ores. The most unusual of these yielded the extremely high-arsenic (23% by weight) copper-arsenic alloys referred to in chapter 2.

The artifact designs that mark Period 2 appear at approximately the same time at a number of sites in West Mexico and areas immediately adjacent to it. As the map in figure 5.1 indicates, artifacts representing this technological complex are found in Michoacán, at the Tarascan centers of Tzintzuntzan (Cabrera C. 1988; Grinberg 1989; Hosler 1986; Rubín de la Borbolla 1944) and Urichu[1] in the Pátzcuaro basin, at Tres Cerritos and Huandacareo in the Cuitzeo basin (Macías G. 1989), and at Milpillas near Zacapu.[2]

These developments also occurred outside of the Tarascan cultural sphere. Elements of this complex appear at Apatzingán (Michoacán), along the Río Balsas at the Infiernillo sites and La Villita, and in coastal Guerrero at Bernard (Brush 1962). Period 2 designs have also been recovered to the northwest, at Lo Arado in Jalisco and at El Chanal in Colima (Kelly 1980). Undated artifacts from Calixlahuaca, the Matlazinca center in the Valley of Toluca conquered in 1476 by the Aztecs, also reflect this technical complex. Period 2 objects were also excavated by Michael Smith in assemblages from the rural Aztec towns of Cuexcomate and Capilco in Morelos (see chapter 7 for an extended discussion). Although these artifacts occur at many sites, scant physical evidence exists for ore processing anywhere in West Mexico. The evidence we do have indicates that processing and production probably took place over a wide area. Archaeologists have reported slag, the by-product of the smelting process, from Bernard and La Villita. Objects identified as pieces of copper ore were found at Apatzingán. Several pieces of material that seem to be partially processed ingots were recovered in Jalisco in the Lago Chapala region.[3] Ethnohistoric evidence (Pollard 1987) from Tarascan Michoacán suggests that smelting and production took place at a number of different centers, although no archaeological research has yet been carried out to investigate this possibility.

The new technical complex was subsequently exported to more distant Mesoamerican centers. To date, the only other Mesoamerican area where there is strong evidence for production of bronze alloys and objects is the Huastec region, at the sites of Platanito and Vista Hermosa (see figure 7.1) (Hosler and Stresser-Péan 1992). However, certain objects recovered at Lamanai in Belize, the Cenote de Sacrificios in the Yucatán Peninsula, Chiapa de Corzo in Chiapas (see figure 5.1), and elsewhere were imported from the West Mexican metalworking zone. They have usually been recovered in contexts that also contain objects representing local metallurgical traditions. The exportation of the technology from West Mexico is treated fully in chapter 7.

The property of metal that had most interested Period 1 West Mexican smiths was its sound, manifest in a variety of bells differing in pitch and fashioned by lost-wax casting. During Period 2, metalsmiths continued to experiment with design and with pitch, using the new alloys

The Florescence of West
Mexican Metallurgy

to invent a plethora of new bell types. These were thinner and more intricately decorated than bells made from copper and were fashioned in a range of sizes, shapes, and material compositions. During Period 2, however, a second property of metal became a focus of metalworkers' technical experiments: color. West Mexican artisans employed the new alloys—copper-tin, copper-arsenic, and copper-silver—in status objects, varying the concentrations of the alloying elements enough to perceptibly, and sometimes dramatically, alter artifact color. Bells, ornamental tweezers, open rings, and sheet metal body ornaments were most frequently made from these alloys, although metalworkers also used them to craft other ritual and status items.

Smiths also took advantage of the utilitarian properties of these new materials. The two bronzes, copper-tin and copper-arsenic, and the ternary copper-arsenic-tin alloy are optimal materials for tools. When cold-worked to shape, they become harder than tools made from copper. Metalworkers fashioned a variety of hard, cutting, carving, and piercing implements, such as axes, unipointed and bipointed awls, awls with narrow blades, and needles, from the new alloys. Yet despite the multiple utilitarian possibilities of the bronzes, the principal focus of the metallurgy was not on tools and work implements but on the brilliantly colored sumptuary objects that expressed fundamental religious and ritual concerns. Bells continue to be the most numerous of these objects.

West Mexican metalsmiths' penchant for technical diversity, like their interest in sound, is already visible in the early assemblages at Amapa and Tomatlán, and this sets the subsequent pattern. From Period 1, they formed metal objects using two distinct approaches: casting the liquid metal into a closed mold, so that the object's form was determined by the final shape of the wax model; and hammering the object to shape from a cast blank. West Mexico was the only region in the Americas where the smiths fully developed both forming possibilities simultaneously.

This same proclivity for technical variety persisted throughout Period 2. Metalworkers continued to employ both methods, and added a third, hot work, to manage the properties of certain alloys. The interest in variety is also clearly apparent in the ways the new alloys were used. For example, metalsmiths employed several bronze alloys concurrently: copper-arsenic, copper-tin, and copper-arsenic-tin. What is more, artisans often used these bronze alloys interchangeably to manufacture the same artifact types, and in some instances, such as at Cuexcomate and Milpillas, assemblages containing some artifacts made from each of the three alloys appear at a single site. Even though both alloy types were introduced from the Andean zone, there they were generally restricted to separate and distinct geographical regions. Copper-arsenic alloys were employed primarily in central and northern Peru and southern Ecuador (Hosler 1986, 1988b; Lechtman 1979; Owen 1986; Scott 1988). Copper-tin alloys were used in the southern highlands of Peru and in Bolivia (González 1979; Mathewson 1915; Mead 1915). West Mexican smiths incorporated, experimented with, and built upon techniques belonging to very different metallurgical traditions.

Archaeological Evidence for Period 2 Metallurgy

During Period 2, metal objects become more common in all regions of the metalworking zone. Artifact types, fab-

5.1
Archaeological sites in the metalworking zone and southeastern Mesoamerica. Period 2 metal objects have been reported at all but Amapa and Tomatlán.

rication methods, and metals and alloys resemble one another, even where cultural affiliations, as seen in other aspects of the material record, seem minimal. Both archaeological and documentary evidence indicates that the primary center for metalworking was the Tarascan empire, whose capital, Tzintzuntzan, lay in the basin of Lago Pátzcuaro in highland Michoacán. The empire eventually extended over 75,000 square kilometers, reaching from the Balsas to the Lerma river system, and included all of the modern state of Michoacán. Tzintzuntzan encompassed slightly more than six square kilometers and had a population of 25,000 to 35,000 people. The city itself consisted of a ceremonial precinct as well as low-, middle-, and high-status residential zones. The surrounding Pátzcuaro basin is estimated to have supported a population of around 80,000 people at the time of the European invasion (Pollard 1993). Numerous metal objects have been recovered at Tzintzuntzan, primarily from burials.

Metal objects also come from other sites in highland Michoacán: for example Urichu,[4] Milpillas,[5] and Huandacareo (Macías G. 1990). Most were excavated from burials. Urichu was one of eight Tarascan communities in the Pátzcuaro basin that served as a local administrative center. Tarascan nobility governed Urichu, reporting directly to the royal dynasty at Tzintzuntzan (Pollard 1993). There, Pollard excavated four burials that contained a total of 19 metal objects.

Milpillas, a Postclassic center to the north near Zacapu, covered at least 54 hectares. Twenty-two metal objects were recovered from burials, a test trench, and a midden deposit, all of which date to after A.D. 1200. Cultural affiliations are unclear. Tarascan ceramics are present but infrequent. Domestic ceramics resemble those Pollard has excavated at Urichu.[6]

A large assemblage of metal artifacts was excavated at Huandacareo, a major Tarascan administrative center in the Lago Cuitzeo basin. Huandacareo was founded by the Tarascans when they conquered the basin in about A.D. 1440. Tarascan elite pottery appears at this site associated with metal objects; 117 sumptuary artifacts and 17 tools have recently been excavated there, primarily from burials and test pits (Macías G. 1990).

The adjacent Valley of Toluca seems also to have been an important metalworking center. Virtually no datable metal objects have been recovered, but numerous items are reported from the site of Calixlahuaca, a few kilometers to the northwest of the modern city of Toluca. Calixlahuaca was conquered by the Aztecs in 1476.

During Period 2, metal artifacts have also been recovered at a number of sites in the Pacific coastal lowlands. Some assemblages are associated with Tarascan materials, others are not. Kelly (1947) recovered 107 metal objects from burials, refuse mounds, and test trenches at sites in the Apatzingán region, located in the Balsas depression of Michoacán. She also recovered a few examples of what she thinks is copper ore. Kelly believes that the metal objects represent local production and use (1947: 143). The ceramic complex at the site is unrelated to Tarascan state material, the basis for Kelly's contention that there is no evidence for a Tarascan presence there, and extremely little evidence for even casual trade relations.

Many objects from the Infiernillo sites, discussed in chapter 3, and from La Villita on the lower Río Balsas also date to this period. The ceramics from Infiernillo represent ties with both the Tarascan state and other regions of West Mexico, specifically northwest Jalisco, Nayarit, and Colima.

Kelly (1949) also extensively surveyed a number of sites in Tuxcacuesco (Jalisco), then carried out limited excavations at six of them, recovering 42 objects from burials, surface collections, and test pits. Several pieces of what seems to be partially processed metallurgical material appear among these items. The predominant ceramic material at Tuxcacuesco is a local style showing some relationships with Colima, to the south.

Metal objects have also been found at Lo Arado, a large Postclassic site in the coastal lowlands of Jalisco. Lo Arado has not been scientifically excavated (Covarrubias V. 1961; Mountjoy 1970). Covarrubias, who wrote a brief report on the site, observed burial mounds lined up in rows along a north-south axis and ceramics that he describes as Postclassic (A.D. 900 to A.D. 1521). He also describes a few of the many metal objects recovered there as made from copper, silver, and gold. Artifacts ascribed to Lo Arado appear in the RMG collections and were analyzed in the course of this study.

Numerous metal objects have been reported from the site of El Chanal (figure 5.1) in Colima, a large Late Postclassic ceremonial center 45 kilometers from the coast (Kelly 1980). El Chanal, which has been extensively looted, dates to A.D. 1250 or later and encompasses at least 5 acres. El Chanal seems to have remained outside of the Tarascan empire, judging from materials recovered in surface collections, seen in private collections (Kelly 1980), and described in historical records.[7] The ceramic material from the Chanal phase in Colima represents a local West Mexican base (Kelly 1980), but contains certain elements that also relate it to Central Mexico. Many El Chanal metal objects also made their way to the RMG collection.

Large numbers of undatable metal artifacts have been reported to the southeast, in Guerrero. The only reasonable temporal assignments at the moment are from the Bernard site on the coast, where 26 metal objects were recovered from limited test excavations; all date to after A.D. 1250. The excavators also identified pieces of material that they interpret as slag, and suggest that metal production was probably taking place at the site (Brush 1962).

In Sinaloa, the northwestern limit of the metalworking zone, metal objects have been excavated from a burial mound at the site of Guasave (Ekholm 1942). Guasave lies on the coastal plain of Sinaloa a few kilometers from the shore (see figure 6.1) and dates to about A.D. 1200 (Meighan 1974). Excavations revealed 166 complete burials; 134 metal objects were recovered, all in burial contexts. Multiple examples of elaborate polychrome ceramics, identified as Aztatlán, were excavated at Guasave. In addition, 19 metal artifacts were recovered from burials and middens at Culiacán (see figure 6.1), at a series of sites surveyed along the Río Culiacán valley south of Guasave (Kelly 1945). Some are associated with polychrome pottery similar to that found at Guasave; the assemblage dates to the same period.

Evidence for Period 2 metallurgy appears primarily at the sites discussed here. However, artifacts from these sites constitute only a small proportion of the metal items known from the metalworking zone; many others come from casual finds and looters' activities.

The Metallurgical Technology of Period 2: New Materials and New Designs

During Period 2, metalsmiths explored the properties of the alloys to make major changes in the formal designs both of lost-wax cast bells and of objects that were hammered to shape: open rings, tweezers, needles, axes, and awls.[8] They also used the alloys for new artifact types. The most abundant of these were sheet metal ritual and status objects and axe-monies. Both these new types required the strength, toughness, and ductility of these alloys. In the Tarascan region, copper-silver sheet metal ritual items—shields, neck pieces, pendants, breastplates—became so common that the Spaniards referred to the alloy as the metal of Michoacán (Warren 1985). From it, smiths crafted extremely thin, delicate designs with highly reflective, silvery surfaces. Axe-monies, thin, T-shaped objects, were made primarily from copper-arsenic alloys. They appear in Oaxaca and in some areas of the West Mexican metalworking zone (Hosler 1986; Hosler, Lechtman, and Holm 1990). The West Mexican variety is found in Guerrero and Michoacán and has its closest counterparts on the coast of Ecuador.

We have too few dated assemblages to know the extent to which Period 2 smiths continued to make Period 1 designs from unalloyed copper. The evidence from Cuexcomate, Tzintzuntzan, Urichu, Milpillas, Bernard, Huandacareo, and Calixlahuaca[9] suggests that perhaps they did not continue to do so; at these sites most metal artifacts are designs optimized by using the alloys. However, in other regions, for example at Apatzingán, Tuxcacuesco, and La Villita, the impact of this second wave was more limited. People living in these areas probably were marginal with respect to obtaining either finished objects or raw materials. At these sites, as well as at the more distant sites in southern Mesoamerica, only certain artifacts have been recovered that represent the new technical complex. To the north and northwest, in the state of Sinaloa at the sites of Guasave and Culiacán, these developments were barely felt, and the Period 1 tradition persisted.

Lost-Wax Casting: Bells

Bells were the focus of a great deal of technical experimentation during this period. West Mexican metalsmiths used the tin and arsenic bronze alloys to create at least five new bell designs (4, 7, 8, 9, 10) and numerous variations on them, as well as new versions of earlier types (1b). Bell designs that first appear during Period 2 are illustrated in figure 5.2; all types illustrated here are found

Table 5.1 Period 2 RMG Bell Types (Lost-Wax-Cast): Composition and Number Analyzed in RMG Collection, and Archaeological Sites of Appearance

		Specimens from RMG Collection		
RMG Type	Datable Archaeological Sites*	Number Made from Alloy	Number Analyzed	Number in Collection
1b	Milpillas (Cu-Sn)	—	—	3
4a	Tzintzuntzan	4	5	10
7a	Cuexcomate (Cu-Sn) Milpillas (Cu-Sn) Tzintzuntzan	6**	9	30
7b	Tzintzuntzan			
7c	Tuxcacuesco			
7d	Milpillas (Cu-Sn)			
8a	Tzintzuntzan	12**	12	41
8b	Tzintzuntzan			
8c	Tzintzuntzan			
8d	Tzintzuntzan			
9a	Bernard (Cu-Sn) Tzintzuntzan	6	11	185
10b	Milpillas (Cu-Sn) Tzintzuntzan	18	18	27

* Composition, if available is indicated in parentheses.
** The numbers found in the columns for 7a and 8a encompass 7a–c and 8a–d respectively.

5.2

Period 2 bell types identified in RMG collection and present in datable archaeological contexts.

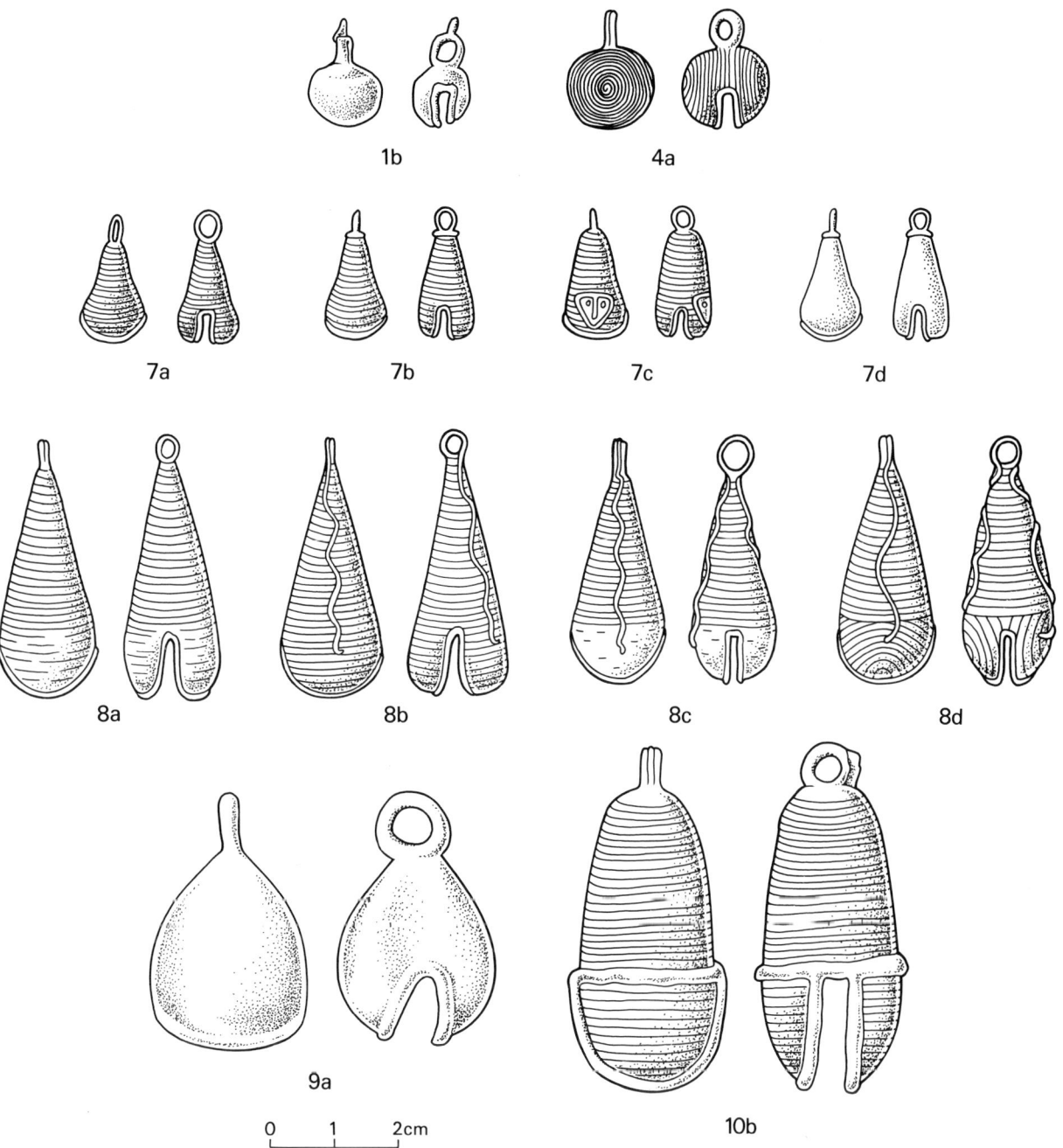

both in archaeological contexts and in the RMG collection. The RMG examples were chemically analyzed in all cases but one (1b). Table 5.1 lists them and the Period 2 sites where they are found. Not all specimens belonging to a particular type are made from an alloy, but those that are not differ in formal characteristics from their bronze counterparts; they are invariably thicker, for example.[10] This probably means (1) that not all metalworkers had access to alloy metal, and (2) that some of the RMG examples (for instance 9a) may have been fashioned in Period 1 and from copper. What we do know is that metalworkers continued to craft several Period 1 designs (1a, 5b, and 11b), which are found at some Period 2 sites. Table 5.2 lists those sites. They used tin bronze for at least one of these types (1a). Bells listed in these tables are not the only varieties to appear during Period 2, but were types present in the RMG collection in sufficient numbers for an appropriate sample for laboratory studies (see appendix 1).

None of these new bell designs has equivalents in lower Central America or Colombia. By contrast, the alloy systems, tin and arsenic bronze, had been developed earlier in South America. There, however, metalworkers tended to use these alloys for cold-worked implements and tools. Also, the concentration of these two elements in the South American alloys is typically low (circa 0.05–3.00% by weight); tin generally does not exceed 5 weight percent (González 1979; Mathewson 1915). In West Mexico, artisans used the two bronze binaries for bells, which they cast by the lost-wax method. Furthermore, they frequently added the alloying elements, most often tin but sometimes arsenic, in exceptionally high concentrations. The highest concentration yet reported is in a bell containing 30% tin, from Bernard (Brush 1962); in the RMG collection, alloy element concentrations in bells reach to 19.98% tin and 23.47% arsenic. This unusual means of managing these alloys, primarily for their color,

Table 5.2 Period 1 Bell Subtypes Found at Period 2 Sites

RMG Type	Archaeological Sites
1a	Apatzingàn
	Bernard (Cu-Sn)
	Cuexcomate (Cu-As)
	Culiacán
	Guasave (Cu)
	Infiernillo
5b	Guasave
11b	Guasave
	Infiernillo

Note: Composition, if available is indicated in parentheses.

is one of the original contributions of West Mexican metallurgy to the metallurgies of the Americas.

Period 2 bells, like their Period 1 counterparts, have been recovered most frequently from burials: at Tzintzuntzan, Huandacareo, and Milpillas in Michoacán. Bells have sometimes been recovered in household debris, for example at Cuexcomate in Morelos and in Guerrero. The most unusual contexts in which bells have been found is in caves, one in Guerrero where "literally hundreds of bells" were reported (Weitlaner J. 1964: 530). Caches of bells have also been reported from caves elsewhere in Mesoamerica (Bray 1977).

Two properties of the bronze alloys were essential to achieve the new bell designs: their solidification characteristics and their strength. Table 5.3 shows the relation between design characteristics (dimensions) and alloy composition for RMG varieties containing five or more specimens. The average height-to-thickness ratio for the Period 2 alloy bells is 39.6; for Period 1 copper bells the ratio is 19.2. Even using this very general measure, it is clear from table 5.3 that the strength of these bronze

Table 5.3 RMG Bells: Design Characteristics and Compositions

Type	Period*	Metal/Alloy	Average Wall Thickness (cm)	Average Height** (cm)	Ratio Height:Thickness
1a	1, 2	Cu, Cu-Sn	0.05	0.61	12.2
4a	2	Cu, Cu-As	0.05	1.30	26.0
7a	2	Cu, Cu-As, Cu-Sn	0.06	1.90	31.7
8a–d	2	Cu-As, Cu-Sn	0.05	3.20	64.0
9a	2	Cu-As, Cu-Sn	0.08	2.00	25.0
10b	2	Cu-As, Cu-Sn	0.07	3.60	51.4
2a	1	Cu	0.06	0.90	15.0
5b	1	Cu	0.09	1.60	17.8
6a	1	Cu	0.06	1.20	20.0
6b	1	Cu	0.07	1.40	20.0
11b	1	Cu	0.13	3.00	23.1

* Period 1: A.D. 600–1200/1300; Period 2: A.D. 1200/1300–1521.
** Height: distance from base of bell to top of resonator.

alloys allowed the smiths to reduce bell thickness by nearly one half.

The other striking feature of these new designs is the unusual surface elaboration known as wirework (see figure 5.2). The resonator cavity walls of these wirework bells (types 4a, 7a–c, 8a–d, and 10b) appear to have been constructed from coiled threads of wire, oriented circumferentially. Many variations occur: large numbers exhibit complex vertical or horizontal zigzag patterns; sometimes, more rarely, the bells have recognizable anthropomorphic or zoomorphic designs. Artisans made the original models for the bells entirely in wax, wrapping individual wax threads around a clay core, and building the bell using a technique similar to coiling pottery. They then placed additional strands of wax over areas requiring reinforcement: around the base of the resonator chamber and at the midpoint, or on areas of the chamber where they elaborated other designs. The metallographic evidence shows that all of these bells, however intricate the design, were cast in one piece. Figure 5.3 shows a cross section through the midpoint reinforcement of a wirework high-arsenic copper-arsenic bronze bell. The photomicrograph shows a continuous cast structure from the bell wall across the reinforced region.

All but two of the new, Period 2 bell designs found in the RMG collection are of the wirework type; those sampled are, in most cases, made from either a copper-tin or a copper-arsenic bronze (see table 5.3 and appendix 2). As table 5.4 indicates, all but three of the Milpillas bells were made from tin bronze alloys. Data for Cuexcomate objects appear in chapter 7.

The enhanced properties of the bronze alloys allowed metalsmiths to achieve thin, intricate castings. The presence of tin or arsenic in copper lowers the melting temperature of the metal. Copper melts at 1053° C, copper alloyed with 15% tin melts at 825° C, and copper a bell mold after the wax has been melted out, the molten metal progressively advances, filling in the complex mold cavity detail as it gradually solidifies. This form of solidification was crucial in casting the crisscrossing, threaded wirework designs of the resonator cavity walls.

Longitudinal sections cut from wirework bronze bells show that wall thickness in the wirework designs varies at each point where the threads of metal join to form the bell walls, sometimes as much as three or more times (see figure 5.3). The strength and fluidity of these bronze alloys were essential in designs with such abrupt changes in wall thickness, especially in these extremely thin-walled objects. Some wirework bells, such as a few belonging to type 7a, were cast in copper, but those bells were either deformed in casting or were smaller than their counterparts made from bronze. Table 5.3 shows that copper was successfully used to cast the designs of very small bells (type 1a) and bells whose walls were thick relative to their height (type 11b).

Alloys are desirable for thin-walled castings for other reasons. When the alloy solidifies across a range of temperatures, microporosity often results, especially in thick-walled objects. Microporosity develops less readily in cast objects with thin walls.

By taking advantage of the strength of copper-tin and copper-arsenic bronze, West Mexican metalsmiths not only optimized bell designs but enlarged the range of pitches the bells produced by increasing bell size. The alloys enabled them to cast larger bells: three of the five new bell types in the RMG collection are larger than 2 cm. Period 1 copper bells tended to be smaller. Larger bells produce lower pitches. The increased strength provided by bronze indirectly permitted greater tonal variety. In addition, bronze alloys, unlike copper, dissipate gases more readily during solidification, so they inhibit macro-

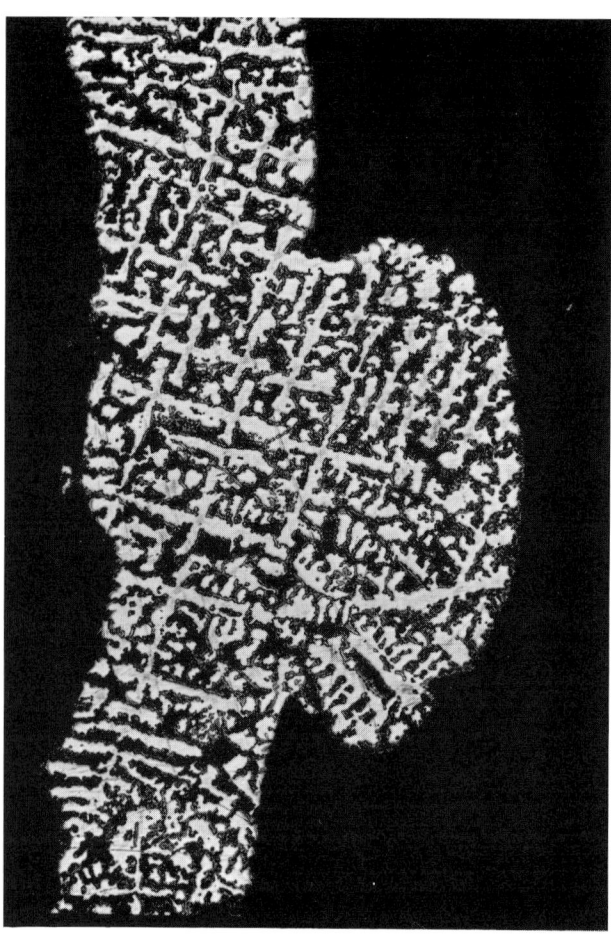

5.3

Cross section through the wall and wirework reinforcement of RMG Period 2 Cu-As alloy (12.28% As) bell. Sample etched in potassium dichromate and hydrochloric acid (mag.: 100).

alloyed with 15% arsenic melts at 800° C. These alloys not only melt at lower temperatures, but they are more fluid than pure copper. They flow easily through thin mold cavities since, unlike copper, they solidify over a range of temperatures. When bronze metal is poured into

Table 5.4 Chemical Analyses of Period 2 Artifacts from Milpillas

	ID No.	Composition (weight percent)								
		Ag	As	Au	Fe	In	Ni	Pb	Sb	Sn
Bells	M-3	0.18	0.3	—	0.03	—	0.008	0.004	0.08	9.56
	M2-1	0.1	0.21	0.01	0.02	0.007	0.01	0.01	0.03	1.74
	M2-2	0.16	0.02	—	0.04	0.02	0.01	—	0.03	4.94
	M2-3	0.06	0.02	—	0.06	0.03	0.01	0.006	0.05	7.48
	M2-4	0.16	0.02	—	0.04	—	0.01	—	0.03	4.85
	M2-5	0.09	0.04	—	0.14	—	0.006	0.007	0.11	—
	M2-6	0.05	—	0.003	0.03	—	0.01	0.007	0.02	3.12
	M2-7	0.27	—	—	0.02	—	0.03	—	0.01	3.13
	M2-8	0.06	—	—	0.05	0.007	0.007	0.003	0.01	3.72
	M2-9	0.1	—	—	0.04	—	0.01	0.003	0.01	0.91
	M2-10	0.27	0.01	—	0.08	—	0.02	—	0.01	4.06
	M2-11	0.13	—	—	0.03	0.01	0.02	0.002	0.007	1.86
	M2-12	0.27	—	—	0.02	0.02	0.02	—	0.004	2.12
	M2-13	0.27	—	—	0.05	—	0.02	0.003	0.006	2.92
	M2-14	0.14	—	—	0.07	—	0.04	0.002	0.005	2.57
	M2-15	0.19	—	—	0.08	—	0.02	0.002	0.006	2.68
	M2-16	0.03	—	—	0.17	—	0.01	0.005	—	—
	M2-17	0.13	—	—	0.13	—	0.009	0.009	0.03	—
	M2-18	0.2	—	—	0.02	—	0.006	0.004	—	2.64
Open rings	M-4	0.15	—	0.01	0.003	—	0.05	—	0.01	—
Needles	M-1	0.12	—	—	0.004	—	0.006	—	0.08	—

Note: Analyses carried out by atomic absorption spectrometry. A dash indicates element not detected; "na" indicates element not analyzed quantitatively because not detected in qualitative analysis.

pore formation in the solid metal. Macroporosity, as noted earlier, influences pitch and tonal quality by damping vibrations.

The most exceptional characteristic of these bells is their composition. As appendix 2 indicates, the concentration of tin or arsenic in the alloys varies greatly among artifacts. Even at low concentrations (1–2%), color changes become apparent in the cast metal, especially in the copper-arsenic system. Bells fashioned from copper-arsenic alloys contain arsenic in concentrations from 0.49% to 23.47%. These alloys are a pale silver-pink color at low arsenic levels. At higher concentrations the metal takes on a silvery hue. When the arsenic concentration reaches about 10%, the color becomes a brilliant silvery-white. Alloy composition in the tin bronze bells ranges from 2.8% to 19.98%. As tin concentration increases, the color of the metal gradually changes from the pinkish hue of copper to an increasingly golden color.

West Mexican smiths were using these alloys to create an array of colors ranging from pale pink to silvery white or to a brilliant gold. We do not yet know how the metal for the extremely high-arsenic copper-arsenic alloys was smelted. From the widely varying concentrations of arsenic in bells made from these alloys, shown in table 5.5, it is clear that at least in ranges above about 5% arsenic, metalsmiths did not control alloy concentration in any systematic fashion. Lechtman's extensive experiments in producing copper-arsenic alloys by direct-smelting a variety of copper and arsenic ores indicated that, in that alloy system, the concentration of arsenic is virtually impossible to predict (Lechtman 1985; Rostoker and Dvorak 1991).[11]

On the other hand, metalworkers did exercise greater alloy control in the production of tin bronze, although the data indicate that standard alloys were never developed in West Mexico. Objects worn and used by religious functionaries, such as ornamental tweezers, consistently contain tin in concentrations varying between about 8% and 11%. Copper-tin bronze lent itself to greater compositional control than the copper-arsenic alloy because of the way it was produced. This involved either (a) smelting cassiterite (tin oxide ore) to obtain metallic tin, then melting together the batches of tin metal and copper metal (these batches could easily be weighed or separated by volume so that certain proportions of tin and copper could be maintained), or (b) co-smelting in a single furnace charge quantities of cassiterite ore and copper ores whose relative weight or volume could be controlled. Smiths must have rapidly developed a sense of the amount of ore or metal that had to be added to produce a low- or high-tin alloy.

During Period 2, by creating a number of new bell designs, metalworkers simultaneously created tonal and visual variety by casting bells in different sizes and shapes, and from a range of bronze alloys that gave them a broad spectrum of new colors. The high-arsenic and high-tin alloy bells looked as if they were fashioned from silver and from gold. However, pure gold or silver, like copper, lacks

Table 5.5 Arsenic Concentration in Selected RMG Copper-Arsenic Bronze Bells

ID No.	As Concentration (wgt. %)
201	22.12
204	12.82
207	23.47
1474[n]	8.69
1475	12.90
1485	13.8

[n] Analysis performed by neutron activation.

the solidification characteristics and strength to accomplish these designs. Unalloyed gold and silver also solidify at fixed temperatures, making complex castings difficult although not impossible. When thin-walled objects are cast from these metals, they are mechanically fragile. Bells had to be sturdy; they were incorporated into rattle boards and rattlesticks, and they were worn by dancers in bunches around the ankles and waist. Bell metal had to meet various requirements: the alloy had to allow rugged, thin-walled, intricate castings, and bells needed to look golden or silvery. It is not surprising therefore that West Mexican metalsmiths cast very few objects from pure gold or silver, although they did use these metals when possible for brilliantly colored sheet metal ornaments. Instead, to make these particular bell designs look silvery or golden, the artisans resorted to the unusual technical expedient of casting their bells from high-tin and high-arsenic bronze. West Mexico was the only region in the Americas where these alloys were used in this fashion.

Cold and Hot Work: Sumptuary Items

This same ingenuity in managing the new materials is evident in the objects manufactured by working metal to shape: principally tweezers, open rings, axes, needles, awls, axe-monies, and sheet metal ornaments. These were fashioned from a cast ingot, or blank, by hammering the material either at ambient temperatures (cold work) or in the heated condition (hot work). When the alloying element, tin or arsenic, was present in high enough concentration to cause brittleness, the objects were worked hot (forged). Otherwise the metal was worked cold, with intermittent rounds of annealing. Metalworkers took advantage of the strength and hardenability of these alloys for new artifact designs. They worked extensively with the binary copper-tin and copper-arsenic alloys, but they also used alloys of copper-silver and the ternary copper-arsenic-tin system. Metalworkers also used alloys of copper-arsenic-lead and copper-arsenic-silver, but infrequently.

Using these alloys, metalsmiths crafted thinner status objects such as open rings and tweezers, and tougher, harder, and finer implements. Table 5.6 shows the kinds and numbers of RMG worked objects analyzed, the numbers made from alloys, and the archaeological sites where they have been excavated.[12] The compositions of the objects from archaeological sites are indicated where known. In some cases these new designs were identical to objects made in lower Central America and Colombia, or in Ecuador, Peru, and Bolivia, and prototypes were introduced from those regions (see chapter 6). In other cases they were unique to this metallurgical tradition.

Only two new classes of worked objects were numerically significant during this period, axe-monies and large, ritual and ornamental sheet metal objects. Metalworkers also crafted a few other new object types, however. One of these was the Tarascan *tarquea,* or hoe. They also fashioned composite ornaments consisting of loop eye needles with small, lost-wax cast bells suspended from the loops (figure 5.4), and small U-shaped bar ornaments, also with small, suspended, lost-wax cast bells. However, the important achievements during Period 2 had to do with metalworkers' interest in new materials and in the design possibilities these materials allowed, rather than in creating a range of new kinds of objects.

Rings. West Mexican smiths continued to make large numbers of open rings, but the formal design characteristics changed, the materials were new, and in some cases fabrication methods changed. These rings have been recovered primarily from burials, as in Period 1. They are most frequently found in groups, adjacent to the cranium

5.4

Period 2 composite Cu-Sn alloy ornament. The object is a small (3 cm) loop eye needle with a bell suspended from the loop.

of the skeleton. Table 5.6 gives the sites where such open rings have been excavated. They appear in burials in the Infiernillo area, at Urichu and Tzintzuntzan, in primary burials at Huandacareo (Macías G. 1990), and at Milpillas. They have also been excavated from burials at Bernard (Guerrero), Culiacán, and Tuxcacuesco.

At Bernard, rings are made from a high-tin bronze (10% or greater) (Brush 1962); at Tzintzuntzan (Grinberg 1989) from copper-arsenic bronze and copper-silver alloys, with silver concentrations sometimes reaching as high as 55%. My macroscopic examinations of rings from Urichu burials suggest that some are made from high-tin bronze alloys and others from alloys of copper and silver. In all cases, the alloying elements, tin, arsenic, and silver, are present in high enough concentrations to noticeably alter the color of the copper. This is true both for objects from the sites just mentioned and for their counterparts from the RMG collection. Forty-six of the 68 rings analyzed from the RMG collection are made from copper alloys: four of arsenic bronze, and the remainder of tin bronze. Several rings are made of ternary copper-arsenic-silver alloys (see appendix 2, ring no. 2421). The concentration of arsenic in the arsenic bronze rings is generally less than 7%. The tin bronze rings, by contrast, contain tin in quite high concentrations, as appendix 2 indicates.

Metalsmiths making high-tin bronze rings had to do so by hot-working the metal. It is impossible to shape a high-tin bronze alloy by cold work, because the metal is inherently brittle. Figure 5.5 shows a longitudinal section cut through a tin bronze ring containing 16.86% tin. Strain markings are visible in some but not all the equiaxed grains, indicating that the metal was worked hot but continued to be hammered lightly as it cooled. The grains also exhibit annealing twins. Elongated pools of can be identified because the eutectoid becomes plastic at temperatures between 600° C and 800° C, when the brittle delta phase converts to the plastic beta phase. If the metal is worked within that temperature range, the eutectoid elongates, as it has in this case. The presence of strain markings indicates that shaping continued below the recrystallization temperature, and this cold work introduced strain in some of the crystal grains. Microhardness readings of this sample range between 131 and 201 VHN; the harder regions were worked after recrystallization.

Metallographic studies of the high-tin bronze, round cross section open rings show that they were made in the same way as their earlier counterparts in copper: by folding the metal over along its longitudinal axis, leaving a central fissure. Open rings with a rectangular cross section, by contrast, were made from solid, cast-rod stock. The bronze alloys allowed smiths to reduce the thickness of the rectangular cross section. From the data in table 5.7 we see that their average band thickness measures 0.16 cm when made from copper and 0.06 cm when made from tin bronze. The ratios of band width to thickness for the copper and tin bronze artifacts provide the best illustration of design modifications. The bronze alloys allowed metalsmiths to fashion thinner, wider bands than when making these rings from copper.

In the case of the round cross section open rings, the bronze alloy also allowed thinner designs, although the overall trend is not as dramatic. Bronze was mechanically necessary to reduce thickness only in open rings with large diameters, as table 5.7 shows. As diameter increases, use of the alloy allows the metal to be fashioned thinner yet still maintain its shape. Nonetheless, metalsmiths were making these open rings from bronze, even in objects whose design did not require it, adding tin to copper in the very high concentrations shown in appendix 2. If these artisans were interested solely in a thinner design,

5.5

Longitudinal section through RMG Cu-Sn alloy ring containing 16.86% Sn by weight. Elongated eutectoid reflects hot work (at 600° C to 800° C) while brittle delta phase has undergone transition to plastic beta phase. Sample etched in potassium dichromate followed by ferric chloride (mag.: 200).

eutectoid are strung out in the direction of plastic flow of the metal. A hot-worked structure is generally indistinguishable from one that has resulted from cold work followed by annealing; both are characterized by equiaxed grains and annealing twins. However, hot-worked copper-tin alloys containing the eutectoid microconstituent

Table 5.6 Period 2 Worked Objects: Composition and Number Analyzed in RMG Collection, and Archaeological Sites of Appearance

		Specimens from RMG Collection		
RMG Type	Datable Archaeological Sites	Number Made from Alloy	Total Analyzed**	Total in Collection
Rings, round cross section	Bernard (Cu-Sn) Infiernillo (Cu-Sn)* Tuxcacuesco	29	45	499
rectangular cross section	Bernard (Cu-Sn) Infiernillo (Cu-Sn)* Milpillas (Cu-Sn) Urichu (Cu-Sn, Cu-Ag)*	17	23	188
undetermined cross section	Huandacareo Tzintzuntzan (Cu-Ag, Cu-As)***	—	—	—
Tweezers, shell	Apatzingán Cuexcomate (Cu-Sn) Huandacareo Infiernillo (Cu-Sn)* Tzintzuntzan (Cu-Ag, Cu-Sn)* Urichu (Cu-Sn)* La Villita	28	29	33
Axes	Palos Blancos (Cu-Sn) Tzintzuntzan	18	35	38
Needles, perforated eye	Apatzingán Cuexcomate (Cu-As-Sn) Infiernillo Milpillas (Cu-As-Sn) La Villita	0	13	31

continued

Table 5.6 (continued)

RMG Type	Datable Archaeological Sites	Specimens from RMG Collection		
		Number Made from Alloy	Total Analyzed**	Total in Collection
loop eye	Cuexcomate (Cu-As-Sn)	10	17	21
	Huandacareo			
	Tres Cerritos			
	Tzintzuntzan			
	Urichu (Cu-Ag, Cu-Sn)*			
	La Villita			
Awls, unipointed	Apatzingán	1	4	6
	Bernard (Cu-Sn)			
	Infiernillo			
	Tzintzuntzan			
bipointed	Apatzingán	1	7	9
	Cuexcomate (Cu-As-Sn, Cu-Sn)			
	Infiernillo			
	Tzintzuntzan			
blade	Cuexcomate (Cu-As-Sn, Cu-Sn)	2	5	7
	La Villita			
Fishhooks	Apatzingán	1	3	3
	La Villita			
	Tzintzuntzan			
Axe monies (type 1a)****	No archaeological context	11	17	103
Ornaments, sheet metal	Bernard (Cu-Ag)	22	24	76
	El Chanal (Cu-Ag)			

* Judged to be alloys based on macroscopic examination. (In all other cases, alloys were identified by chemical analysis.)
** Identification of alloys was based primarily on quantitative determinations. Not all artifacts were analyzed quantitatively. In those cases, determinations about alloy type were based on qualitative emission spectrographic results available in Hosler (1986).
*** Some Tzintzuntzan objects have been analyzed for chemical composition (Grinberg 1989); I was able to examine others macroscopically; still others have no chemical compositional information but are illustrated in Rubín de la Borbolla (1944).
**** See also Hosler (1986) and Hosler, Lechtman, and Holm (1990).

Table 5.7 Comparison of Dimensions and Compositions of Rectangular Cross Section and Round Cross Section Open Rings

Rectangular Cross Section

ID No.	Metal/Alloy	Diameter (cm)	Band Thickness (cm)	Band Width (cm)	Ratio Width: Thickness
620	Cu-Sn (8.50% Sn)	2.5	0.05	0.6	12.0
2313	Cu-Sn (10.30% Sn)	2.6	0.05	0.4	8.0
635	Cu-Sn (12.44% Sn)	2.6	0.07	0.7	10.0
36a	Cu	3.0	0.15	0.4	2.7
36b	Cu	3.0	0.13	0.5	3.8
633	Cu	3.1	0.20	0.2	1.0

Round Cross Section

Number of Objects	Metal/Alloy	Average Diameter (cm)	Average Thickness (cm)
11	Cu-Sn	2.3	0.15
5	Cu	2.3	0.14
4	Cu-Sn	2.7	0.13
4	Cu	2.7	0.23

low-tin bronze could easily have imparted the required strength or toughness. Metalworkers systematically added tin in very high concentrations for a range of golden colors, just as they had in lost-wax cast bells. They made silvery-colored open rings by using high-silver copper-silver alloys; they occur at Tzintzuntzan, and Urichu, and also in material recovered from the Infiernillo project.

Silvery colors thus were realized in two ways by West Mexican smiths. In bells, they used high-arsenic copper-arsenic alloys; in the cold-worked open rings, sheet metal objects, and some tweezers, copper-silver alloys accomplished the same end. Copper-silver alloys are ideal for worked or hammered items because, during the sequences of hammering and annealing required to shape the metal, copper oxidizes and is gradually lost at the surfaces, thereby enriching the surfaces in silver. During fabrication, these alloys take on a characteristically silvery color, even when silver is present in concentrations of about 10–15%. Furthermore, copper-silver alloys are both malleable and tough, resisting brittle fracture due to their unusual, intermeshed lamellar microstructure. They are perfect alloys for shape retention in objects that need to be thin. By contrast, when these alloys are cast, they do not appear silvery until they contain about 40% silver.[13] Thus copper-arsenic bronze is the more appropriate alloy for casting a silver-looking bell. Copper-arsenic alloys begin to look silvery when arsenic concentrations reach about 7 weight percent.

Open rings were items made to be worn to display social rank, as I have indicated elsewhere. At Urichu,[14]

Huandacareo, and Tres Cerritos (Macías G. 1989) they appear to be ear ornaments; in Infiernillo burials and at other sites they may form part of a composite hair ornament. The fact that copper-tin bronze, a traditionally utilitarian alloy, was used so systematically and with tin in such high concentrations for these items clearly illustrates how West Mexican societies perceived the alloy. What most interested the metalsmiths about the tin bronzes were the golden colors they conferred, not only in bells and open rings but for all of the artifact classes I subsequently describe.

Tweezers. The tweezers smiths crafted during this time were also objects of adornment, and worn suspended by cords around the neck. They are highly intriguing, both because ethnohistoric sources indicate that they became key symbols of rank and office and for the exceptional technical skill required to manufacture them. Metalsmiths took advantage of the properties of the two bronze alloys, the ternary copper-arsenic-tin alloys, as well as the alloys of copper and silver to fashion an entirely new tweezer design, various iterations of which are illustrated in figure 5.6. The new design, which in engineering parlance is termed a shell, exhibits three-dimensional curvature below the tweezer hinge, giving the blades a shallow domelike appearance. RMG tweezers range in length from 3 to 7 cm; tweezers in other collections sometimes measure as long as 12 cm. This design has Colombian and coastal Peruvian prototypes, although West Mexican tweezers are often larger. Type c, however, the version that is technically most difficult to execute, is unique to the Tarascan region of West Mexico. These tweezers are distinguished by four absolutely symmetrical spirals, one emerging from each side of each blade.

Tweezers, at least the spiral variety, represented sacred power and priestly office in the areas dominated by the Tarascans. They are found most frequently in burials and are also mentioned and illustrated in the *Relación de Michoacán*. That document's description of the activities of the Tarascan chief priest mentions tweezers as items of priestly regalia. One priestly duty was to judge wrongdoers:

Pues venido el día desta justicia general, venía aquel sacerdote mayor llamado *Petámuti*, y componaíase. Vestíase una camiseta llamada *ucata-tararénguequa* negra, y poníase al cuello unas tenazillas de oro y una guirnalda de hilo en la cabeza . . . [When this day of general justice came, that chief priest named Petámuti came, and he arrayed himself. He got dressed in a black shirt called a *ucata-tararénguequa* and he put around his neck some golden tweezers and a fiber wreath on his head . . .] (Tudela 1977: 13)

Another priest, "Zurumban," is described as wearing a wreath of fiber around his head and some gold tweezers around his neck (Tudela 1977: 47). The *Relación* also relates that gold tweezers were tendered as gifts to visiting Chichimec leaders. Several illustrations show the Tarascan chief priest carrying out his official duties wearing a large pair of type c spiral tweezers around his neck (see figure 3.16). One illustration shows him wearing a tweezer while judging wrongdoers; in another he wears a tweezer while discoursing on the history of the Tarascan ancestors.

These ethnohistoric data suggesting that tweezers, especially the large spiral variety, were key cultural symbols among the Tarascans are paralleled by the archaeological findings. At Infiernillo, Urichu, and Tzintzuntzan these tweezers appear in high-status burials, located in the chest area of the deceased, as if worn around the neck. Large numbers of the spiral tweezers have been excavated recently from Huandacareo (Macías G. 1989). West Mexican tweezers were also status items elsewhere in Mesoamerica, and are found in elite burials as far south as Lamanai, Belize.

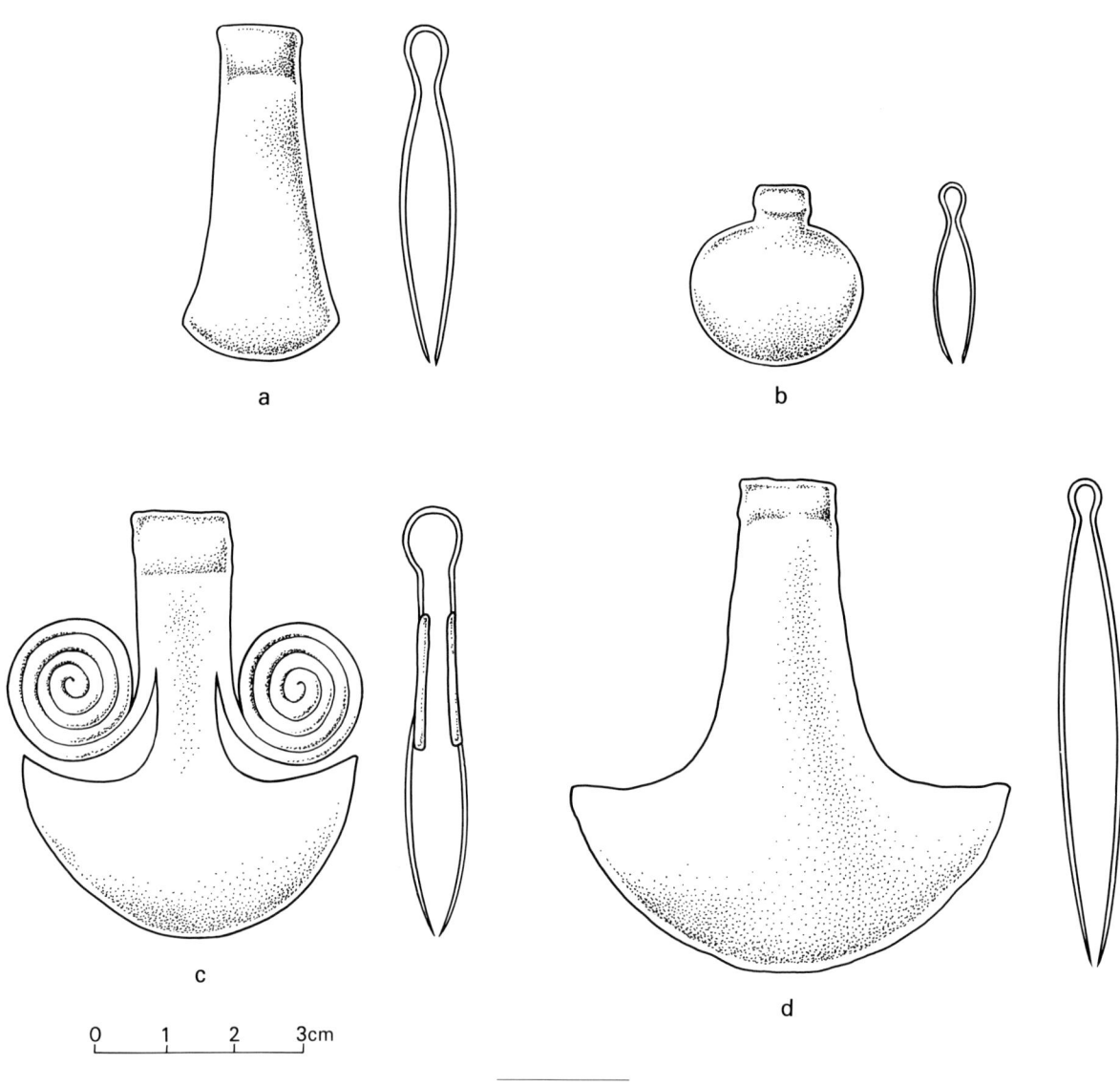

5.6
Shell tweezer designs; front and profile.

Documents written at the time of the Spanish invasion suggest that in certain regions, at least some tweezers were also functional depilatory implements. The Indians of Yucatán are reported by Cervantes de Salazar to have removed beard hairs with pincers (Tozzer 1941); Dahlgren de Jordan (1979) reports that golden tweezers were used by Mixtec chiefs for the same purpose. Other data, including the ethnographic observations cited in chapters 3 and 4 from throughout the Americas, indicate that depilation of beard and other facial hairs was, and is, a widespread practice.

All shell design tweezers were made from alloys; as I will show, the characteristics of this design require the strength and stiffness of these materials. The tweezer metal was made very thin, the blades sometimes measuring as little as 0.01 cm in thickness. The design is such that stress is distributed across the domed surface of the blade rather than concentrating in one small area, as is the case with Period 1 beam design tweezers. As appendix 2 shows, tin concentration ranges from 2.54% to 14.15% by weight, and arsenic concentration ranges from 2.7% to 4.43% in tweezers made from arsenic bronze. Two tweezers made from copper-silver alloys contain silver in concentrations above 20%. Grinberg (1989) reports a tin bronze tweezer from Tzintzuntzan containing 8.48% tin, and each of two examples from Cuexcomate (Morelos) contains slightly more than 10% tin (see table 7.2).

How were these tweezers manufactured? The metallographic studies show that shell tweezers containing more than 6% tin were hot-worked to shape. Others, fashioned at ambient temperatures, were left in either a cold-worked or annealed condition. The design that is technically the most difficult to execute is type c (see figure 5.6), the spiral tweezers illustrated in the *Relación de Michoacán*. These tweezers consist of two symmetrical dome-shaped blades joined by a hinge. The blades and hinge were hammered to shape from one continuous piece of metal. A rectangular strip of metal coiled into a spiral emerges at approximately the midpoint of each side of each blade. All four spirals are strikingly symmetrical; they are made from the same piece of metal as the blades and the hinge.

Several aspects of this design raise practical questions about how these tweezers could have been made from a single continuous piece of metal. The most perplexing characteristic is the length of the rectangular strip that forms each spiral; in the artifact studied, these strips would measure more than 15 cm if unwound. Fifteen centimeters is greater than the entire length of the tweezer were it to be opened and laid flat. Also, the spirals are considerably thicker than the blade metal immediately adjacent to them. It is difficult to visualize a manufacturing sequence that would result in such marked and local differences in thickness. The metallographic study was intended to determine whether these tweezers could have been made from more than one piece of metal, with the spirals soldered to the blades. The tweezer examined (no. F8 in the RMG collection) is a copper-arsenic-tin alloy with tin present at a concentration slightly above 10%. A cross section of metal was removed for metallographic study close to the juncture of the spiral and blade to determine whether or not there is a join at this location. The microstructure (figure 5.7) is typical of a hot-worked copper-tin alloy, containing equiaxed grains with annealing twins. Strain markings are present in some but not all grains. The microstructure is continuous, showing no evidence of a join. The presence of pools of elongated eutectoid visible in figure 5.8 provides evidence that the metal was worked hot, between approximately 600° C and 800° C. The microstructure of this section, cut even closer to the spiral-blade juncture, remains continuous across the juncture zone. This cross section was ground down through a series of successive planes to search for any evidence of a join. The photomicrograph in figure

5.7

Spiral tweezer. Cross section cut inward from blade edge just above juncture of blade and spiral. Note equiaxed grains with annealing twins; strain markings are present in some grains. The continuous structure indicates that blade and spiral were formed from a single sheet of metal. Sample etched in potassium dichromate followed by ferric chloride (mag.: 100).

5.9 represents the microstructure of the section precisely at the juncture of the spiral arm and the blade. The metal again is structurally continuous across the two; there is no evidence of solder or of any other method of joining. Electron microbeam probe analysis across the blade and spiral portions of the section provides substantiating data;

5.8

Cross section cut below section in figure 5.7 and closer to but above juncture of tweezer blade and spiral. Note elongated eutectoid indicating hot work while the beta phase was stable. Sample etched in potassium dichromate followed by ferric chloride (mag.: 200).

the chemical compositions in the two regions are nearly identical, representing a single, original piece of metal.

The information gained from the metallographic studies of this section, of a second section from the blade itself, and of still another from the spiral indicates that the entire tweezer was made from a single piece of metal. The metal was initially cast flat. The original thickness and shape are unknown, but they had to be able to provide sufficient metal to fashion each of the four spirals. The form of the cast blank might have approximated the one illustrated in figure 5.10, diagram (1). The smith cut away the reserve metal for the spirals from the blade, as diagram (2) suggests. Next the region to become the tweezer hinge was flattened by hot-working, and shaping of the blades began (3). The blades were shaped before work on the spirals began, because the spirals had to remain in precise alignment with one another; once the metal was bent at its midplane to form the hinge, the spirals had to coincide laterally and front-to-back.

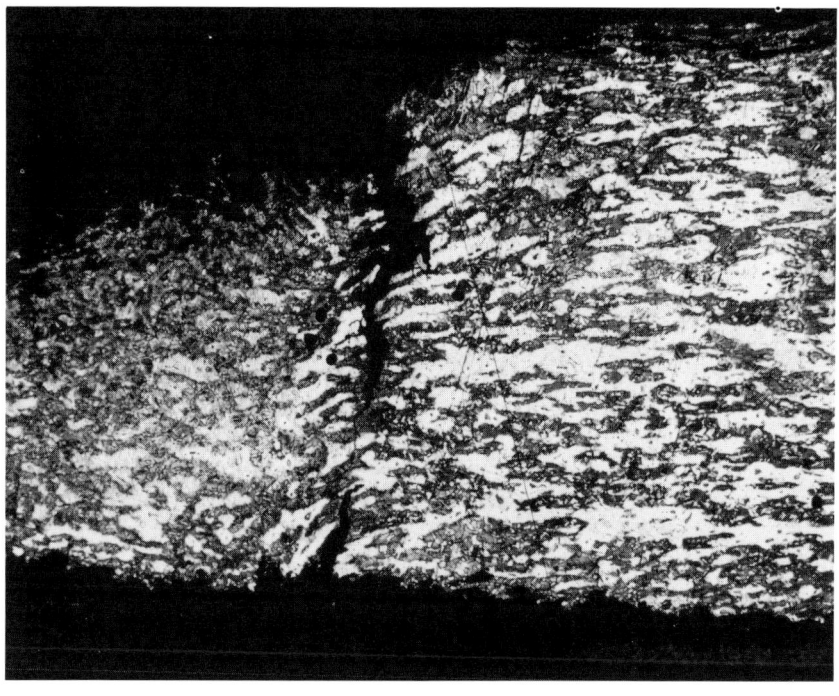

5.9

Cross section cut through juncture of tweezer blade and spiral. The continuous cored and highly oriented structure indicates that the blade and spiral were formed from one piece of metal sheet. Sample etched in potassium dichromate followed by ferric chloride (mag.: 100).

The most technically challenging task was to make the spirals. Each of the four rectangular metal strips was hammered to its desired length (4), then bent into a tight coil (5). This operation is difficult even when performed on a detached piece of rectangular cross section wire made from ductile metal and using a pair of pliers. Since information about ancient West Mexican metalworking tools is unavailable, we cannot reconstruct the process precisely. After the spirals were fashioned, the tweezer was folded over at the midpoint and the hinge was hot-worked to final shape. The precise and elegant symmetry of the tweezer was ultimately achieved through cutting and abrading or filing of the metal around its perimeter, especially along the arc of the blade tips.

Two questions arise concerning the shell tweezers. One is whether shell design tweezers, even this very large spiral variety, were functional tools. The other is why the shell design was consistently made from alloys: copper-tin, copper-arsenic, copper-arsenic-tin, and copper-silver. Computer simulation studies of the shell design, like the studies of the Period 1 beam design tweezers, were carried out to determine whether shell tweezers could have

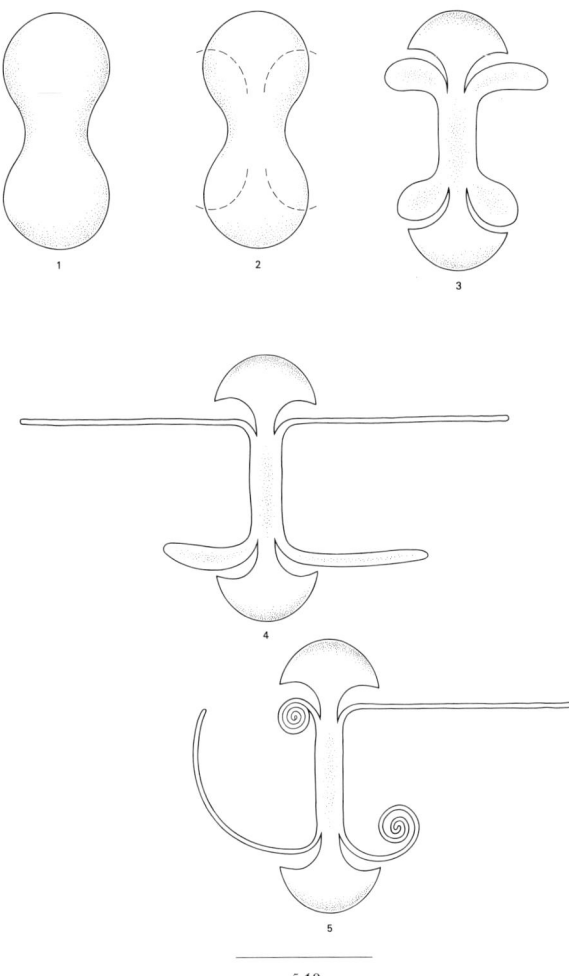

5.10

Reconstruction of spiral tweezer fabrication sequence: (1) supposed shape of cast bronze blank; (2) reserve metal for spirals cut away from tweezer body; (3) hinge hammered to a flat band and shaping of lunate blades begun; (4) reserve metal hammered into long, rectangular-section strips; (5) each strip bent into a tight spiral.

functioned as depilatory tools, given their particular dimensions, material compositions, and fabrication techniques. A model of one shell tweezer was constructed as a prototype and its stress and deformation behavior simulated. Figure 5.11 shows the model.

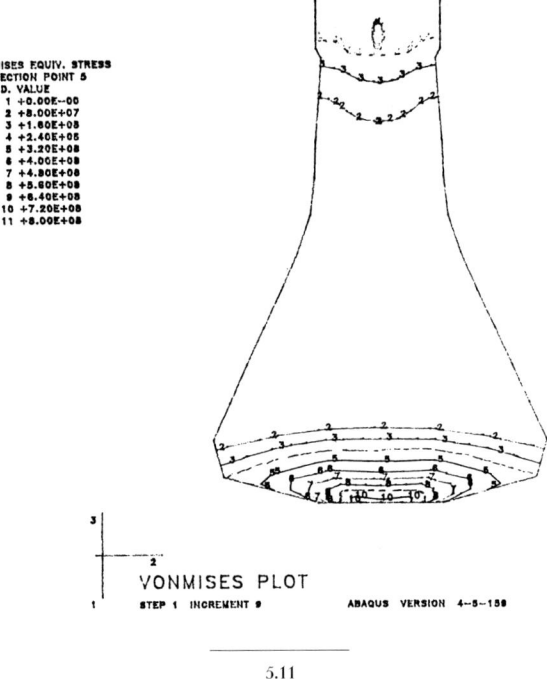

5.11

Shell tweezer: computer model exhibiting stress concentration contours developed during a simulation trial of the tweezer in use. Stress concentrates at the blade tips and at the hinge. Finite element analysis carried out with ABAQUS simulation program.

Table 5.8 gives the composition, design, and fabrication techniques of the prototype, as modeled. The table also shows the parameter values that were changed (in composition and fabrication technique) to assess the functionality of the other RMG shell design tweezers. The data demonstrate that all the West Mexican shell design tweezers studied were fully functional depilatory implements given the particulars of their individual designs. The discussion of the beam tweezers (see chapter 3) indicates that for any tweezer to operate, the stresses generated in the tweezer hinge cannot exceed the value for the yield strength of the material. The yield strength values for copper-tin and copper-arsenic alloys are significantly higher than those for unalloyed copper, as table

Table 5.8 Simulation Data for Shell Tweezer Prototype with Parameter Changes

		Parameter Changes				
Variables	Prototype	1	2	3	4	5
Composition	5% Sn	10% Sn	3.5% As	"	"	"
Fabrication technique	cold-worked	annealed/hot-worked	annealed	"	"	"
Length (cm)	3.5	"	"	"	"	"
Hinge width (cm)	0.5	"	"	"	"	1.2
Blade width (cm)	1.7	"	"	"	"	"
Blade thickness (cm)	0.1	"	"	0.05	0.01	"
Ratio: length to thickness	35	"	"	"	"	"
Tip opening (cm)	0.18	"	"	"	"	"
Contact force (g)	126	123	123	211	141	113
Yield strength (psi)	377	418	345	377	377	377
Stress excursion (MPa)	256	256	256	133	440	233

Note: " indicates same value as prototype.

5.8 reveals. As alloy concentration increases, yield strength values also increase, and the tweezer can tolerate greater stresses at the hinge. The stress excursion for the prototype is 256 MPa. This value does not nearly approximate the yield strength of the 5% tin alloy, and this tweezer was an operable depilatory tool. Because the blade metal of the prototype tweezer was the thickest with respect to blade length of any of the shell tweezers examined, I simulated the effect of reducing thickness so that the length-to-thickness ratios would approximate the ratios of other tweezers. All other dimensions were held constant. As table 5.8 indicates, when blade thickness was reduced to values even as thin as 0.01 cm, these shell tweezers were functional as long as certain length-to-thickness ratios were maintained. When the metal is cold-worked, stress excursions can exceed yield strength within the limits discussed previously without failure.

I also explored the effect of altering hinge width in the model, because hinges of West Mexican shell tweezers are typically wider than that of the prototype. Increasing hinge width decreases stress because the overall stress is distributed over a wider area. Table 5.8 shows that the tweezer model with a hinge measuring 1.2 cm has a stress excursion of 233 MPa, whereas higher stress excursions (256 MPa) characterize the object whose hinge measures 0.5 cm. Data from the simulation testing indicate that all of the shell tweezers that maintained a certain length-to-hinge-width ratio were functional when made from alloys.

All of the 29 shell tweezers in the RMG collection, including the large elaborate spiral design tweezers worn by priests, could function as tools. The next question I addressed was why they consistently were made from copper alloys. I did so by simulating the behavior of a tweezer with typical shell dimensions but made from

Table 5.9 Simulated Performance of Shell Tweezer Made from Copper

Composition	Fabrication Technique	Length (cm)	Blade Thickness (cm)	Ratio Length:Thickness	Hinge Width (cm)	Tip Opening (cm)	Contact Force (g)	Yield Strength (psi)	Stress Excursion (MPa)
Cu	cold-worked	4.5	0.01	450	1.0	0.38	100	265	400

copper. Table 5.9 shows the behavior of this tweezer model. The tweezer is nonfunctional when made from copper because the stress excursions far exceed the yield strength. However, when I increased thickness from 0.01 to 0.10 cm, stress excursions became far lower and the tweezer became fully functional. All but five of the 29 shell tweezers in the RMG collection measure less than 0.10 cm in thickness, and all of these are made from alloys (thicknesses range from 0.01 to 0.05 cm). The strength of these alloys was essential to achieve the extremely thin shell design. It is probably also true that to make a very thin functional tweezer, a shell design was necessary to offset the inherent fragility of the material by distributing the stress over a larger area.

West Mexican metalsmiths took selective advantage of certain properties of the tin and arsenic bronzes and copper-silver alloys in fashioning shell tweezers. Tin concentration in these tweezers, which ranges between 2.54% and 11.76% (one measured 14.15% by neutron activation; see appendix 2), are higher than necessary to accomplish the shell design and for the tweezer to operate. These tweezers, like the rings and bells, looked golden when made from high-tin copper-tin alloys and silvery when made from copper-silver alloys. The two copper-silver alloy shell tweezers in the RMG collection contain silver in concentrations above 20%. Metalworkers also used copper-arsenic alloys for three RMG shell tweezers, but never incorporated arsenic in high enough concentrations to produce a strong silver color. Arsenic was present at levels below 5% in all three cases. This alloy imparted the strength necessary to achieve the shell design but avoided the brittleness that characterizes worked objects when arsenic is present at levels above about 7%. Nevertheless, the arsenic concentration in these three tweezers—ranging between 2% and nearly 5%—did alter the metal color to a pinkish silvery hue.

These elaborate shell tweezers represent a far greater investment in materials and labor than the Period 1 beam tweezers, and mark the Tarascan area as the only region in the world where these aesthetically compelling tools became symbols of priestly office and social rank.

Sheet Metal Ornaments. Period 2 smiths developed the superior properties of the new alloys to produce yet another set of cold-worked status display objects. They made these thin and delicate ornamental breastplates and shields, headbands, neck pieces, pendants, earrings, disks, and bracelets from copper-silver sheet metal, taking advantage of this material's special set of qualities: malleability, unusual toughness, strength, and ability to develop a silvery-looking surface during working and annealing cycles. Some artifacts were made from silver, when the design allowed it; others were sometimes made from gold or from copper-gold and copper-silver-gold alloys. Metalsmiths very rarely used these metals and alloys for lost-wax castings. Some of these alloys apparently were used for castings in Oaxaca and Central Mexico, from the illustrations in Torres and Franco (1989).

The earliest silver artifacts reported (nose rings, open rings) are attributed to the Early Postclassic Period (prior to A.D. 1250) and come from the Infiernillo region (Maldonado C. 1980). Although these artifacts are reported as silver, my macroscopic examinations indicate that they probably were made from copper-silver alloys. The best evidence thus far that after A.D. 1200 smiths began to employ alloys of gold and silver with copper more widely comes from the vast numbers of sheet metal artifacts looted from the cemetery at the El Chanal site in Colima. As mentioned, many El Chanal artifacts appear in the RMG collections, and another El Chanal group has been described and analyzed by Kelly (1985). These artifacts do not derive from controlled excavations, but Kelly's observations of the cemetery clarify their temporal associations and their provenience; she believed the El Chanal metal dated after A.D. 1200.

All but a few of the El Chanal artifacts are made from extremely thin metal sheet. Kelly (1985) had eight objects analyzed qualitatively; they were made of silver, of silver-copper alloys, and, in five cases, of a ternary copper-silver-gold alloy. One is a binary alloy of gold and silver. Ternary copper-silver-gold alloys have been identified only twice before from West Mexico: in two small cast bells from Nayarit, illustrated by Lumholtz (1973) and analyzed by Root (in Lothrop 1952), and in four examples of metal sheet from Bernard (Guerrero), analyzed by Brush (1962). The binary gold-silver alloy had never before been identified in Mesoamerica.

The 60 gold and gold alloy objects and 76 silver and silver alloy objects in the RMG collection (see table 5.6 and appendix 1) are made from thin (0.1 cm) sheet metal in all but a few cases. All RMG gold and gold alloy objects are said to have come from El Chanal. I was unable to analyze these artifacts, but it is clear from their color and from surface corrosion characteristics that they are made from gold alloys, rather than from pure metal. Five of the silver and copper-silver alloy artifacts are from El Chanal, and 34 are from Lo Arado in Jalisco. The sites from which the remainder were obtained are unknown. Objects made from silver and silver alloys include disks (figure 5.12), neckpieces, nose and lip rings, crescent pendants, and coils of metal that I believe are bracelets. Chemical analyses (see "sheet metal," appendix 2) indicate that they are made from copper-silver alloys with silver in concentrations ranging from virtually pure silver to 14% silver in copper. All silver and copper-silver objects are made from sheet metal, and all were cold-worked to shape.

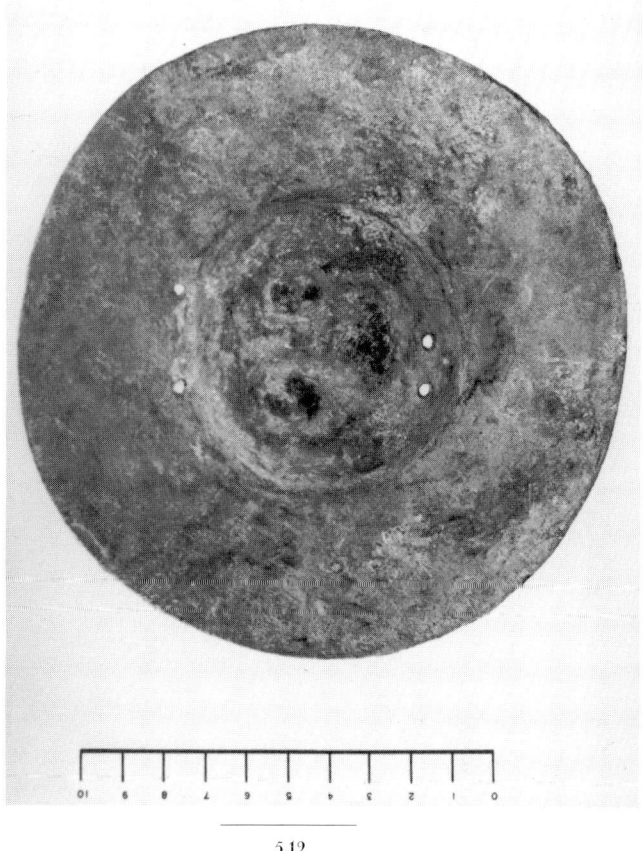

5.12

West Mexican disk in RMG collection made from a silver-copper alloy.

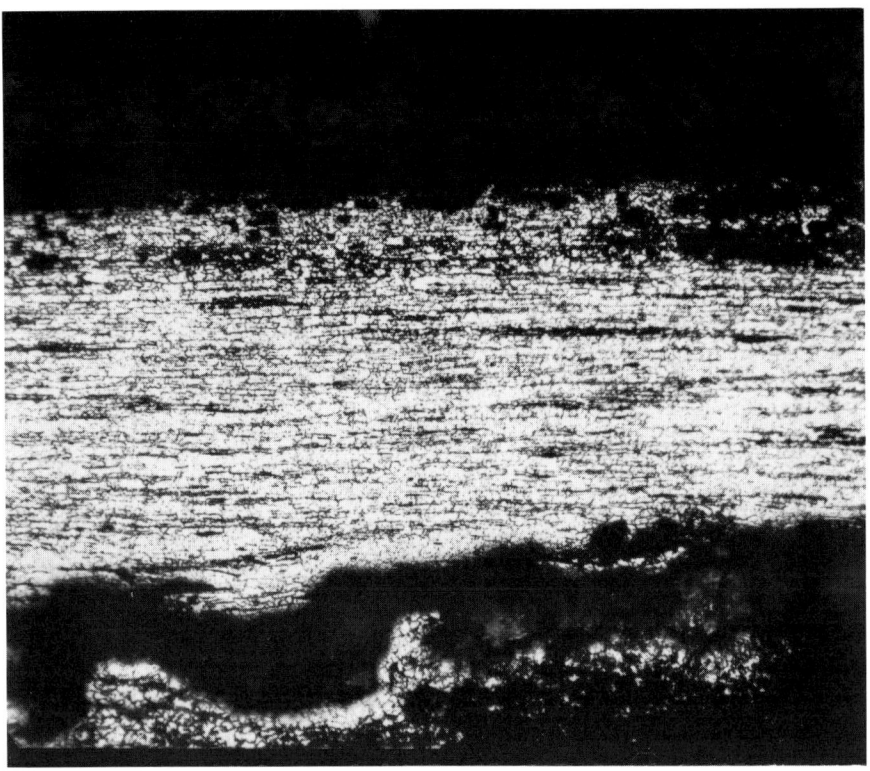

5.13

Cross section through Ag-Cu alloy disk illustrated in figure 5.12. Lamellar microstructure typical of these alloys develops even at low concentrations of the alloy element, here 5% Cu. The object was shaped through cycles of cold work and annealing. Sample etched in potassium dichromate with hydrochloric acid (mag.: 50).

The *Relación de Michoacán* reports that the ruler wore a silver disk or shield (called a *rodela*) at his back when he led his people into battle, and was interred with a gold disk on his chest. Crescent pendants were worn by dancers. One of the large disks in the RMG collection, illustrated in figure 5.12, came from a burial where it was found adjacent to the cranium of the skeleton and associated with other copper-silver alloy artifacts. This object contains 95% silver and 5% copper. The photomicrograph in figure 5.13 represents a cross section of metal removed from the disk and illustrates a fully cold-worked structure. The metal underwent repeated cycles of cold work followed by annealing; it was left in the cold-worked condition. The copper-silver alloys are particularly useful in fabricating thin sheet metal objects due to the toughness imparted by the laminated microstructure, which inhibits the propagation of cracks across lamellae. What is more, silver enrichment effects ensure that the color of the metal will always be silver, even in alloys containing primarily copper.

Documentary sources (Tudela 1977; Warren 1985) make clear that sheet metal objects made from gold, silver, and their alloys were common in Michoacán, so that the numbers presently found in museum collections do not reflect the situation prior to the Spanish invasion. For example, Cristóbal de Olid, the captain of the first occupying force in Michoacán, delivered Tarascan metal objects to Hernán Cortez that he had looted from Tarascan temples. The documents state that 1,000 marks (or in excess of 60 pounds) were made from a copper-silver alloy, and 5,000 marks were made from an alloy of gold and silver (Warren 1985). In Michoacán, objects made from silver and gold and from the copper-silver alloys were so abundant that Cortez's first expedition to the Orient, headed by captain Alvaro Saavedra, took with it Tarascan copper-silver alloy objects to trade, including 122 shields, 100 diadems, 11 bracelets, and other objects (Warren 1985). Reports of Tarascan treasure sent to Nuño de Guzmán, who became governor of New Galicia in 1531, describe three shipments: one that included 600 shields of gold and 600 of silver, a second containing 400 gold objects and 400 of silver, and a third that included 200 shields of gold and 200 of silver in addition to gold earrings, bracelets, and pendants. A conflicting account reports that the second shipment contained 400 objects of silver and gold including bracelets, earplugs, shields, miters, and small vessels (Warren 1985). Regardless of the particulars, a very large quantity of gold, silver, and copper-silver alloy sheet metal was being made in Michoacán. I discuss the significance of these highly reflective golden and silvery objects in chapter 8.

The use of gold and silver objects and the mining of gold and silver metal were probably under state control. This may partly explain why we do not find the objects at the sites recently excavated in Michoacán and elsewhere. Access to gold and silver must have been highly restricted. The systematic and extensive looting carried out by the Spaniards probably also accounts for the relatively few items made from these metals present in collections and excavated assemblages.

The *Relación de Michoacán* reports that Tarascan kings stored gold and silver objects in royal treasure houses dedicated to the gods. These were located at Tzintzuntzan and on several islands in Lago Pátzcuaro. Silver objects were placed there in honor of the moon deity, and according to the Spaniards were made of silver varying in quality (Warren 1985). The latter assessment probably refers to the copper-silver alloy. The Europeans describe shields and breastplates that were thin enough to be cut in half with a sword (Warren 1985), suggesting that they were probably identical or very similar to the large disk illustrated in figure 5.12. Spanish friars report finding gold and silver objects in temples, made as offerings to idols. Sometimes the idols themselves were made from metal. Friars also report seeing figures of felines and other animals made from these metals, although none of these figures has survived (Warren 1985). The Tarascans considered gold and silver property of the gods. A Tarascan ruler is reported to have said to the Spaniards, as they looted the royal treasure houses of gold and silver objects: "What was here was not ours, but it belonged to you who are gods, and now you are taking it for yourselves because it was yours" (Warren 1985: 61).

Metal was used most often between A.D. 1200 and the Spanish invasion for three artifact classes: bells shaped by lost-wax casting, cold-worked sheet metal ritual items and ornaments, and hot- and cold-worked open rings. Rings and tweezers symbolized and communicated social power and rank. Bells and sheet metal were ritual items and also status markers. In all three cases, metalsmiths employed the alloys, copper-tin, copper-arsenic, and copper-silver, to optimize artifact designs. They also used the new materials to create golden and silvery reflective colors associated with the concepts of the sacred discussed in chapter 8, and with the solar and lunar deities. They did so for objects whose design characteristics often disallowed the use of pure metals. Golden colors were achieved with high-tin bronze alloys in bells, open rings, tweezers, and occasionally sheet ornaments; silvery colors with high-arsenic bronze in bells, and with copper-silver alloys for objects that were cold- and hot-worked to shape, such as open rings, tweezers, and sheet metal ornaments. Copper-silver alloys are difficult to cast, and copper-arsenic alloys become brittle when worked if the alloying element is present in high concentrations. The metalworkers' solution to creating silvery-looking metal was precisely in keeping with the properties of these particular alloys.

Cold Work: Tools and Axe-monies

Nearly all other objects made using these new materials are implements for woodworking and woodcutting, cloth production, metalworking, and related activities. They include needles, axes, unipointed and bipointed awls, and awls with narrow blades. Tools used for subsistence activities—such as hoes, fishhooks, and digging stick points—while not abundant, also figured in the technical repertoire. One artifact type, axe-monies found in Guerrero, Michoacán, and Oaxaca, pertains to a wholly different functional category. Documentary sources suggest that the variety found in the West Mexican metalworking zone was used for tribute.

Axes. Datable axes are found infrequently in this region, although they are mentioned in many ethnohistoric sources from the Late Postclassic Period. The documents indicate that in Michoacán metal axes were used for woodcutting and woodworking. In fact, the *Relación de Michoacán* relates that a guild of woodcutters represents one of the craft specialties supported by the king of Michoacán. The guild is depicted in the *Relación*, and their leader holds a hafted metal axe, the symbol of their vocation (figure 5.14). The primary task of the guild was to gather wood for temple fires. Wood gathering itself apparently could be a ritual act. For example, the *Relación* states that a man who remarried was required to spend four days gathering wood beforehand as a kind of penance. The *Relación* illustrates such an individual, burdened with wood, wielding a hafted metal axe (figure 5.15). In Central Mexican sources, metal axes are shown being used as tools, dissociated from ritual contexts. A

5.14

Tarascan woodcutters holding a hafted, metal axe. (From Craine and Reindorp 1970, plate 3.)

5.15

Wood gathering as a ritual act among Tarascan peoples. (From Craine and Reindorp 1970, plate 12.)

5.16

The production of metal axes, from the Florentine Codex (Sahagún 1950–1982 book 11, plate 796).

carpenter depicted in the Codex Mendoza holds a hafted axe (Clark 1938, vol. 3, fol. 70r), which he uses to trim a branch. Durán (1967, vol. 2, plate 61) illustrates individuals using metal axes to construct wooden boats for Hernán Cortez.

The Florentine Codex illustrates some aspects of axe production in the prehispanic era. Figure 5.16 shows smelting in a crucible-type furnace, from which the molten metal flows into an axe-shaped mold. Next to the crucible lies a finished axe. Metallographic studies show that axes such as these were then cold-worked to achieve their final form and to increase hardness.

Some of the most revealing ethnohistoric information makes clear that the metal axe form was itself invested with power. Most of this information does not come directly from the metalworking zone but rather from sources from Central Mexico and the Highland Maya and Mixteca Puebla regions. One version of the *Popol Vuh,* the creation story of the Maya peoples, relates that the sky was held up by an upturned copper axe (Jansen 1990). Metal axes also are illustrated in ritual and calendrical codices from the Mixteca Puebla region. In the Codex Cospi (Kingsborough 1964–1967, vol. 4, plates 18, 22), Mictlantecuhtli (figure 5.17) and Tlaloc, two of the Nine Lords of the Night, are each illustrated grasping a hafted metal axe. In the Codex Laud (Kingsborough 1964–1967, vol. 3), from the same region, Tlaloc is also depicted with a hafted metal axe. Also in the Codex Laud a hafted axe appears in a sequence showing which days are appropriate for penances. Portrayed with it is a sacrificial knife, a bundle of wood, and a tool used for bloodletting. In the Codex Selden (Kingsborough 1964–1967, vol. 2, plate 87), hafted axes are shown among objects identified as offerings. The Codex Fejervary Mayer (Kingsborough 1964–1967, vol. 4, plate 38) shows the Lord of the Earth brandishing an axe over a captive who cowers at his feet (figure 5.18).[15]

The idea that the axe form represents power is ancient in Mesoamerica. The ethnohistoric sources suggest that metal axes like axes made from other materials were also accorded those meanings. The *Relación de Michoacán* mentions that the king had a subordinate whose primary task was to carry the king's copper axes. Although the *Relación* does not explicitly say so, these axes must have figured among the objects that represented sacred, royal power.

During Period 2, smiths cast their axes using the bronze alloys (figure 5.19). Although few metal axes from this West Mexican region derive from secure archaeological contexts, one was excavated at Tzintzuntzan, and Brush (1962) analyzed a tin bronze celt from Palos Blancos, a Late Postclassic site on the coast of Guerrero. The 18 axes made from bronze alloys in the RMG collection range in length from 7.8 to 15.8 cm. As appendix 2 indicates, three are made from a ternary copper-arsenic-tin alloy, ten from tin bronze, and the remainder from copper-arsenic alloys. Appendix 2 also gives the range of concentration of the alloying element in these axes: tin is present in amounts from 1.26% to 8.72%; arsenic varies from 0.71% to 5.67%.

The metalworkers employed these bronze alloys to enhance the performance of the tools. Table 5.10 shows that by using bronze they could fashion implements that were both harder and thinner than axes made from copper. Period 2 axes are approximately three-fourths the

5.17

Codex Cospi: Mictlantecuhtli holding a metal axe. (From Kingsborough 1964–1967 vol. 4, plate 18.)

5.18

Codex Fejervary Mayer: Lord of the Earth brandishing an axe over a captive. (From Kingsborough 1964–1967 vol. 4, plate 38.)

5.19
West Mexican tin bronze axe, RMG collection.

thickness of their copper counterparts; the length-to-thickness ratio increases by about 50%. These new and stronger alloys allowed the metalsmiths to cast thinner designs, but the axes are also far harder. Table 5.10 indicates the respective Vickers microhardness values (VHN) for axes made from copper and its alloys.

All RMG bronze alloy axes were studied metallographically, and the results show that fabrication techniques were identical for all of them. These axes were initially cast as a blank, and their final form achieved through a sequence of stages of cold work followed by annealing. All axe blades were cold-worked as the final operation to shape and to harden them. The butt ends of some were also cold-worked after the final anneal.

Metallographic data, artifact chemistries, and hardness values demonstrate that smiths maximized blade hardness in all cases. However, hardness was controlled and increased through fabrication methods (by work hardening) rather than by any careful control of the alloying element concentration (solid solution hardening). Table 5.10 and appendix 2 indicate just how much alloy concentration varies in these objects. The difference between 0.77% and 8.06% tin in a copper-tin alloy represents very significant differences in the hardness of the cast metal but also in the working properties, including hardenability, of the alloy.

Certain characteristics of the final product were more easily managed once the axe was cast, since crucial properties were controlled through working rather than by using standard alloys. Standard alloys, at least of copper-arsenic bronze, are difficult to prepare. The amount of oxygen present is unpredictable, yet it too affects working

Table 5.10 Copper and Copper Alloy Axes: Comparison of Average Dimensions and Hardness Values

Metal/Alloy	No. of Objects	Concentration Range of Major Alloying Element (wgt. %)	Average Length (cm)	Average Thickness (cm)	Ratio Length: Thickness	Vickers Hardness Number (kg/mm^2)*
Cu	17	—	10.8	1.0	10.8	80–135
Cu-Sn	5	0.77–8.06	11.5**	0.7**	16.1**	131–274
Cu-As	9	0.71–4.80				70–195
Cu-Sb***	1	—				101–171
Cu-As-Sn	3	Sn: 1.33–5.30 As: 0.64–1.30				123–297

* Values represent range of maximum hardness readings on blades.
** The averages for axes made from alloys are for all 18 objects.
*** Qualitative determination only.

properties. These factors probably explain why metalsmiths preferred to manage hardness through final cold work and shaping of these tools. The issue is more complex in the case of tin bronze axes, because metalworkers did control tin concentrations in some object classes although they did not standardize the alloys.

If we compare the composition and hardness data for a group of these axes (table 5.11), the extent to which hardness was controlled through cold work becomes apparent. The axe containing 2.48% tin is somewhat harder than the axe that contains 8.06% of that element. Metallographic studies of the blades of both axes reveal that the 8.06% tin bronze was cold-worked to the point where toughness was compromised, as figure 5.20 indicates. This photomicrograph shows the blade section of that axe; fracture lines are clearly visible in the eutectoid. The eutectoid microconstituent of a copper-tin bronze alloy is inherently brittle and will undergo brittle fracture if severely cold-worked. This artifact contained such high concentrations of tin that the smiths stopped at this point. Maximum hardness measured 210 VHN. Experimental data show that a 7% tin bronze reaches 170 VHN when reduced in thickness by only 25%, but can reach 248 VHN if reduced by 75%.[16] By contrast, the axe containing only 2.5% tin was so severely cold-worked that it reached maximum hardness.

Table 5.11 RMG Axes: Concentration of Alloying Element and Hardness

ID No.	As (wgt. %)	Sn (wgt. %)	Vickers Hardness Number (kg/mm2)*
367	—	2.48	188–253
2249	—	1.94	159–213
374	—	7.92	185–256
403b	—	8.06	156–210
351	0.71	1.33	148–202
369	0.64	5.31	122–297
386	1.31	3.10	132–177

* Values represent range of hardness readings on blades.

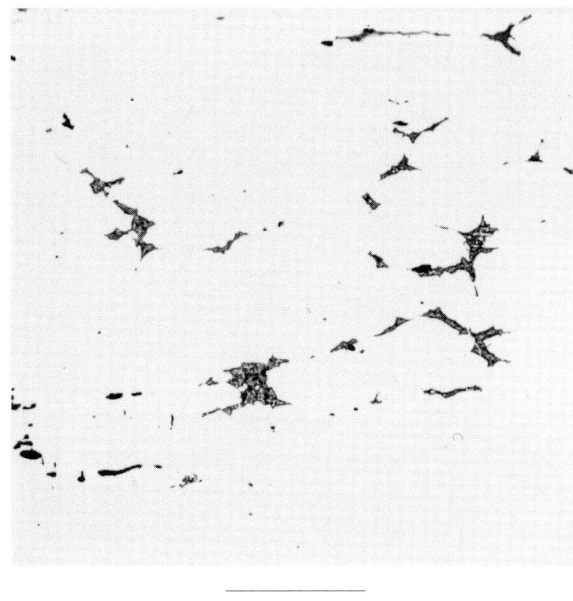

5.20

Section from the blade of an 8.06% tin bronze axe, as polished; the brittle eutectoid exhibits fracture lines resulting from excessive cold work (mag.: 200).

Awls: Unipointed, Bipointed, and Bladed. The design of other tools also took advantage of the superior mechanical properties of these bronze alloys. Awls of both types present in Period 1 (see figure 3.26), as well as an awl with a blade (figure 5.21), were made during Period 2 from alloys of copper-arsenic, copper-tin, and copper-arsenic-tin (table 5.6).

The uni- and bipointed awls were made as before, but a new type appears distinguished by a blade at one end. Intact examples in the RMG collection range from 6.6 to 9.4 cm in length. Figure 5.22 shows a longitudinal section from the tip of such a blade. The object has been severely cold-worked; in addition the metal at the tip end is doubled over from use.

The awls made from alloys, including the type with the blade, were fashioned in the same basic way as the

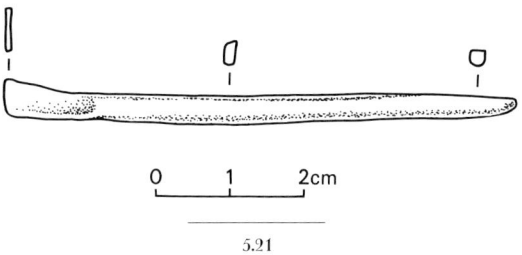

5.21

Awl with narrow blade, a type new to Period 2; top view.

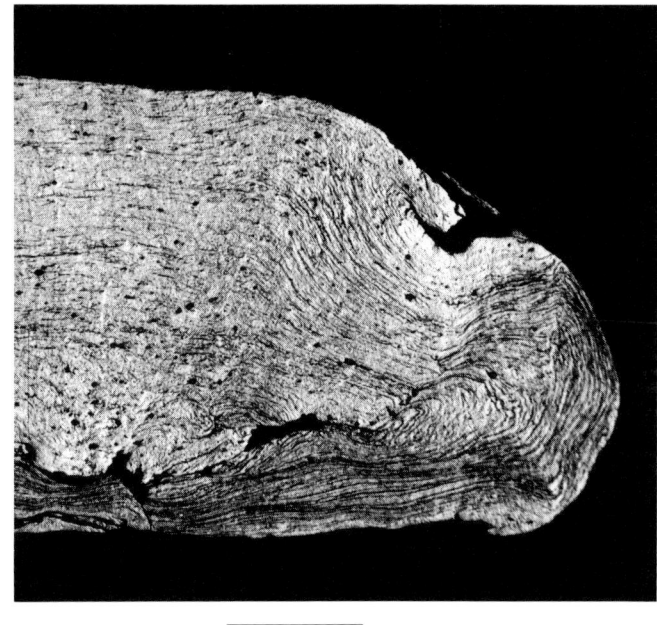

5.22

Longitudinal section from tip of awl with blade. The heavily cored metal has undergone severe plastic deformation. The working tip is blunted and doubled back from use. Sample etched in potassium dichromate (mag.: 100).

Period 1 copper awls. Metalworkers cast a long rod, then hammered it around to achieve its final shape. Those with a round cross section tended to be folded longitudinally leaving a central fissure; square or rectangular cross section objects were cast closer to their final shape, then trued up by hammering. Figure 5.23 shows a photomi-

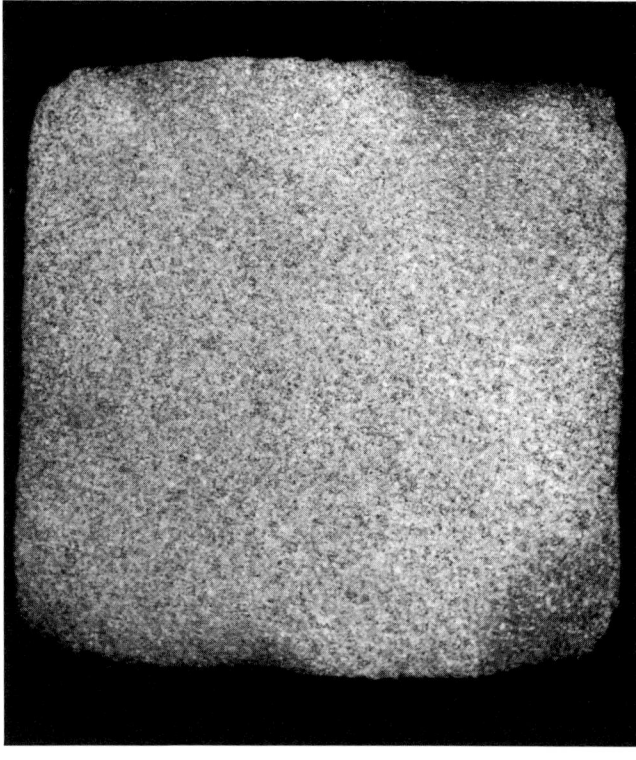

5.23

Cross section of a Cuexcomate awl with narrow blade (6.99% Sn, 1% As). The section has been hammered almost square, with corners most heavily worked, perhaps from use; no central fissure is visible in the metal. Sample etched in potassium dichromate (mag.: 21).

crograph of the cross section of a rectangular cross section copper-arsenic-tin awl from Cuexcomate containing 6.99% tin and 1% arsenic. It has no central fissure. Few examples of the round cross-sectional type are thus far known from this period; all Cuexcomate specimens, for example, are made with rectangular cross sections.

Although these tin and arsenic bronze tools were capable of working far harder materials than was copper alone, they required a sturdy, tough shank capable of absorbing impact. Those made with an internal longitudinal fissure could have been used only to work soft materials, such as soft woods, certain fibers, leafy materials, skins, and hides. Sahagún mentions that people making lost-wax castings used a metal tool to carve clay and charcoal cores. These awls would have been appropriate for such tasks, as well as for sculpting or smoothing the wax itself, or perhaps for chasing designs in gold objects. Sahagún (1950–1982, book 9, plates 54–56) shows metalworkers using an incising tool to chase gold designs.

All three varieties of these small hand tools appear at numerous sites in West Mexico (see table 5.6). Tin bronze alloy awls have been recovered at Bernard in Guerrero (and at Calixlahuaca in the state of Mexico). Michael Smith's excavations at Cuexcomate unearthed 14 of these objects in household debris,[17] all alloys either of copper-arsenic-tin or of copper-tin (see table 7.2). These probably were not metalworking tools; no convincing evidence for metal production has been found at Cuexcomate.

Metallographic studies show that the Cuexcomate implements have been cold-worked to shape. They are extremely hard as a result of their high tin content and final work hardening, and they may have been used for carving wood. Only two Cuexcomate implements are intact. Four have blades, and one is bipointed; the fragmentary condition of the others does not retain evidence of the form of their working ends. Thus it is impossible to determine whether, overall, the design of this group of tools was altered to take advantage of the properties of the alloys. Evidence from the RMG collection and from awls excavated from burials at Urichu suggests that they were made longer than their counterparts in copper, but the design differences are not as dramatic as in bells, tweezers, and axes. Table 5.12 documents these differences.

Table 5.12 Awls (Unipointed, Bipointed, Blade): Comparison of Dimensions and Compositions

Provenience	Metal/Alloy	Length (cm)	Midpoint Thickness (cm)	Ratio Length : Thickness
RMG	Cu	5.00	0.30	17
RMG	Cu	6.80	0.39	17
RMG	Cu	7.90	0.30	26
RMG	Cu	10.00	0.70	14
RMG	Cu	10.30	0.25	41
RMG	Cu	11.30	0.40	28
			Average ratio (Cu):	25
RMG	Cu-As	12.80	0.40	32
Cuexcomate	Cu-Sn	6.20	0.30	20
Cuexcomate	Cu-Sn	9.95	0.40	24
Urichu	Cu-Sn?	10.40	0.40	26
Urichu	Cu-Sn?	9.00	0.16	56
Urichu	Cu-Sn?	9.70	0.20	48
			Average ratio (alloys):	34

Needles. Period 2 sewing needles were made from bronze alloys. The alloying element is present in concentrations of about 1 to 2 percent. The most common needle eye design was the loop eye type. These needle eyes are made by doubling over an extremely thin (less than 0.07 cm) tab of metal which tucks into the needle shaft. A variant on the design is made in the same way except that the tip of the tab protrudes. The piercing end of these needles is pointed. Both loop eye needle designs are illustrated in figure 5.24. The strength of the bronze alloys was essential to the extremely thin loop eye metal; this design was only marginally functional when made from copper.

Artisans fashioned these needles by hammering a flat rectangular strip of metal to the desired thickness and length, leaving a thin tab at the eye end. They then hammered the metal around its longitudinal axis to form the shaft, leaving a long internal fissure. The eye was made by doubling over the tab, inserting the free end into the shaft, then hammering the shaft down over the tab to secure it. The metal underwent many sequences of hammering and annealing and was ultimately left in the cold-worked state. Figure 5.25 shows an etched section through such a needle eye; the pronounced flow lines that run circumferentially around the shaft indicate the direction in which the shaft metal was initially cold-worked. The equiaxed grains and annealing twins that result from hammering and annealing are also visible in the photomicrograph. Microhardness readings (150 VHN) are highest at the edges where cold work was most pronounced and are consistent with the metal composition: a copper-arsenic alloy containing 1.89% arsenic.

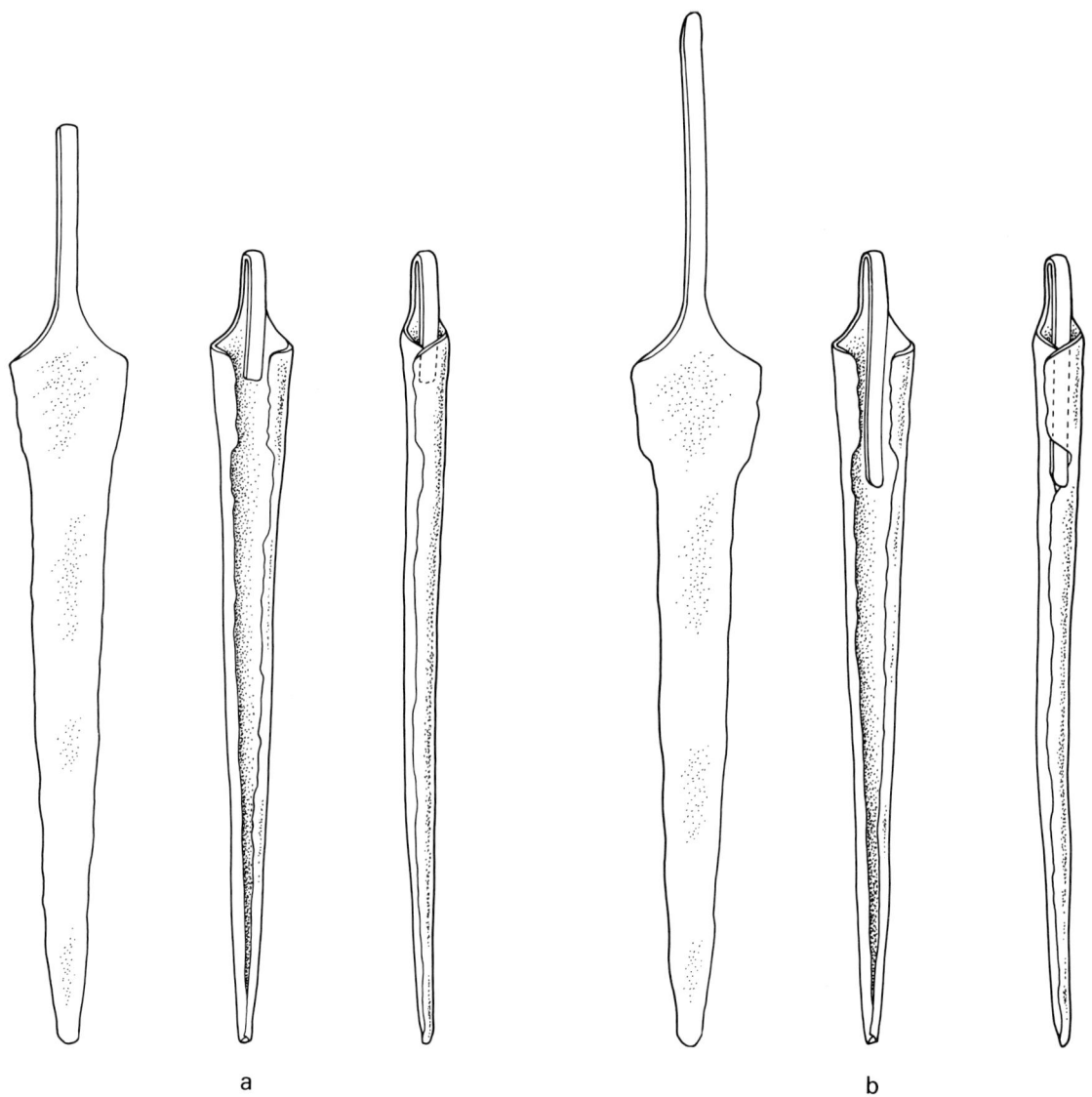

5.24

Loop eye needles. Two designs and their fabrication sequence: (a) loop eye tab tucks into needle shaft; (b) eye tab protrudes from beneath overlapping lateral flaps.

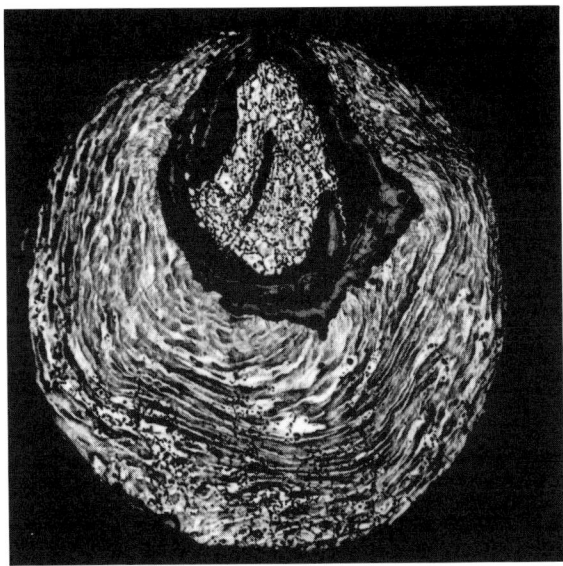

5.25

Loop eye (tucked) needle (1.89% As). Cross section through the eye shows tab end completely enclosed within the hollow of the shaft. Flow lines in heavily cored shaft metal indicate direction of metal flow during rounding of the shaft. Sample etched in potassium dichromate (mag.: 50).

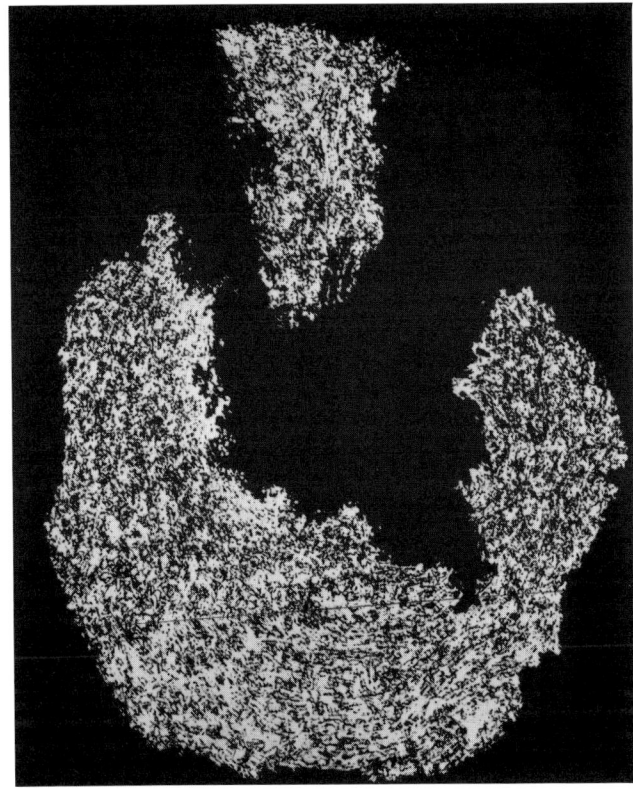

5.26

Loop eye (flaps) needle. Cross section just below the eye shows tab end protruding beyond edges of lateral flaps. The metal is in fully annealed state. Sample etched in potassium dichromate (mag.: 100).

The other iteration of this design was fashioned in basically the same way except that the shaft was hammered flat just below the tab, providing a pair of lateral flaps. The eye tab was then bent over and the two flaps hammered down to secure it, leaving the end free. The section through the shaft just below the eye (figure 5.26) shows the protruding tab. The only example of this needle eye type in the RMG collection is made from copper. Two examples have been excavated at Cuexcomate (Morelos), both made from bronze alloys.

Smith's excavations at Cuexcomate and Capilco produced 11 needles (see table 7.2); the two loop eye needle designs and the variety made during Period 1 with a perforated eye were used at these sites during the same period. All were made from copper-arsenic-tin alloys. At these sites loop eye needles with a protruding tab date to between A.D. 1200 and A.D. 1350, the earlier part of Period 2. (This variety may subsequently have been substituted by the other, sturdier type, which constitutes by far the most pervasive needle type in Mesoamerica.) The design has intrinsic stresses introduced into the metal as the loop is doubled over and the two tabs are hammered around to secure it. Annealing relieves these stresses, but it also makes for a softer needle. We do not know how these distinctive needles were used at Cuexcomate. However, these implements support Smith and Heath-Smith's (1994) contention that Cuexcomate was a zone of specialized cloth production.

Both needle designs, the perforated eye and loop eye, have also been found in the Balsas drainage in burials at Infiernillo and La Villita (see table 5.6) but these examples have not been analyzed. Apart from Cuexcomate, where needles were recovered from household contexts, loop eye needles have been excavated from burials at a number of other sites: at Urichu, Tzintzuntzan, Huandacareo, and Tres Cerritos. The perforated-eye needle has also been found at Apatzingán.

None of the Cuexcomate needles measures longer than 11 cm. The bronze alloys were not being used to make longer needles but to fashion sturdier designs. The chief design requirement was to achieve a very narrow metal loop eye: the loops measure only 0.05 cm in some cases. However, needles excavated at Urichu did take advantage of the properties of the alloys for length and a finer design: for example, one measures 16.5 cm in length and is only 0.1 cm thick, with an extraordinarily narrow (0.03 cm) loop eye.

Axe-monies. Axe-monies constitute a unique artifact class. Axe-monies have been reported most frequently from Oaxaca (Easby, Caley, and Moazed 1967; Hosler 1986; Hosler, Lechtman, and Holm 1990), but the variety found in Guerrero and along the Michoacán-Guerrero border do not appear outside this metalworking zone (figure 5.27). The West Mexican axe-monies measure from 14 to 20 cm in length, are thin (mean thickness measures 0.05 cm), and are shaped like an axe.

Axe-monies have rarely been found in archaeological context in this region, and never with secure dates. The Oaxacan variety dates to after A.D. 1200. Ethnohistoric evidence for the West Mexican type indicates that these objects were tribute items. Objects that closely resemble them are illustrated in the Codex Mendoza as tribute items to the Aztec from two provinces in Guerrero, and similar artifacts, but made from silver, were tribute to the king of Michoacán (Clark 1938; Schöndube B. 1974). *Hachuelas* is a term in Spanish sometimes used to describe the axe-shaped objects that were used as tribute. An inventory of the Casa de Munición in Mexico City, drawn up in 1528, lists among the copper objects stored there eight hundredweight of copper, 500 copper shields, and 113 cases of copper *hachuelas* (Barrett 1981: 12). The probability is good that these *hachuelas* were the type 1a axe-monies described and illustrated here.

In descriptions of archaeological explorations in Naranjo, central Guerrero, Weitlaner reports acquiring a "packet of 13 copper leaves [*láminas*] in the form of an axe but with the thickness of heavy paper, about whose use we were unsure" (Weitlaner 1947: 79). Fragments of *láminas muy delgadas* [very thin leaves] were excavated at La Villita on the Michoacán-Guerrero border (Cabrera C. 1976), and villagers at the site of Xochipala, Guerrero, also report finding them, again referring to them as *láminas* (Hosler 1986; Hosler, Lechtman, and Holm 1990).[18] A group of 30 were collected from the Balsas drainage in Guerrero and are housed in the regional museum of Cuernavaca in the state of Morelos. Examples from Guerrero are also found in the British Museum, the Museo Nacional de Antropología e Historia in Mexico City, and the American Museum of Natural History in New York.

The distinctive characteristics of these objects, determined from studies of the RMG collection, is that they are usually made from arsenical copper and often from an alloy of copper and arsenic (see appendix 2). Arsenic concentrations range from 0.05% to 6.35% by weight. Axe-monies were fashioned from an original cast blank; successive sequences of cold work and annealing pro-

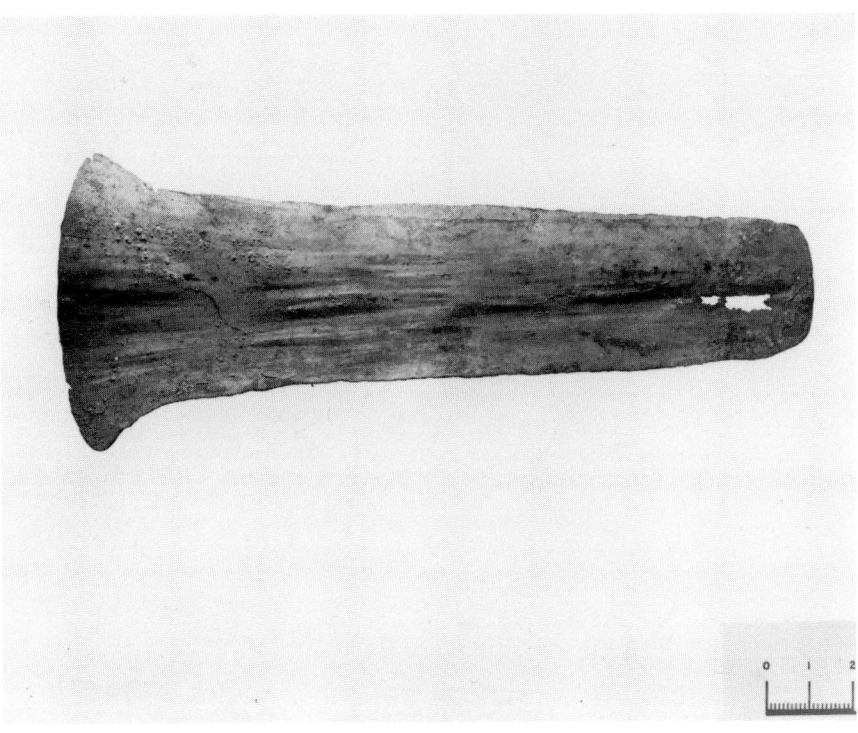

5.27
Axe-money (type 1a), found in Guerrero.

duced the extremely thin metal sheet used to shape them. The photomicrograph in figure 5.28 of a longitudinal section through a typical Guerrero specimen shows a fully annealed structure with some very slightly elongated inclusions reflecting prior cold work. An alloy optimizes the design of these leaflike objects because the metal is so thin. Most axe-monies were left annealed, although some were left in the cold-worked condition. If metalsmiths consistently had made them this thin from copper, they would have retained their shape with difficulty. Even so, the concentration of arsenic is highly variable and clearly was not systematically controlled.

The use of the copper-arsenic alloy for this West Mexican design and for all other axe-monies suggests that they, like their Ecuadorian counterparts (see chapter 6), may have served as a repository for copper-arsenic metal. Lengths range from 14 to 20 cm, as I have noted; thickness varies from 0.02 to 0.06 cm. The fact that lengths are fairly uniform is consistent with their use as tribute items. Axe-monies are made so that they can easily be stacked on top of one another, and Weitlaner's description of a "packet" of 13 leaves suggests that in this zone, as in Ecuador, these objects were packaged and bound in lots. They were easily portable repositories of copper-

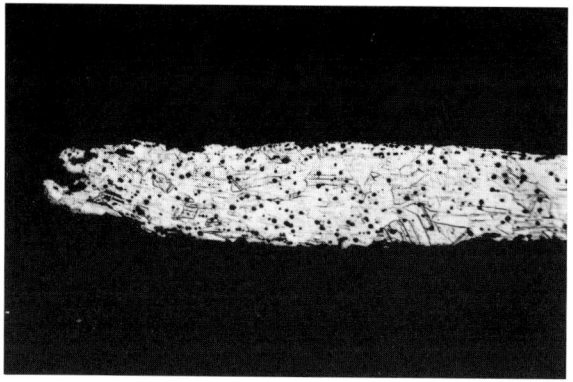

5.28

Longitudinal section through the tip of a type 1a axe-money (0.60% As). Microstructure characterized by equiaxed grains with annealing twins; the cold-worked and annealed metal has been left in the annealed state. Sample etched in ammonium hydroxide plus hydrogen peroxide (mag.: 100).

arsenic metal, and their axe shape reflects the traditional importance of the axe form in Mesoamerican societies.

The artifacts described here in some detail—bells, open rings, tweezers, needles, axes, awls, and axe-monies—are the primary metal object classes produced during Period 2 in the metalworking zone. They are by no means the only ones. Fishhooks, hoes, digging stick points, arrow points, nose rings, lip plugs, beads, buttons, hand-held rattles, finger rings, and assorted small ornaments (bells attached to needles, and so forth) were also fashioned during this time and from these same metals and alloys (Hosler 1986, 1988a). However, they appear relatively infrequently in museum collections and in archaeological excavations. Sheet metal gold and silver ritual items, sometimes made from alloys of those metals, are the only significant artifact class that cannot be thoroughly treated here. They do not appear either in museum collections or in archaeological contexts in the numbers suggested by the documentary sources.

The Focus of Period 2 Metallurgy

During the period after A.D. 1200, a new technological complex emerged in the area of the metalworking zone described here. That area includes the Tarascan region of Michoacán, the Valley of Toluca, northwest Guerrero, Colima, and parts of Jalisco. However, apart from occasional references to several mines in Michoacán where ore processing occurred (Hosler 1986; Pollard 1987; Warren 1968, 1989), we do not know where mining and production took place. The few sites mentioned here from which slag or ore have been reported do not constitute primary processing centers. We also have little idea of how the production of these bronze and copper-silver objects may have been controlled. Laboratory data offer little evidence for standardization either in alloy types or in artifact design, except at the broadest levels of property control, suggesting that processing and manufacture were carried out at multiple centers.

A defining characteristic of the Period 2 technological complex is that metalsmiths used copper alloys extensively: the two binary bronzes, copper-silver, and ternary alloys of these metals. New smelting regimes (described in chapter 2) were required to produce this range of alloys. These alloys allowed artisans to optimize the design of objects that had previously been made in copper, employing new fabrication techniques, such as hot work, to manipulate them when necessary. Objects made from bronze alloys were fashioned in only one other region, the Huastec area in eastern Mexico (Hosler and Stresser-Péan 1992), and the data so far indicate that this occurred just before the Spanish invasion. Elsewhere, for example to the north in Sinaloa and at sites along the lower Balsas, smiths continued to work in copper as before. To the south, in Oaxaca, a very different technology emerged centering on the production of lost-wax castings from copper-gold alloys. Although West Mexican metallurgy underwent radical changes during Period 2 in materials and in processing methods, little change took place in the constellation of objects made from metal, or in their numbers relative to one another. The emphasis or orientation of the technology remained the same.

The artisans' interest in the resonant properties of metal continued to develop during this period. Bells were the most significant artifact class with respect to sheer numbers and the varieties of new designs. Smiths used the alloys to optimize these new designs, making thinner, larger, and more complex castings taking advantage of alloy strength and solidification characteristics. During Period 2, these artisans also became extremely interested in metallic colors, which they achieved by raising the tin or arsenic concentrations to levels such that arsenic bronze bells looked like silver and tin bronze bells like gold.

The focus on metallic colors is also evident in status display objects made by hot or cold work. The computer simulation studies make clear that the strength of the bronzes and of the copper-silver alloy were required to accomplish the design of the shell tweezers and the open rings. However, metalsmiths used the alloying elements in higher concentrations than necessary, in order to achieve the golden and silvery colors these alloys conferred on worked objects: they were choosing among technical alternatives. They accommodated their procedures to the alloy concentrations, working high-tin bronzes hot when necessary, so that neither formability nor color was compromised. These alloys also allowed the metalsmiths to craft a range of new artifact types from extremely thin sheet metal: ornaments, neck pieces, breastplates, and ornamental shields.

At the same time, metalworkers used bronze to improve the design and function of tools and implements, which they made harder, tougher, and finer. Axes made from tin bronze, arsenic bronze, and the copper-arsenic-tin alloy were nearly twice as hard and half as thick as the Period 1 examples in copper. The bronze alloys also permitted a new needle design, with a thin, loop eye. In addition, metalworkers used arsenic bronze for axe-monies, the extremely thin leaves or *láminas* that served as tribute items from Guerrero.

In summary, the chemical analytical data (appendix 2) indicate that Period 2 smiths were making several basic alloy types: (a) a low-tin or low-arsenic bronze (with less than about 6% of the alloying element); (b) a high-tin or high-arsenic bronze; (c) a high-silver copper-silver alloy. Despite this general pattern, alloy concentration often varies greatly within a single artifact class. In open rings, for example, tin concentration ranges from 8% to 19%. West Mexican metalsmiths understood and controlled the properties of all these alloys, but they did not exercise tight control of alloy concentrations. When possible, they optimized properties through processing techniques.

These data and the data in chapter 3 establish that the fundamental orientation or pattern of West Mexican metallurgy became established very early and did not change significantly. Nonetheless, as the technology developed—at least in the regions and at the sites discussed here—smiths greatly amplified the range of designs and fabrication methods to optimize certain properties of the new alloy materials. Some of these properties, such as color, were new. Others, particularly the resonant qualities of metal, were unquestionably valued throughout. Further, while the laboratory data make clear that these artisans fully understood and manipulated the properties and manufacturing regimes required for tools, utilitarian artifacts never became a primary concern. These societies defined metal as a material whose principal use was for objects that communicated sacred power—through sound and through color—and for objects used in ceremony and worn by elites.

6

PERIOD 2: ORIGINS AND TRANSFORMATIONS

Some Period 2 developments had their roots in the metallurgies of lower Central America and Colombia or southern Ecuador and northern Peru, the same two regions from which elements of the technology were introduced earlier. Others derived from metallurgical traditions even farther to the south, from the south-central coast of Peru and the adjacent highlands of Peru, Bolivia, and northwest Argentina. Still others arose independently in West Mexico. Period 2 metalsmiths integrated these diverse elements according to their own precepts concerning how metal should be used and the particular native metals and ores available to them.

Alloying

The three copper binary alloys had been developed in South America hundreds of years before they were first used in West Mexico. All three were introduced to West Mexico around A.D. 1200 or slightly before; copper-arsenic and copper-silver alloys seem to predate the alloys of copper-tin. These alloys then were elaborated locally using local resources. A few of the new designs crafted from these copper alloys do derive from South or Central American prototypes. However, as I have pointed out elsewhere, Andean peoples did not export the raw materials, either as ores or in ingot form; nor were the objects themselves ever imported to West Mexico on a large scale.

Southern Ecuador and northern Peru contributed two major alloy systems to the metallurgy of West Mexico, copper-arsenic and copper-silver, as well as prototypes for objects made from them. Copper-arsenic and copper-silver alloy objects are extremely common throughout this Andean zone but are largely absent in the area between Ecuador and West Mexico.

Arsenic Bronze. Arsenic bronze artifacts date to as early as A.D. 300 in coastal Ecuador. Metalworkers on the north coast of Peru were also experimenting with this alloy at about the same time. By A.D. 950, and perhaps earlier, low-arsenic copper-arsenic alloys served as stock material in southern Ecuadorian metallurgy. This same alloy, made locally, was also a stock material in northern Peru. Andean smiths used arsenic bronze to fashion open rings, tweezers, bells, awls, axes, needles, and axe-monies.

We do not yet know where in Ecuador the arsenic bronze alloy was produced. The highland provinces are the likely possibility, since arsenopyrite, as well as enargite and other arsenic-bearing copper ores, occur there. Shimada (1985) has argued that most arsenic bronze metal used in Ecuador was imported as ingots from primary smelting centers, like Batán Grande, on the north coast of Peru.[1] Batán Grande (figure 6.1) became a major center for arsenic bronze production by about A.D. 900, and it continued as a primary north coast supplier of that alloy to the Sicán and Chimú until about A.D. 1400 (Shimada 1985; Shimada and Merkel 1991).

In West Mexico, metalsmiths used arsenic bronze alloys extensively only after A.D. 1200. Two Period 2 artifact types made from this alloy most directly link the region with the southern Ecuadorian and northern Peruvian zone. One is the paper-thin axe-money and the other the versions of the distinctive loop eye needle. Metallographic studies show that the fabrication sequence for the axe-monies is identical in both regions; methods used to manufacture the two needle varieties likewise are identical. More generalized artifact forms—axes, awls, and small hand tools—also were made from copper-arsenic bronze in both regions, and their manufacturing methods and design characteristics were also the same. Nonetheless, axe-monies and loop eye needles provide the least ambiguous evidence for contact due to their unusual forms and specialized use. We know that axe-monies were used for similar purposes in West Mexico and Ecuador.

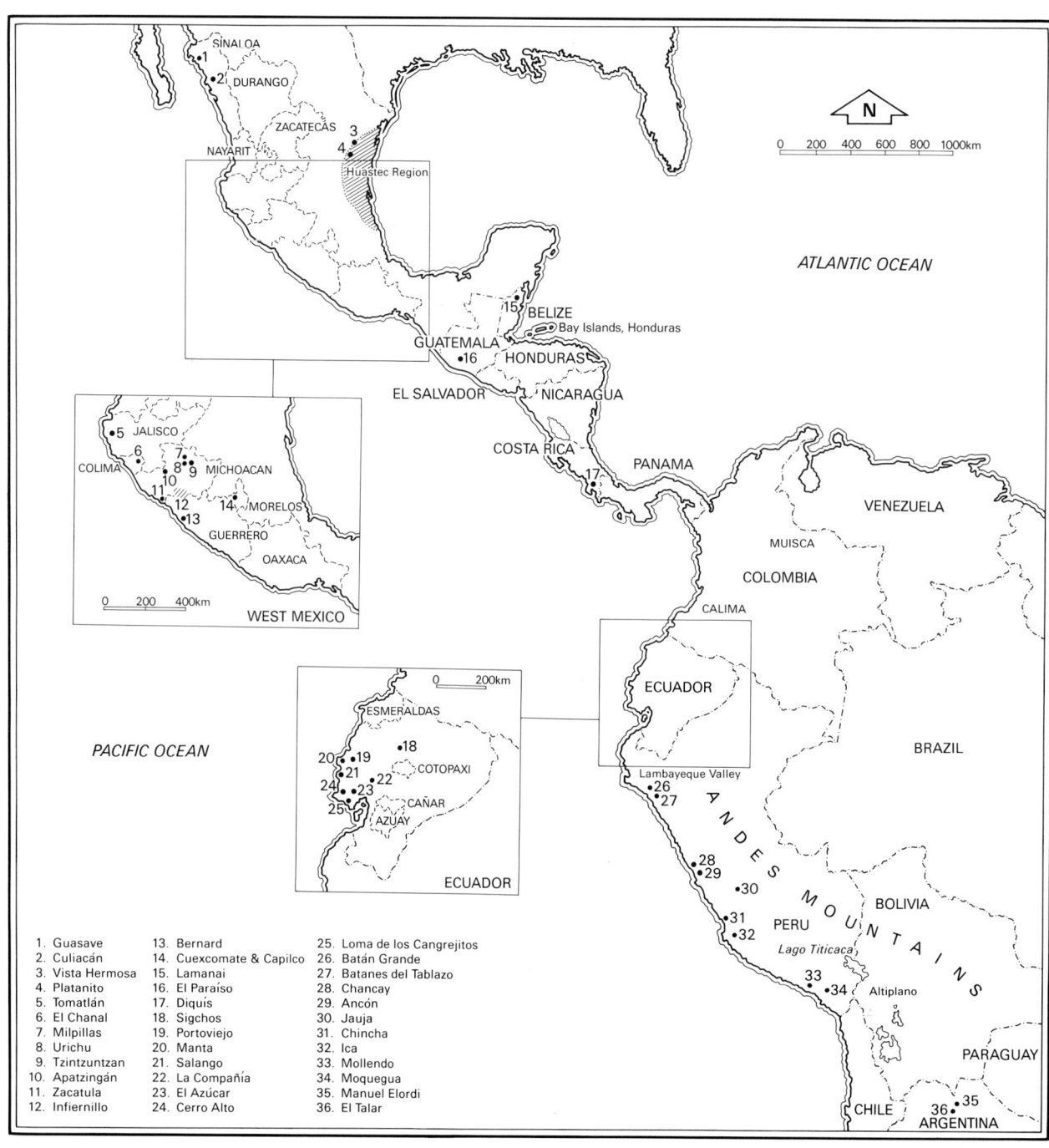

PERIOD 2: ORIGINS AND
TRANSFORMATIONS

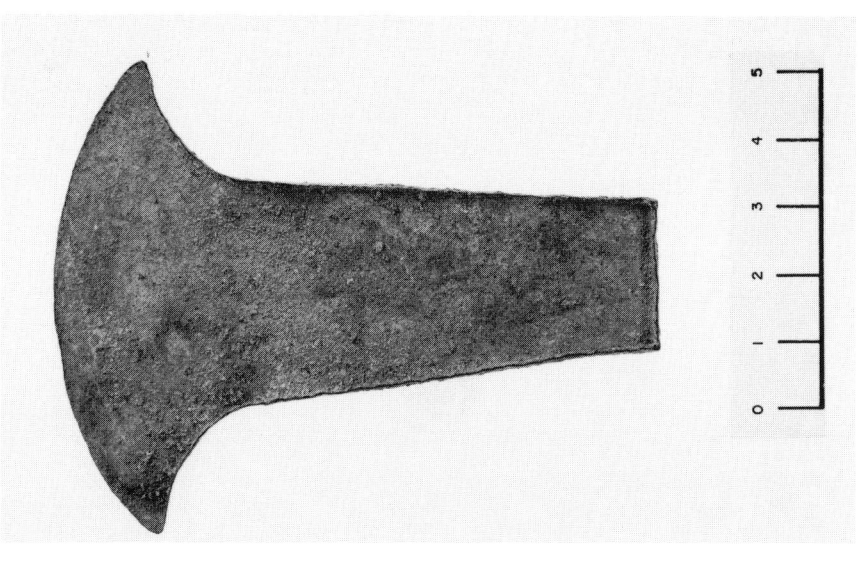

6.2

Axe-money from Ecuador (type 2). Photograph by Jeanne Mandel.

The same probably holds true for the loop eye needles; the design precludes some tasks and facilitates others.

Axe-monies are common in southern Ecuador, and appear in West Mexico (and Oaxaca) after A.D. 1200. Both the Andean and the Mexican varieties are usually made from copper-arsenic alloys. Among the various Mesoamerican, Ecuadorian, and Peruvian designs, three are closely related with respect to dimensions, fabrication techniques, and alloy composition: the Ecuadorian type 2 axe-money (figure 6.2); the variety common to Michoacán and Guerrero, type 1a (see figure 5.27); and type 2a (Hosler 1986).

6.1

New World archaeological sites and regions associated with Period 2 West Mexican metallurgy.

In the coastal polities of Ecuador these thin, axe-shaped objects circulated as a form of wealth. In West Mexico, the type 1a axe-money was a tribute item, and, as I have suggested here, may also have served as a repository for copper-arsenic metal. Both the Ecuadorian and West Mexican varieties are T-shaped, thin to paper-thin; the thinnest varieties have been found stacked and bound in packets. Arsenic, the major alloying element, is present in low concentrations, ranging from 0.1 to 6.4 weight percent. The extreme thinness of the metal (less than 0.05 cm) makes an alloy a requirement of the design. In general, mean values for arsenic concentration in Mexican specimens are somewhat lower than in objects from Ecuador (Hosler, Lechtman, and Holm 1990). Fabrication methods were nearly identical. The Ecuadorian variety (figure 6.3), like its West Mexican counterpart (figure 5.28), was cold-worked to shape, annealed, then left in

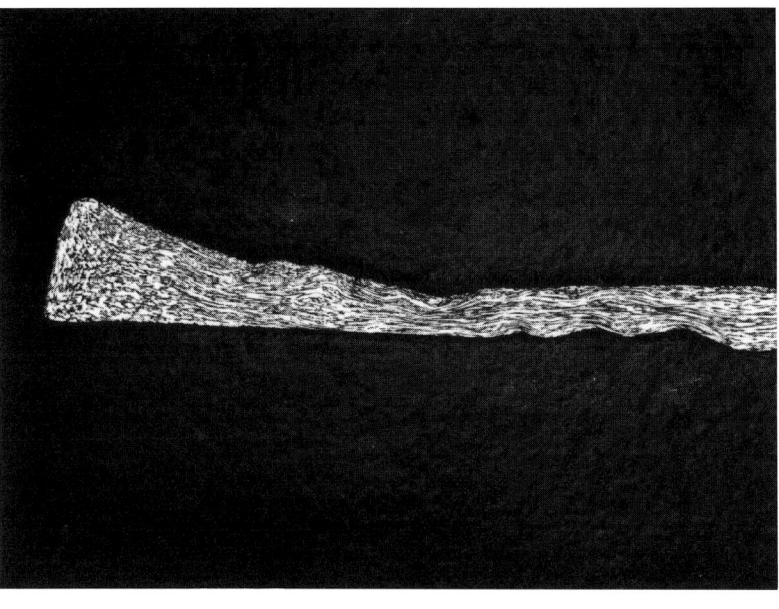

6.3

Cross section of Ecuadorian axe-money (1.81% As), severely cold-worked to shape but left in the annealed condition except locally beneath surface indentations. Sample etched in alcoholic ferric chloride (mag.: 11).

an annealed condition. The pronounced evidence of prior cold work in the Ecuadorian example results from its higher arsenic content: 0.60% versus 1.81%.

During Period 2, metalworkers in West Mexico also used copper-arsenic alloys to optimize the loop eye needle design (see figure 5.24). The same design was fashioned from this alloy significantly earlier in southern Ecuador and on the Peruvian north coast. The RMG collection has two loop eye needle varieties and the two also appear at Cuexcomate. Remarkably, both also occur in Ecuador, and at some Ecuadorian sites such as Salango both appear in the same depositional context. Figures 6.4 and 6.5 show the two Andean versions of this design.

West Mexican smiths and their counterparts in southern Ecuador and northern Peru fashioned these two versions of the loop eye needle in exactly the same way. Chapter 5 documents the fabrication sequence for the two West Mexican specimens (see figure 5.24). Photomicrographs of cross sections through two Ecuadorian needles appear in figures 6.6 and 6.7. The Ecuadorian needles were made by folding and hammering a rectangular strip of metal around its longitudinal axis, creating a round shaft with an internal fissure. The loop portion for the eye was flattened, then bent over. In the needle of type a, this tab was tucked into the shaft; for type b needles it was bent back against the shaft and two flaps of metal were hammered around and over to secure it. The tab of the loop protrudes. The final step in both cases was to anneal the metal. The photomicrographs illustrate sections of the needles where the tab tucks into the shaft

6.4

Loop eye needle, tucked (type a), from northern Peru.

6.5

Loop eye needle, flaps (type b), from coastal Ecuador.

(type a; figure 6.6) or at the position of the securing flaps (type b; figure 6.7). In the latter case, the two flaps that encircle the tab have been hammered down upon it to secure the loop.

The tucked version of this design (type a) is widely distributed in the Andean region: on the coasts and in the highland areas of Ecuador and Peru, and in the *altiplano* of Bolivia and northwest Argentina. In the northern Andes, metalworkers made these needles from copper-arsenic alloys. In the southern Andean highlands they used alloys of copper and tin. In both cases, the alloying element appears in concentrations of about 1–2%. A few examples of this needle type are known from Colombia. They are probably made from alloys of copper and gold, although we have no analytic data for them.

The earliest specimens of this needle type appear in the southern Andean highlands and date to between 400 B.C. and A.D. 200 (González 1979). Analyses of a group from northwest Argentina show that they are made from a copper-arsenic-tin alloy (Fester 1962) deriving from an unusual local ore body. These date to between 200 B.C. and A.D. 650. Other needles from the same highland region are fashioned from the more typical southern Andean tin bronze alloy (Nordenskiöld 1921). Needles of

6.6

As-polished transverse section of Cu-As alloy north Peruvian loop eye needle (type a), showing rectangular tab and needle shaft enclosing it. Sample etched in potassium dichromate (mag.: 100).

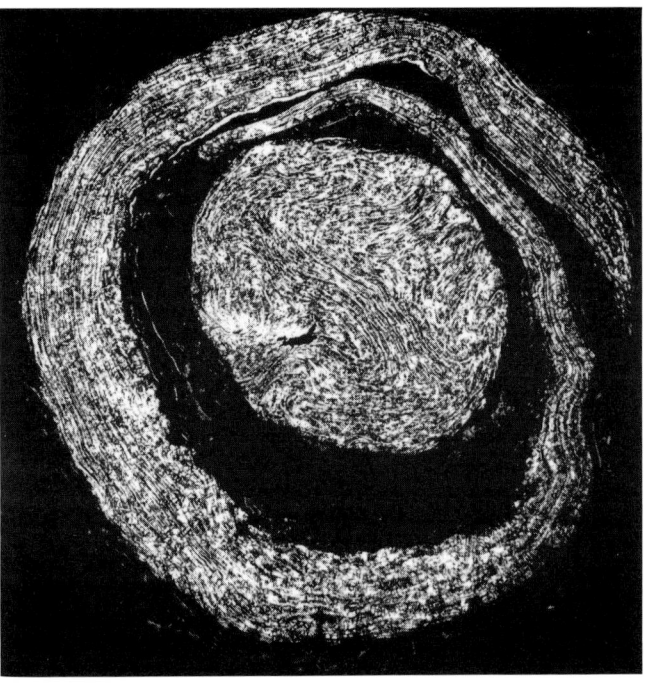

6.7

Transverse section of loop eye needle (type b) from Ecuador. The oval tab is secured by the overlapping lateral flaps hammered down onto its surface. Sample etched in potassium dichromate (mag.: 100).

type a appear around A.D. 900 on the north coast of Peru and in southern Ecuador (Lechtman 1981). At Loma de los Cangrejitos, they date between A.D. 900 and A.D. 1150 and are made from a copper-arsenic alloy; at Salango and Cerro Alto, from copper-arsenic and copper-silver alloys. The version of the design with a protruding loop tab and flaps (type b) is restricted in South America to Ecuador. As I noted, both designs appear at Salango, where they are made from a low-arsenic copper-arsenic alloy.

Copper-Silver Alloys. The second major contribution to West Mexican metallurgy from this central-north Andean region is the copper-silver alloy system. Two production centers for copper-silver alloys existed in the New World: one in southern Ecuador and northern Peru, where they appear around A.D. 500, and the other in the Tarascan region of Michoacán, where they become common after A.D. 1200. Metalworkers used these alloys in both regions for ritual and status objects—pendants, shields, disks, and other display items—which they cold-hammered from metal sheet. They also fashioned open rings and, occasionally, tweezers from these alloys, again by cold-working the metal to shape. Copper-silver artifacts are rare in the intervening regions of Colombia, Central America, and southern Mesoamerica. Some undatable copper-silver alloy objects are found in collections from Oaxaca.

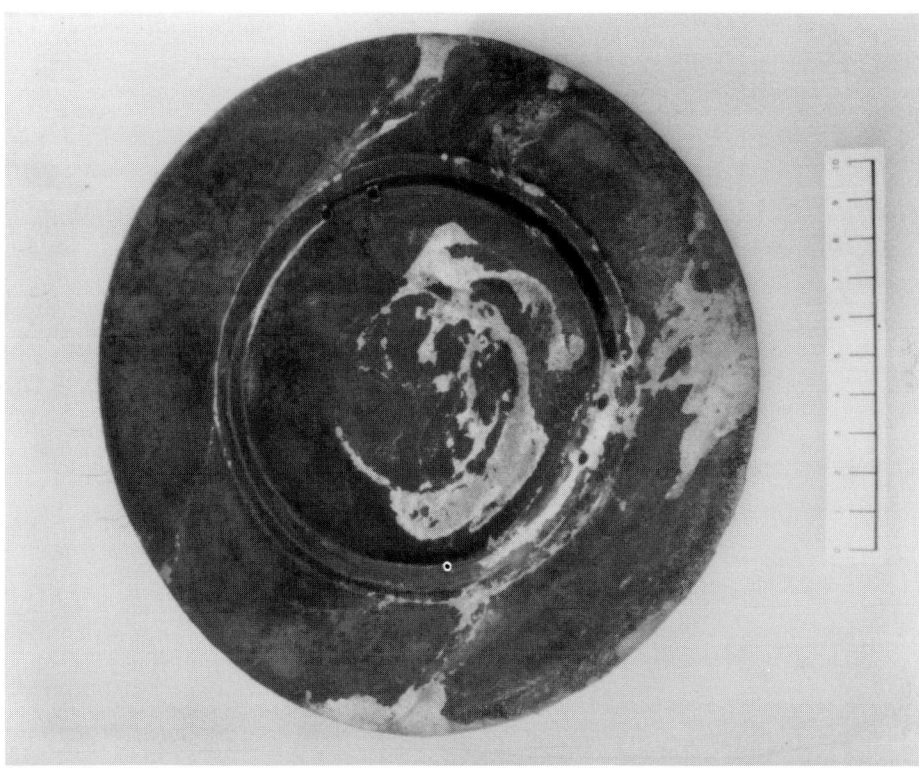

6.8

Copper-silver alloy disk, from MAG collections, Ecuador.

Ecuadorian metalsmiths first used these alloys in the middle Guangala period (see chapter 4) at such coastal sites as El Azúcar (circa A.D. 280). The assemblage at the much later site of La Compañía (circa A.D. 800 to A.D. 1530) contains a variety of copper-silver sumptuary items, including pendants, breastplates, and bowls. Copper-silver sheet metal ritual objects were equally early on the north coast of Peru. There, these alloys were initially used by the Moche (200 B.C. to A.D. 800) and subsequently were widely employed by the Chimú (A.D. 1000 to A.D. 1470). On the south coast of Peru, metalworkers also used copper-silver alloys, but somewhat later: first at Ica, before A.D. 1000, and then at Chincha (see figure 6.1), between A.D. 1000 and A.D. 1100 (Kroeber and Strong 1924). Although most West Mexican copper-silver alloy objects date to after A.D. 1200, some artifacts from the Infiernillo sites look like copper-silver alloys, and associated material suggests that their dates are slightly earlier. Copper-silver artifacts were recovered in Guerrero at Bernard, at El Chanal in Colima, and at Urichu and Milpillas in Michoacán. As I have already noted, Tarascan metalsmiths employed the alloy so ubiquitously that the Spaniards called it "the metal of Michoacán" (Warren 1985).

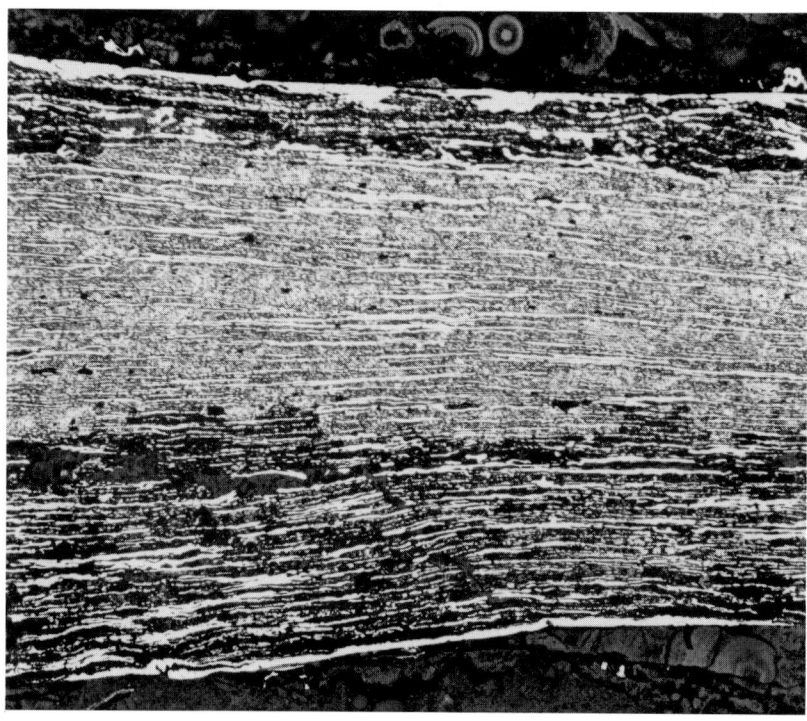

6.9

Transverse section from Cu-Ag alloy disk illustrated in figure 6.8. The microstructure exhibits the lamellae formed by alternating copper-rich and silver-rich phases, so typical of Cu-Ag alloys hammered into sheet. Sample etched in potassium dichromate followed by ferric chloride (mag.: 200).

The properties of copper-silver alloys that metalsmiths most appreciated in both regions were their malleability, toughness, and color. These alloys can be greatly reduced in thickness without embrittlement, permitting metalworkers to elaborate extremely thin sheet metal that will hold its shape. Unlike on the copper-gold *tumbaga* alloys, surface enrichment effects are difficult to accomplish on copper-silver castings. Not surprisingly, lost-wax cast copper-silver artifacts are unusual in West Mexico.

In addition to handling this material in a similar way, West Mexican Period 2 metalworkers and their Andean counterparts sometimes used it to make the same kinds of objects. One example is open, round cross section rings, measuring about 2.5 cm in diameter and found in burials. In West Mexico these rings are most abundant in highland Michoacán. They also have been recovered in great numbers at coastal Ecuadorian sites. In addition, metalsmiths in both regions used copper-silver alloys to fashion large, round shields or disks (approximately 15 cm in diameter) through repeated sequences of hammering and annealing. The disk illustrated in figure 5.12, from a burial in Michoacán, is a virtual replica of certain disks made in Ecuador. Figure 6.8 illustrates an example from coastal Ecuador, presently in the MAG collections in Guayaquil. The West Mexican disk and its Andean analogue are essentially identical in fabrication techniques and alloy composition. The photomicrograph of the Ecuadorian disk in figure 6.9 shows the characteristic microstructure of the heavily worked copper-silver alloy, a semilaminated structure that develops as the metal is reduced in thickness. The resulting material is strong, malleable, and tough.

Copper-silver alloys were used in West Mexico and the Andes for similar ends. Metalworkers fashioned thin, silvery-looking ritual and status objects from them by cold-working them to shape. In addition to disks, this alloy was used in both regions for sheet metal pendants, open rings, and shell design tweezers, although copper-silver shell design tweezers in the Andes are generally confined to the central and southern area.

Copper-Tin Bronze. The smiths of the southern highlands of Peru, Bolivia, and northwest Argentina contributed a third major alloy system, copper-tin bronze, to West Mexican metallurgy. Tin bronze objects were manufactured in only two zones in the Americas, West Mexico and the southern Andean highlands. West Mexican copper-tin bronze objects date to after A.D. 1200. In the Andean region they date to before A.D. 850 and continue through the Spanish invasion. Tin bronze artifacts in South America were restricted to the southern portion of the Andean region until the Inka consolidation of the central Andean zone in about 1480, after which they sometimes appear associated with Inka material in conquered Andean territories to the north. Several isolated examples have recently been analyzed from Colombia.[2]

In Mesoamerica tin bronze objects were fashioned in West Mexico and were sometimes exported to other areas (see chapter 7). The alloy metal itself was also smelted in the Huastec region (Hosler and Stresser-Péan 1992) in the century before the Spanish invasion. We do not yet know for certain whether bronze objects also were fashioned there.

The most abundant sources of tin in the New World are found in the southern Andean highlands. The earliest tin bronze objects come from this region (circa A.D. 300) and belong to a technological complex that probably derived from the *altiplano* of Bolivia (González 1979). By A.D. 850, the tin bronze alloy was employed in northwest Argentina (see, for example, Ventura 1985), about the same time (A.D. 600 to A.D. 1000) that it spread to the southern Peruvian highlands. Andean metalworkers used low concentrations of tin (on average, about 5%) in knives, tweezers, axes, awls, needles, and other tools. They also occasionally incorporated tin in higher concentrations, sometimes to 10%, for ritual paraphernalia and symbols of state: star-shaped mace heads, ceremonial axes, rectangular and round plaques, plumes, pendants, and figurines (Ambrosetti 1904; Boman 1908; Gordon 1985; Mathewson 1915; Mead 1915; Rutledge and Gordon 1987). Cassiterite, the primary raw material, is so abundant that southern Andean metalsmiths used copper-tin bronze alloys as an all-purpose metal, in the same way that southern Ecuadorian and northern Peruvian smiths used alloys of copper and arsenic. However, chemical analyses show that, in contrast to arsenic, when tin appears in South American artifacts it almost always is found in concentrations high enough to alter the working properties of the artifact metal. The different smelting regimes used to produce these two bronze alloys may be responsible for these differences. Tin bronze alloys are made by adding either the molten metal or cassiterite separately to the crucible, giving the artisans greater control over alloy concentration. Andean metalworkers obviously controlled alloy composition in copper-tin objects, using tin in low concentrations (about 2–7%) for tools. Tin very occasionally appears in higher concentrations for sumptuary objects. Mead (1915) reports two cases of tin levels to 12%, presumably for its coloring effects. Other analyses show only few objects with tin present in concentrations substantially higher than 12%. Two open rings from northwest Argentina dating to about A.D. 950 contain 20% tin by weight (Ventura 1985). Generally, however, artifacts containing more than 10% tin are very rare in the Andes.

The systematic use of high-tin bronze alloys for ritual and status objects seems to be confined to West Mexico. There, metalsmiths used the alloy, with tin concentrations of 10% and higher, for color, and to optimize the design in shell design tweezers, open rings, lost-wax cast bells, and miscellaneous ornaments. Low-tin copper-tin alloys usually were reserved for tools.

Three artifact types characteristically made from tin bronze appear in the southern Andean region and later in West Mexico: rectangular cross section open rings, loop eye needles, and round-bladed shell design tweezers (see figure 5.6, type b). These objects are fashioned in the same way in both areas. Open rings with rectangular cross sections occur in burials in northwest Argentina; their dimensions are identical to their counterparts in West Mexico. A few round cross section rings similar to those appearing in Ecuador and West Mexico are also known from the southern Andes, but there the rectangular cross section design is most frequent. Few chemical analyses have been performed on these open rings. Two for which we do have analytical data are the high-tin bronze rings from northwest Argentina; they come from burials at the sites of El Talar and Manuel Elordi (see figure 6.1). The loop eye needle of type a is also made from tin bronze when found in the southern Andean area. Unfortunately, few shell design tweezers have been analyzed, although they are common in this Andean region. Examples from the southern highlands (in the American Museum of Natural History collections) seem to be made from the bronze alloys, based on macroscopic observations.

Alloys and Artifacts

Evidence that the copper-tin alloy system was introduced to West Mexico is not as strong as that for alloys of copper-arsenic and copper-silver because we have so few chemical analyses of Andean tin-bronze objects. Nonetheless, the presence of shell design tweezers in this Andean region unquestionably links this zone with West Mexico. Three of the four shell tweezer types found in West Mexico (types a, b, and d; see figure 5.6) have earlier prototypes in the southern Andean area. The design characteristics of these Andean tweezer specimens are identical to those of artifacts representing the three types in West Mexico; they were fabricated in the same way, and they were sometimes made from the same materials. Southern Andean smiths made these tweezers from alloys of copper and silver and, apparently, from copper-tin bronze. In West Mexico this tweezer design sometimes is made from copper-silver alloys, but more commonly from alloys of copper-arsenic, copper-tin, and copper-arsenic-tin. Shell design tweezers are atypical of Ecuador, although they do appear there sporadically. They are also reported occasionally from Colombia, where they are made from copper-gold alloys.

Shell design tweezers with a round blade, identical to the West Mexican type b tweezers, are most abundant in coastal Peru and the southern Andean highlands. They appear first in northwest Argentina, where metalworkers made them between 200 B.C. and A.D. 650; metalsmiths continued to fashion the same design into the later peri-

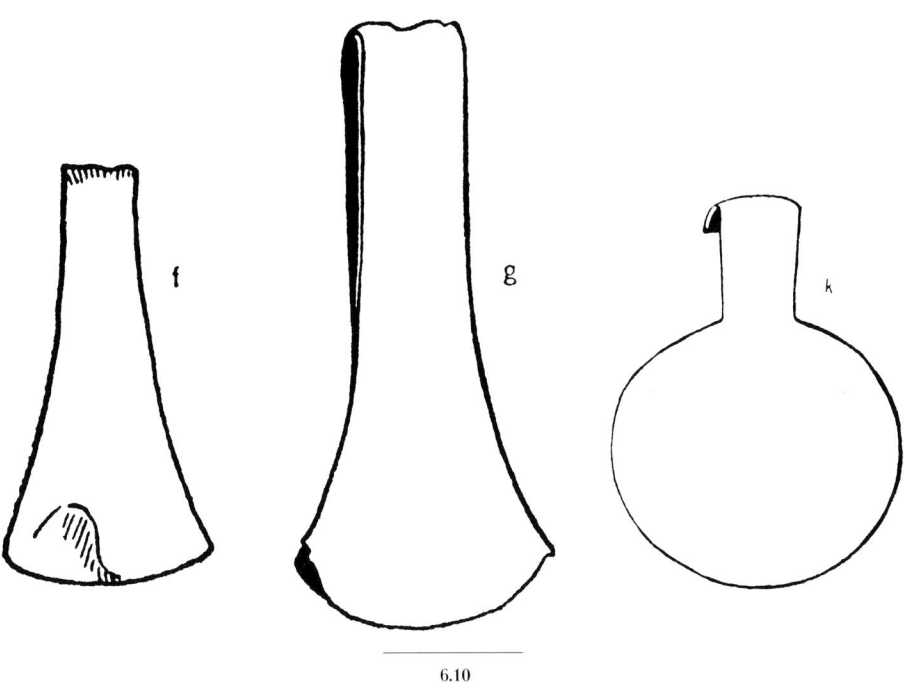

6.10

Tweezers from Chincha, Peru, identical to types a and b from the RMG collection (see figure 5.6). Photograph from Kroeber and Strong 1924, plate 22.

ods. The probability is high that most type b tweezers from the southern highlands were made from tin bronze, given the early dates and pervasive use of tin bronze in the area, and that museum specimens apparently are made from the alloy.[3]

The West Mexican type a shell tweezer design has its earliest counterpart on the south coast of Peru, where such tweezers appear at Chincha. Tweezers of type a excavated at Chincha date to between A.D. 1000 and A.D. 1300 (Kroeber and Strong 1924; Root 1949); they are identical, even in dimensions, to the West Mexican specimens. These Chincha tweezers are made from copper-silver alloys. Round blade (type b) tweezers also excavated at Chincha have slightly later dates (A.D. 1300 to A.D. 1400). They also are made from copper-silver alloys. The third West Mexican tweezer design, a variation of type d, also appears on the coast, specifically at Ancón and Chancay (Hosler 1986, 1988b). The Chincha tweezers are illustrated in figure 6.10.

The West Mexican tweezer design with a semilunar or anchor shaped blade (type d) is also reported from sites in northwest Argentina and in Colombia. Dates for all South American specimens are problematic. One example from northwest Argentina dates after A.D. 1400 (González 1979). In northwest Argentina, this tweezer design also is most likely made from copper-tin bronze alloys. However, Colombian specimens are the most similar to the West Mexican type d design, but they are made from

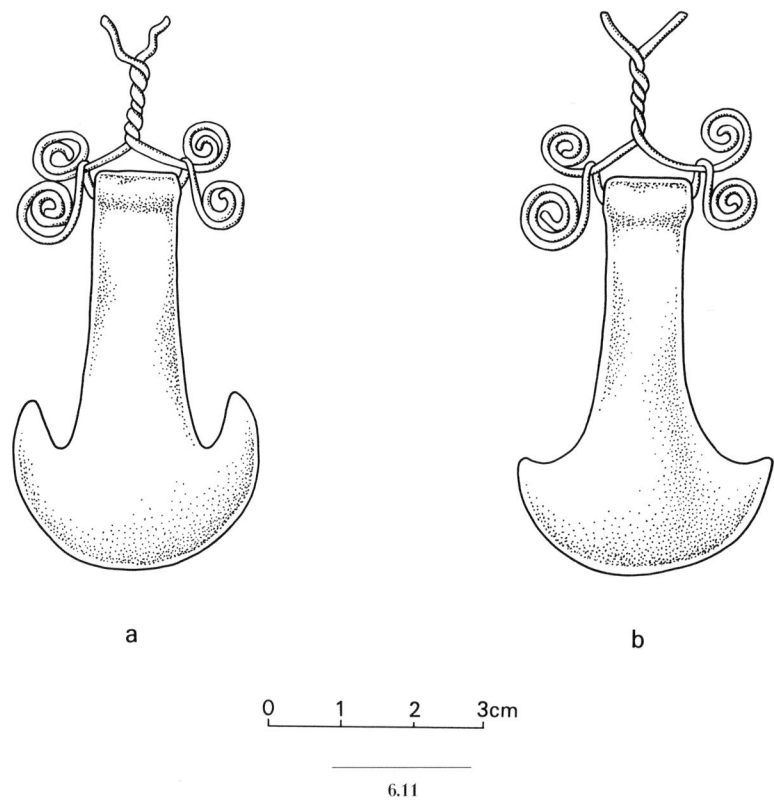

6.11

Tweezer type d: examples from RMG assemblage (a), and from Museo del Oro, Bogotá, Colombia (b).

alloys of copper and gold. One unusual West Mexican example, made from copper-tin bronze, has an exact replica in Colombia, made from a copper-gold alloy. Figure 6.11a shows the West Mexican design, 6.11b the Colombian analogue.

All three of these shell design tweezer types appear in the South American collections of the American Museum of Natural History, acquired from coastal Peruvian sites and from the *sierra* of northwest Argentina and Bolivia. Macroscopic observations indicate that those from coastal Peru are made from a copper-silver alloy; as mentioned, highland examples appear to be made from alloys of copper and tin.

Major elements of Period 2 West Mexican metallurgy thus derive from the sheet metal-oriented metallurgies of Ecuador, Peru, Bolivia, and northwest Argentina. Although the distances are great, the best explanation for the Period 2 appearance of these tweezer designs in West Mexico, and for the tin bronze alloy, is that they were introduced from the southern Andean region. Prototype copper-tin bronze objects, probably including shell tweezers, were moved from the southern highlands to ports on the south coast, where they were transshipped north. A more northerly South American point of origin for the tin bronze alloy is highly unlikely. Smiths used copper-arsenic and copper-silver alloys until the Inka ex-

A few West Mexican artifacts may have been imported from the southern and/or central Andes; we know of at least one Andean object that could have been imported from Mexico. Apart from two gold or gold alloy Nazca-style sun disks found in West Mexico (see figure 6.15) and identified by Furst (1965b), an Inka-style tweezer is exhibited among the West Mexican collections at the Museo Nacional de Antropología e Historia in Mexico City. Perhaps most intriguing is a drawing (figure 6.12) by Reiss and Stübel (1880–1887) of a Tarascan-style spiral tweezer but made with characteristic Inka dimensions; it was found on the coast of Peru. It suggests that Andean smiths actually saw the elaborate, elite Tarascan spiral tweezers. This artifact could represent an Inka copy of a West Mexican spiral tweezer.

The copper-gold casting technology of lower Central America and Colombia contributed far less to Period 2 metallurgy than did the metallurgies of the central and southern Andes. Although lost-wax casting, especially for bells, continued to flourish, Period 2 metalworkers cast bells using copper-tin and copper-arsenic bronze, and they independently invented a host of new bell designs. A few other Period 2 metal objects probably derive from Colombia and lower Central America, one of which is a buttonlike, lost-wax cast ornament (figure 6.13) with

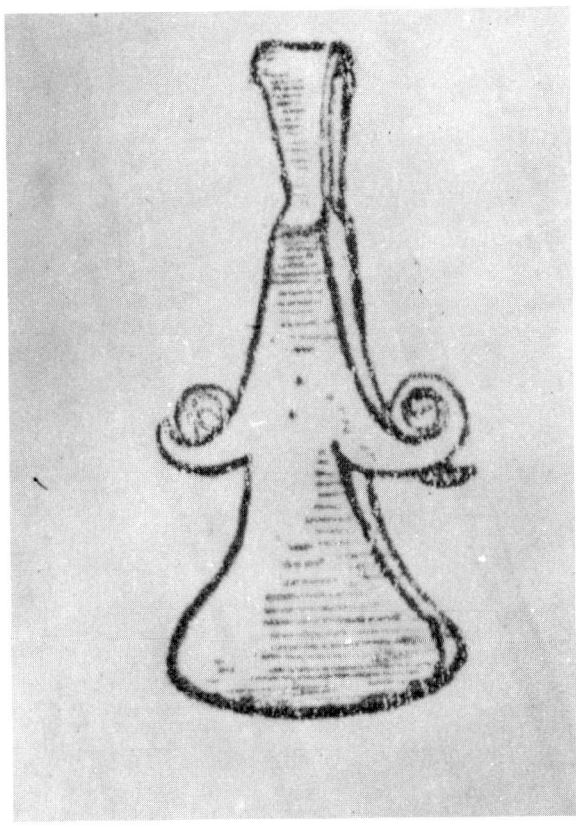

6.12

Tweezer with spirals, found in coastal Peru. Photograph of drawing by Reiss and Stübel (1880–1887, plate 264).

pansion even in the central Andes, for example at Jauja, in the upper Mantaro valley in Peru. After the Inka conquest, local metalsmiths continued to make objects from copper-arsenic alloys, but added tin to them as a marker of Inka hegemony (Owen 1986; Costin et al. 1989). In southern Ecuador, by contrast, although Inka-style copper-tin bronze objects do appear after the Inka expansion (Escalera U. and Barriuso P. 1978), there is no evidence for local production of objects from this alloy.

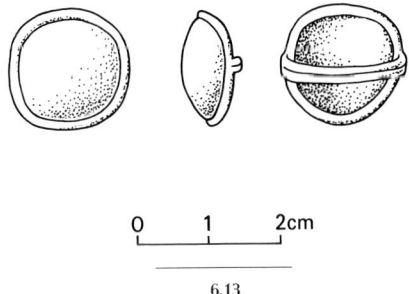

6.13

Lost-wax cast button from RMG collection. Similar objects appear in Oaxaca, Belize, and lower Central America.

many counterparts in lower Central America. These buttons appear at Guasave, Apatzingán, Culiacán, and El Chanal. Also, at least one of the West Mexican tweezer designs (type d) may have been introduced from Colombia, although tweezers were never a major item in Colombian metallurgy.

Mechanisms of Introduction

Some elements of Period 2 metallurgy were introduced via the same maritime exchange system operating off the coast of Ecuador that had earlier transmitted the technical know-how and prototype objects of Period 1. Certainly the copper-arsenic and copper-silver alloy objects that served as models for various Period 2 West Mexican artifact designs, including loop eye needles, axe-monies, and sheet metal status objects, were introduced in this way. However, some lower Central American and Colombian components of the technology, such as buttons, may have diffused overland. As noted earlier, similar objects appear by A.D. 1200, or perhaps slightly earlier, at Lamanai in Belize, on the Bay Islands in Honduras, and in Oaxaca.

The appearance in West Mexico of tin bronze and of artifact designs that characterize the metallurgy of the south coast of Peru argues for a more southerly arm to the Andean maritime exchange system, linking Ecuador and southern Peru. A mid-sixteenth-century Spanish document that discusses Chincha as a large and wealthy kingdom on the south central coast of Peru suggests such a link (Rostworowski 1970). Chincha emerged as a powerful coastal state at approximately A.D. 1200 and continued as a major economic force into the Late Horizon (A.D. 1476 to A.D. 1532). The Inka allowed Chincha to maintain its status because of its central position in a thriving, long-distance Pacific coast maritime exchange network. According to the document, Chincha was a port where "6,000 merchants" resided, who engaged in long-distance trade to points north using fleets of balsa rafts. The document specifically names Quito as the destination for the goods. (Here, "Quito" refers to the Audiencia de Quito, the territory that corresponds approximately to the modern republic of Ecuador.) The document also names Portoviejo, a town currently located slightly inland from the Manteño seacoast capital of Manta, as one of the ports of call. The document does not explicitly state that *Spondylus* was traded south, from Ecuador to Chincha, although it does mention gold beads and emeralds. Bartolomé Ruiz's description of the sailing raft encountered off the coast of Esmeraldas makes it clear, however, that shell from the north was one of the items shipped from equatorial waters south (Sámano-Xerez 1937).

The Chincha document establishes a maritime network that could move goods from southern Peru to West Mexico through Ecuadorian ports. It also raises the possibility that copper, or copper objects, from the south coast of Peru were transshipped northward. According to the document, Chincha merchants bought and sold in copper, apparently using copper as a value in exchange. Rostworowski (1970) argues that Chincha merchants obtained copper from the southern highlands and the *altiplano*, and that copper was the primary good they then shipped north to exchange for the warm water *Spondylus*. She also believes that this copper might have served as a kind of metal standard. Ongoing excavations at Chincha have not yielded metal ingots or artifacts that might have served as such a standard. They also have furnished no evidence that Chincha functioned as a port (Hosler, Lechtman, and Holm 1990). Nonetheless, artifacts identical to West Mexican types, such as the two shell tweezer designs, have been excavated there. Chincha may have served as an intermediary through which objects made from copper-tin bronze were shipped north to Ecuador, eventually reaching West Mexican ports. Although relatively few tin bronze objects were recovered at Chincha, Root has analyzed what he describes as a "lump" of pure tin from the nearby coastal site of Ica.[4] The tin, perhaps smelted locally from imported highland cassiterite ore, provides unequivocal evidence that tin metal was used in this coastal zone. The fact that Chincha began to emerge as a significant political and maritime entity at just about the same time that copper-tin bronze initially appeared in West Mexico may affirm Chincha's possible role as an intermediary in the maritime network.

We also know that as early as A.D. 200 colonists from the Titicaca basin, in the southern highlands, were living at Mollendo and Moquegua on the far south coast of Peru (see figure 6.1). They participated in an exchange system in which goods moved back and forth between the highland lake zone and the coast. Recent archaeological evidence shows that Tiwanaku (A.D. 100 to A.D. 1000) engaged in a deliberate policy of colonization on the south coast (Goldstein 1989a, 1989b). This arrangement fits the Andean exchange pattern that Murra (1972) has called the vertical archipelago. By the Late Intermediate Period (circa A.D. 1000 to A.D. 1476), or about A.D. 1200, the vertical archipelago system of exchange was a key factor in the growing wealth and power of the Lupaqa kingdom, which dominated the territories bordering the entire western shores of Lake Titicaca. Given this well-established economic pattern, it would be highly unusual if copper-tin bronze artifacts (and tin metal in the case of the Ica lump) were not one of the items that moved from the highlands to the coast. Even if Chincha turns out not to have been a major trading entity, long-distance maritime commerce may have been overseen by other coastal Peruvian polities, or perhaps even by the Ecuadorian merchants who organized and orchestrated the movement of these goods.

The maritime network can explain how objects made from metals and alloys typical of the southern Andean highlands, coastal Peru, Ecuador, and possibly Colombia made their way to West Mexico. Ecuadorian seagoing vessels traveling along the Pacific coast reached West Mexican ports and exchanged these metal objects for local goods. Although no evidence for an Ecuadorian-style metallurgy or metal trade items has been found in Central America, mariners probably did sometimes stop off at Pacific ports en route north. However, exploitable deposits of arsenic-bearing copper ores and of ores of copper and silver do not exist in the Central American zone, so that the technological complex introduced to West Mexico could never have appeared in Central America. The requisite raw materials were unavailable.

It was not only prototype artifacts that were introduced to West Mexico during this time but the technical information required to make them. Metalworkers who recognized the ore minerals of silver, arsenic, and probably tin, and who knew how to process these ores, must have been physically present in the area. The smelting technologies are sufficiently complex that personal contact is essential to communicate information about extractive metallurgy as well as about processing and manufacturing techniques. Either Andean metalsmiths traveled to West Mexico or artisans from West Mexico were present in the Andes, or both. The presence of Andean smiths in West Mexico is the most likely alternative because Andean polities controlled the maritime exchange system. The merchants may have imparted basic information concerning the technology as they exchanged tweezers, bells, and open rings with their West Mexican partners, perhaps for *Spondylus*, peyote, and other items. Andean metalsmiths or miners may have accompanied these seagoing traders. Once in West Mexico, the South American artisans probably set off with their West Mexican companions from a port in coastal Nayarit, Colima, Jalisco, or Michoacán, searching for the ore minerals that resembled the materials they used in Ecuador: malachite, chalcopyrite, arsenopyrite, enargite, and tetrahedrite. The physical presence of Andean artisans in West Mexico is the most plausible way to explain the transmission of the smelting, smithing, and casting techniques.

The document from the port of Zacatula (West 1961), discussed in chapter 4, relates that seagoing traders came in canoes from islands to the south with rich cargoes to trade. Sometimes, because the seas were so rough, they waited as long as six months before returning. (This six-month period coincides with the hurricane season in this region of the Pacific, when sailors currently put into port.) It may have been during such protracted layovers that Andean artisans introduced some aspects of their craft to their West Mexican exchange partners.

The most difficult situation to envision, given the great distances involved, is how knowledge of the ore types needed to produce tin bronze reached West Mexico. Cassiterite ores were only available in the southern Andean highlands and in north-central Mexico. Ecuadorian mariners or merchants voyaging off the coast of Peru might have been acquainted with cassiterite, and Peruvian peoples transshipping copper-tin bronze objects or copper and tin ores from the southern Andean highland region to the coast were certainly familiar with it. Cassiterite is no more difficult to smelt than any other oxide ore. The major problem is to find it. If knowledge of tin bronze production was introduced to West Mexico, people who understood smelting procedures not only went to West Mexico but continued inland, to those areas in Zacatecas and Durango where cassiterite deposits occur. They may have taken the painted jay, *Cyanocorax dickeyi*, along with them, to the mountainous area of Durango now inhabited by this bird (see chapter 4). At least two cassiterite deposits are known in this same region, and they happen to lie closer to the Pacific coast than any other cassiterite deposit in Mexico.

This reconstruction makes us wonder why Period 1 smiths used unalloyed copper almost exclusively, and only later, in Period 2, incorporated the alloys. Ecuadorian metalsmiths were already using copper-silver and copper-arsenic alloys at the time metallurgy was first introduced to West Mexico. Although these two alloys may have appeared in the metalworking zone slightly earlier than the data presently admit, the evidence so far indicates that West Mexican smiths did not use them on any significant scale between A.D. 600 and A.D. 1100. Either South American traders simply did not introduce information about the alloying and smelting technologies of these materials—although they did introduce certain fabrication methods and prototype objects—or West Mexican artisans chose not to elaborate them.

The New Technology: West Mexican Alloys and Smelting Regimes

Three alloys—copper-arsenic, copper-silver, and copper-tin—were introduced to West Mexico from regions to the south. To what extent were the methods used to make these alloys also similar in the two areas? The production of copper-silver alloys is the most difficult to explain, because in West Mexico and in Ecuador and Peru silver deposits appear in many forms: as native or metallic silver, as argentite, and as silver sulfosalts. No reliable method exists to distinguish which of these was used on the basis of artifact chemistry alone.

The picture for the copper-arsenic alloys is also complex. In Peru, copper-arsenic alloys (containing arsenic in concentrations between about 0.5% and 5.0%) may have been made by smelting enargite (Lechtman 1976, 1979, 1981, 1991). Enargite is a copper sulfarsenide ore and is abundant in the central Andean highlands. When smelted, it yields copper-arsenic alloys directly, with arsenic present in fairly low concentrations (Rostoker and Dvorak 1991). Lechtman's experiments with the ore have shown that when pure enargite is co-smelted with atacamite, a copper chloride ore mineral, the resulting alloy can contain up to about 7% arsenic (Lechtman 1985).[5] Pure enargite contains 17% arsenic. However, enargite is a highland ore, and large numbers of copper-arsenic alloy objects were made on the Peruvian north coast. Lechtman (1991) argues that these highland ores were one of the major items brought from the *sierra* to the coast; in fact, she has identified a piece of enargite at the coastal metalworking site of Batanes del Tablazo, in the lower Chancay valley. As chapter 2 elaborates, copper-arsenic alloys can also be produced by co-smelting a copper ore, such as chalcopyrite, with arsenopyrite, an iron ore of arsenic. Arsenopyrite apparently does sporadically associate with chalcopyrite on the north coast of Peru (Lechtman 1976, 1991), and it is also found at several locations in the highlands. In general, however, arsenopyrite is far less abundant than enargite in the highland zone. Shimada and Merkel argue that copper-arsenic alloys were produced at the coastal site of Batán Grande by co-smelting copper oxide ores with arsenopyrite. Shimada has recovered samples of arsenopyrite from that site (Merkel et al. 1994; Shimada 1985; Shimada and Merkel 1991). Some coastal copper-arsenic alloys may have been produced in this way. However, Lechtman suggests that enargite was the major ore mineral utilized, based on the sheer volume of metal objects manufactured on the north coast of Peru in the period after about A.D. 800.

The ore mineralogy of Ecuadorian deposits is helpful in understanding how copper-arsenic alloys may have been produced there. Goossens's studies (1972a, 1972b) show that in Ecuador, arsenic-bearing minerals and copper-arsenic ores are found only in the highlands. They appear in the provinces of Azuay, Cotopaxi, and Cañar (see figure 6.1). At the mine of Sigchos near Quito, for example, arsenopyrite and chalcopyrite occur but enargite and tetrahedrite are also present, suggesting that the raw materials were available to produce copper-arsenic alloys either by co-smelting arsenopyrite with chalcopyrite or by smelting enargite or tetrahedrite. Goossens's data indicate that the arsenic-bearing copper ores of enargite and tetrahedrite are slightly more abundant than are those of arsenopyrite and chalcopyrite. The geological data tend to suggest that, at least in Ecuador, both methods could have been used.

In West Mexico, by contrast, arsenopyrite is the most important arsenic-bearing ore mineral. Furthermore, wherever arsenopyrite appears, it is found associated with chalcopyrite (see chapter 2). Enargite and tetrahedrite, while present in Mexico, are uncommon. West Mexican smiths for the most part produced the copper-arsenic alloy by directly smelting naturally occurring mixtures of chalcopyrite and arsenopyrite, or by co-smelting arsenopyrite with chalcopyrite or their weathered products. This situation suggests that Andean metalsmiths, who introduced the copper-arsenic alloy and the know-how behind its production, had to be familiar with arsenopyrite to assist West Mexican metalworkers in locating the ores.

Some analytical data indicate that, in West Mexico, copper-arsenic alloys may have been produced using somewhat different methods than in Ecuador, because arsenic appears in different concentration ranges in artifacts from the two regions (figure 6.14). In the West Mexican objects, arsenic appears in very low and very high concentrations; between these extremes, distributions are spread relatively evenly across the composition range represented. The limited range of low arsenic concentrations for Ecuador may indicate that the alloy was automatic in most cases; smiths direct-smelted the metal from an arsenic-bearing copper ore, and the arsenic concentrations that resulted were not controlled. The distribution characteristic of Mexico may reflect the fact that arsenopyrite was deliberately and intentionally added to, and co-smelted with, copper ore in concentrations high enough to affect the working properties of the alloy. Arsenic concentrations tend to be higher than in the Ecuadorian objects, although they are not systematically controlled.

West Mexican metalsmiths used the copper-arsenic alloy in many cases exactly as it had been used in the Andean region, but in West Mexico the alloy never became an all-purpose stock material. The kinds of ore minerals smelted may account for this. The Andean varieties, unlike those in West Mexico, are largely arsenic-bearing copper ores. When they are smelted, they inevitably produce copper-arsenic alloys.

The smelting regimes used to produce the copper-tin bronze alloys are the simplest to explain. Cassiterite, the oxide ore of tin, is the most common tin ore in both regions; it is the only tin ore in Mexico. Smelting procedures are uncomplicated: in Mexico, cassiterite could either have been co-smelted with chalcopyrite or smelted separately to win metallic tin, which was then added to molten copper. In the Andean region, chalcopyrite and a number of other copper sulfide ores are common. The alloy could have been produced in the same way using any one of them. Stannite, a tin-bearing copper ore, has occasionally been reported from Bolivia; when direct-smelted, this ore produces a copper-tin alloy. Stannite may have been used for copper-tin alloys in the southern Andes, but these ore minerals' relative abundance makes cassiterite a far better possibility. Furthermore, tin con-

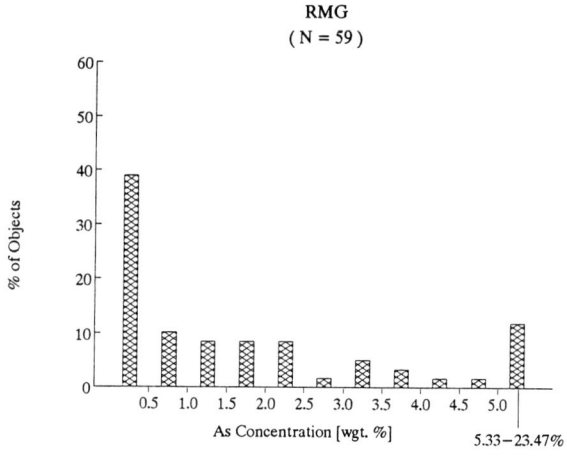

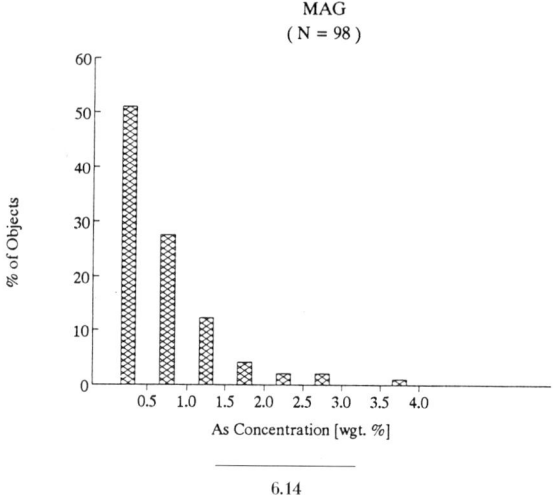

6.14

Ranges of arsenic concentration in Cu-As alloy objects from West Mexico and Ecuador.

centrations in Andean artifacts appear to have been controlled and are always high enough to affect the working properties of the metal. This suggests that the alloy was not the inadvertent product of smelting ores such as stannite, but was the result of a deliberate effort.

The West Mexican Interpretation

During Period 2, West Mexican metalsmiths incorporated new elements into their technical repertoire. In general, these elements pertained to the secular and utilitarian rather than to the elite and sacred spheres of the South American metallurgical traditions. In cases such as axe-monies, West Mexican smiths replicated all components of the source technical complex: object design, the metals and alloys used, and fabrication methods. In other instances they reconfigured those components. In particular, they created an elite sphere of their own technology by reinterpreting and transforming certain secular or more minor aspects of northern and southern Andean and lower Central American and Colombian metallurgies. Their use of the bronzes and their extensive experimentation with lost-wax casting best exemplify this.

We cannot easily reconstruct the historical circumstances that shaped the technical decisions described here. I assume that in West Mexico the process was mediated by political interests and religious ideology, and that, by Period 2, elites probably determined how the new material was used. In South America, traders, elites, or both determined the classes of knowledge and the material suitable for export. Burger (1984), in testing in the New World Renfrew's Old World model (1969, 1975) explaining exchange relations, points out that *Spondylus* and *Strombus* shell are the sorts of nonproductive goods that fit Renfrew's criteria for long-distance exchange items.

(They were traded between Ecuador and Peru.) As exotic materials, they became vested with symbolic meanings by those acquiring them; they are low in information content, limited in variety, and modest in quantity (Burger 1984: 46–49). These criteria nicely describe the volume of exchange and the sorts of metal artifacts moved from Ecuador and Peru to West Mexico.

The most original technical developments in West Mexico during Period 2 occurred in the elite and ritual realms of the technology. West Mexican smiths' sheer inventiveness and the technical variety visible there ultimately and most decisively differentiated this metallurgy from the metallurgies of Central and South America. We see this in the artifact types that became central foci of ritual and elite technology, in the range of fabrication methods used, and in the unusual variety of metals and alloys that characterized this technical enterprise.

Period 2 metalworkers employed both the characteristic Andean approach to fashioning ritual metal objects, working them to shape sometimes into extremely thin metal sheet, and the Colombian method of lost-wax casting. However, as I have observed in chapter 4, West Mexican artisans did not replicate the most esoteric components of those technologies in other respects: ritual artifacts, the variety of casting techniques, and *tumbaga* alloys.

West Mexican metalsmiths likewise did not incorporate the most ritually important objects from the more southerly region: sheet metal plaques, body ornaments, and three-dimensional objects. There is also no evidence that West Mexican smiths ever adopted the complex surface enrichment techniques, including electrochemical replacement plating and depletion gilding described by Lechtman and others.

Period 2 metallurgy most closely approximates the elite spheres of the metallurgies to the south in the ways

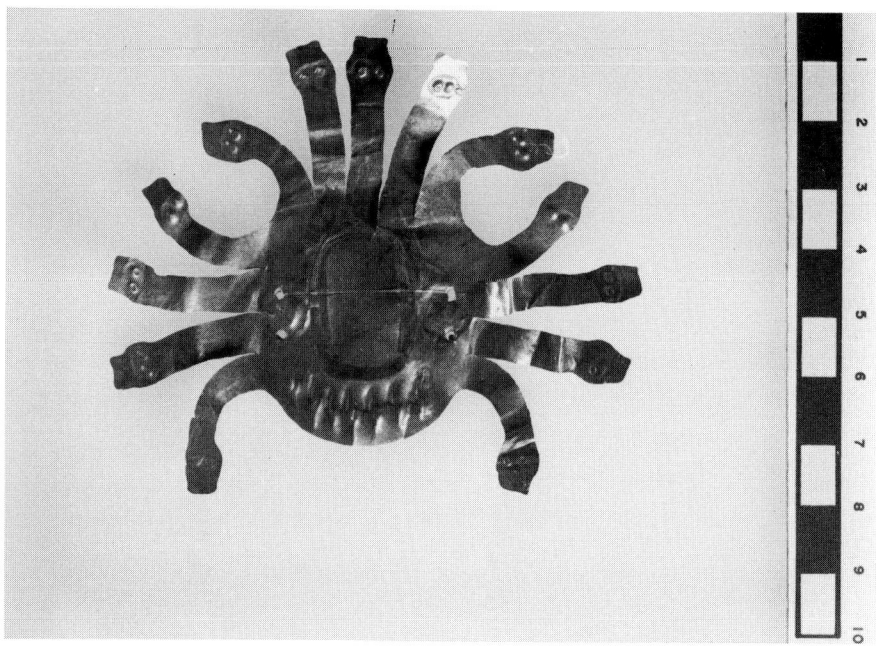

6.15

Sheet metal disk, RMG collection. This disk may have been imported to West Mexico from the south coast of Peru.

copper-silver alloys were used. In Ecuador, smiths made sheet metal status objects from them. They also made objects like tweezers that were cold-worked to shape from an original cast blank. The "silver" objects Bartolomé Ruiz mentions seeing on the native raft he intercepted—diadems, *cintos, tenazuelas,* and so on—were probably copper-silver sheet metal objects that looked silvery because of their processing. Some of these artifacts, especially the sheet metal breastplates and open rings, were incorporated into the ritual and elite component of West Mexican metallurgy. However, West Mexican copper-silver sheet metal objects never display typically Andean iconographies. When we do find designs on these objects (Kelly 1980), the motifs are distinctively Mesoamerican.

What West Mexican peoples appropriated was the *material,* which they used to elaborate the motifs and distinctive artifact designs (like the large, shell design tweezers) that were meaningful in their own belief systems.

Even the few published West Mexican examples of the quintessentially Andean ternary alloy objects (from El Chanal in Colima; see chapter 5) exhibit distinctively Mesoamerican designs. Only the sheet metal disks mentioned previously (figure 6.15) show unmistakable iconographic links to the Andean region, specifically to south coastal Peru. Because they are so similar, they may have been imported. They are made either from copper-gold or copper-silver-gold alloys.[6]

West Mexican metalworkers' most novel experiments were with materials, specifically the bronze alloys, and with fabrication techniques, that is, with lost-wax casting. To use bronze systematically as an elite and ritual material was their most significant contribution to the metallurgies of the Americas. What is more, they designed most bronze objects so that they had to be made from such an alloy; the particular dimensions and/or function required bronze's physical and mechanical properties. They also increased tin or arsenic concentration so that it dramatically affected color, and in doing so created a range of golden and silvery hues. In lost-wax cast bells, the bronze alloys allowed metalsmiths to cast larger, thinner, more intricate designs, which not only displayed a range of ritually significant colors but also produced a variety of new pitches. West Mexico was the only region in the Americas where bronze alloys were used systematically for lost-wax cast bells, and it was also the only region in the ancient world where copper-arsenic bronzes were purposefully elaborated with arsenic present in such exceptionally high concentrations.

The single most unexpected characteristic of Period 2 metallurgy is the widespread use of the tin bronze alloy, because the key material, cassiterite, is scarce in Mesoamerica and rare in the metalworking zone. Tin ore, tin ingots, or perhaps copper-tin ingots must have been transported to West Mexico from deposits in the Zacatecas tin province at least some of the time. The fact that smiths were using bronze so extensively is particularly significant; its scarcity may have contributed to its predominant ritual use.

Apart from using bronze for bells, these artisans also capitalized on this alloy's special physical and mechanical properties for other status objects that were shaped by cold work. They created thinner designs and achieved golden-looking colors in various classes of objects. Shell design tweezers and open rings are the most noteworthy of these. Both had prototypes in South America, but, like

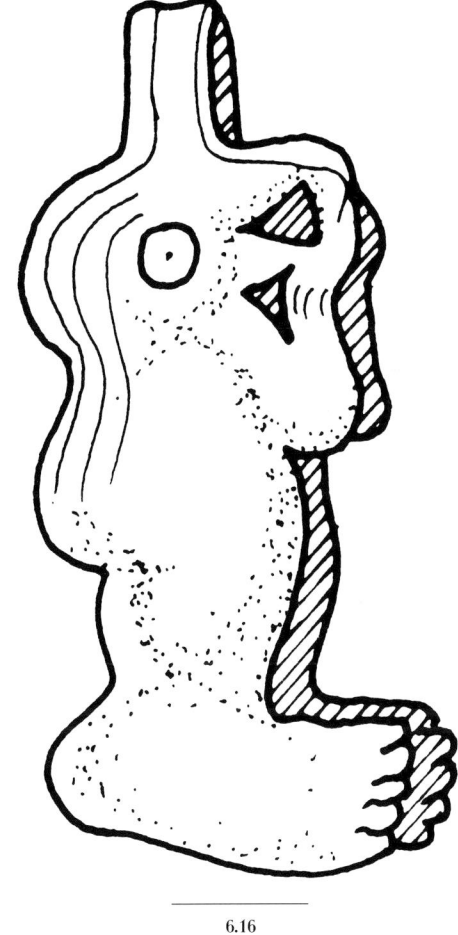

6.16

Tweezer from northwest Argentina. Hair is plucked by closing the feet around the hair. Photograph from González 1979, figure 11.

the bells, they constituted minor, secular, or culturally less significant components of South American technologies. In West Mexico, these object classes became central components of the metallurgical repertoire; technological elaboration and inventiveness focused on them.

Metal tweezers seem to have acquired special significance after they were introduced to West Mexico. The large, elaborate, aesthetically and technically impressive shell design tweezers, particularly the spiral variety unique to the Tarascan region, became central cultural symbols in that area. They were almost invariably fashioned with the alloying element—tin or silver—present in high enough concentrations to alter the color of the metal. Their use in the Tarascan region to communicate priestly office and authority differs from what we can infer about their meaning in certain regions of South America. Contrast, for example, the image of the shell tweezer worn by the Tarascan priest (see figure 3.16) with a tweezer design found in northwest Argentina (figure 6.16). Southern Andean tweezers sometimes are made in the shape of an armadillo; others, such as the tweezers in figure 6.16, have a human shape whose extremely prominent feet form the blades that close around the hair to pluck it. Depilation of facial and body hair was a pan-American concern, but in West Mexico the tool used to perform that activity itself became a critically important symbol. The chief priest wore his ritual tweezers as a mark of office, and, as I have shown, not only were these "symbolic" tweezers golden and silvery in color, but they were all capable of functioning as tools.

West Mexican artisans appropriated some elements of South American metallurgies with little change. One of these was to use bronze for particular kinds of tools and implements. They directly incorporated certain Andean implement types, such as the loop eye needle. They also adopted the idea that metal served as a standard of value, fashioning the artifacts that represented that idea, axe-monies, in precisely the same way as in northern South America. Yet even in this utilitarian sphere of these technologies, there are differences in balance and in scope. The utilitarian component of West Mexican metallurgy is smaller than that of the metallurgies of the Andes; West Mexican metalsmiths made proportionately fewer tools than did their South American counterparts.

West Mexican smiths also fashioned a more limited variety of tool and implement types than the metalworkers to the south.

The most concrete information concerning the relative abundance of certain ancient Andean metal objects comes from Ecuador. There, the MAG's acquisitions policy selects for status objects; but even in view of this, the proportion of status objects to tools is about three to one, or 75%. In West Mexico, 86% of the RMG collection is made up of status ornaments and ritual objects. The RMG acquisition policy gave equal weight to all metal artifacts, including fragments and damaged objects. The Salango assemblage may provide a more realistic Ecuadorian example of the relative proportions of status and ritual objects to tools than the MAG collection. There, half of the metal objects recovered were ornaments and the other half were tools. Salango excavations involved a variety of contexts including middens, house floors, patios, and burials. Excavation procedures were so meticulous that even tiny metal fragments were recovered. The Salango metal objects may constitute as representative a sample, with respect to relative abundance, as can be obtained, given the vagaries of corrosion, excavation strategies, and so on. The proportions of status objects to tools is approximately equal (see Owen 1986)[7] in the southern Andes. However, in Ecuador and Peru metalworkers crafted a broader range of utilitarian objects than in West Mexico. They made hand tools, needles, and awls, but also knives, club heads, and spear points (Mayer 1992). Agricultural tools also are present in northern Peru, including hoes and digging stick blades (Lechtman 1981; Shimada 1985). Agricultural tools are unusual in West Mexico.

The most surprising way in which the utilitarian components of West Mexican technology diverge from those to the south lies in the variety of metals and alloys

West Mexican metalsmiths employed for tools. They used the three bronze alloys—copper-tin, copper-arsenic, and copper-arsenic-tin—virtually interchangeably. In the Andes, alloys of copper and tin were generally restricted to the south, where cassiterite deposits are found, and arsenic bronze alloys to the north, where arsenical copper ores occur. In Mexico, a varied array of mineral deposits—ores of arsenic as well as tin—occur within a reasonably circumscribed geographical region. West Mexican peoples took advantage of these raw materials and Andean technical know-how to develop both bronze binary alloys, and independently invented the ternary copper-arsenic-tin alloy. Objects from Cuexcomate and Capilco, particularly tools, are usually made from this latter bronze, which sometimes contains arsenic in higher concentrations than tin. To date, little evidence exists suggesting that this ternary alloy was produced intentionally in the Andes except at the central highland site of Jauja. There it dates to the Inka presence (A.D. 1460 to A.D. 1532; Costin et al. 1989), certainly later than its appearance in West Mexico. The only other examples of this copper-arsenic-tin alloy, mentioned earlier, are from northwest Argentina, and were probably unintentionally smelted from an unusual local ore that also contained zinc (Fester 1962).

West Mexican smiths took systematic advantage of these three bronze alloys to optimize tool design. They made axes and awls thinner and harder, and fashioned structural designs like loop eye needles, made possible by the strength of these alloys. We do not know if southern Andean metalsmiths managed tin bronze with this same versatility, although it is clear that they were controlling tool hardness by varying tin concentration. In Ecuador, on the other hand, at least in the artifact classes available for study, there is little to suggest that smiths altered artifact designs to accommodate the properties of the alloys they used (see figure 6.14).

Taken as a whole, the fundamental course of West Mexican metallurgy did not alter during its entire history. Only a small proportion of the technology at any time was devoted to utilitarian ends. The configuration and emphasis of Period 2 West Mexican metallurgy, particularly the ways in which tin and arsenic bronze were used, are especially intriguing, given that the environment offered the ore minerals and metals—especially of arsenic—with which these smiths could have elaborated a thoroughgoing utilitarian bronze tool and weapons technology. Metalsmiths did make some of these objects from bronze. However, as we have seen, these peoples employed metal primarily for objects worn by elites and religious functionaries and used in ritual. The properties that interested them were initially sound and then color, which became a central technical focus with the introduction of the alloys in Period 2.

During Period 2, metalsmiths capitalized on the enhanced properties of the new materials to elaborate the pattern they had established in Period 1. They integrated elements of South American metallurgies into the preexisting tradition, expanding its scope and variety while maintaining its emphasis. Certain arenas of the technology were clearly less amenable to experimentation than others. West Mexican smiths were most conservative with respect to object types. Few new artifact types appeared in Period 2, and most Period 1 types had prototypes in Central and South America, and/or had been made in West Mexico but from other materials. These artisans chose to experiment with materials, not with new forms; the fundamental changes that took place from Period 1 to Period 2 came about as they worked with these new materials, taking advantage of alloy properties to streamline and refine design and improve functionality.

This pattern of experimentation in materials and conservatism in artifact types is not simple to explain. It may reflect the fact that metallurgy was introduced as a kind of package, so that people always associated metal with certain kinds of artifacts. It also may reflect broader American or Mesoamerican attitudes, perhaps visible in other materials as well, that constrain changes in form but allow experiments with materials. As I show in chapter 8, the Mesoamerican deities themselves are portrayed this way; they are experimentalists, beings who try various materials until they find those that work best for their human design. The tendency to retain forms but alter materials also may hold cross-culturally, at least in preindustrial societies, providing insight into the conservative role of material forms in human society.

7

THE DISSEMINATION OF WEST MEXICAN METALLURGY

Certain elements of Period 2 West Mexican metallurgy were introduced to other regions of Mesoamerica. They include artifacts, information about processing, and at least one class of raw materials: tin or tin bronze ingots. We know little about how this occurred because archaeological and documentary evidence is sparse. There is no doubt that it did occur, however. The laboratory studies of metal artifacts recovered from sites outside of West Mexico show that some are West Mexican in design and composition and must have been exported from the metalworking zone. By contrast, other objects made from tin bronze and found at sites outside of West Mexico are *not* designs typical of the West Mexican metal assemblage. These artifacts apparently were fashioned at or near those settlements, but the tin metal must have been imported. Strong evidence now also shows that a tin bronze metallurgy developed in the Huastec area of eastern Mesoamerica (see figure 7.1) just before the Spanish invasion and that this technology was closely related to the bronze technology that developed earlier in West Mexico.

In some cases, sufficient archaeological and ethnohistoric evidence exists to infer how certain elements of West Mexican technology were disseminated. Occasionally we can identify the ethnic groups or polities responsible for distributing the objects, information, and raw materials. The Aztec and Tarascan states clearly played primary roles. However, the data are insufficient for such reconstructions for earlier periods.

The amount of tin present in these artifacts provides the key in determining whether certain metal objects were moved over long distances. Virtually any artifact containing tin in concentrations above about 0.40% and recovered in regions other than West Mexico, the Zacatecas tin province, or immediately adjacent areas provides strong evidence for movement of materials or artifacts. The argument is as follows. Chalcopyrite ores often contain tin in disseminated form at low levels of concentration. Copper metal smelted from such ores can be expected to contain low levels of tin. Grains of cassiterite can also sometimes be associated with copper ores, and copper smelted from such ores will also contain tin at low levels. The tin impurity in copper when obtained from such sources virtually never appears in artifact metal in concentrations higher than about 0.30%, and it is usually present in far lower concentrations. On the other hand, when tin is detected in Mesoamerican artifacts at higher concentrations we can safely assume that (a) either a deliberate alloy was made by adding metallic tin, smelted from cassiterite, to copper, or the object was made from remelted bronze artifacts,[1] and (b) the cassiterite was obtained from the geological province that lies on the northeastern border of West Mexico (see chapter 2). In Mesoamerica, cassiterite deposits are restricted to this zone.

West Mexican tin bronze artifacts are one product of Period 2 metallurgy that was moved to other regions of Mesoamerica. The most common exports are ritual or status items, especially the large, cylindrical, type 10b wirework bells (chapter 5) and type a and b shell design tweezers illustrated in figures 7.2 and 7.3. Artisans crafted these particular objects in highland Michoacán, which was under Tarascan control, and in adjacent regions of northern Guerrero, portions of which were incorporated into the Aztec empire after A.D. 1440. These, and other West Mexican metal objects including tools, appear most frequently at sites to the east, south, and southeast of the West Mexican production zones: in Morelos (e.g., Cuexcomate), Tamaulipas, Oaxaca (e.g., Monte Albán), Chiapas (e.g., Chiapa de Corzo), Tabasco (e.g., Madero), the Yucatán Peninsula, Belize, and Honduras (see figure 7.1). They rarely have been reported in the Guatemalan highlands and have not yet been found either on the Pacific coast of Guatemala or the coastal strip of the Mexican

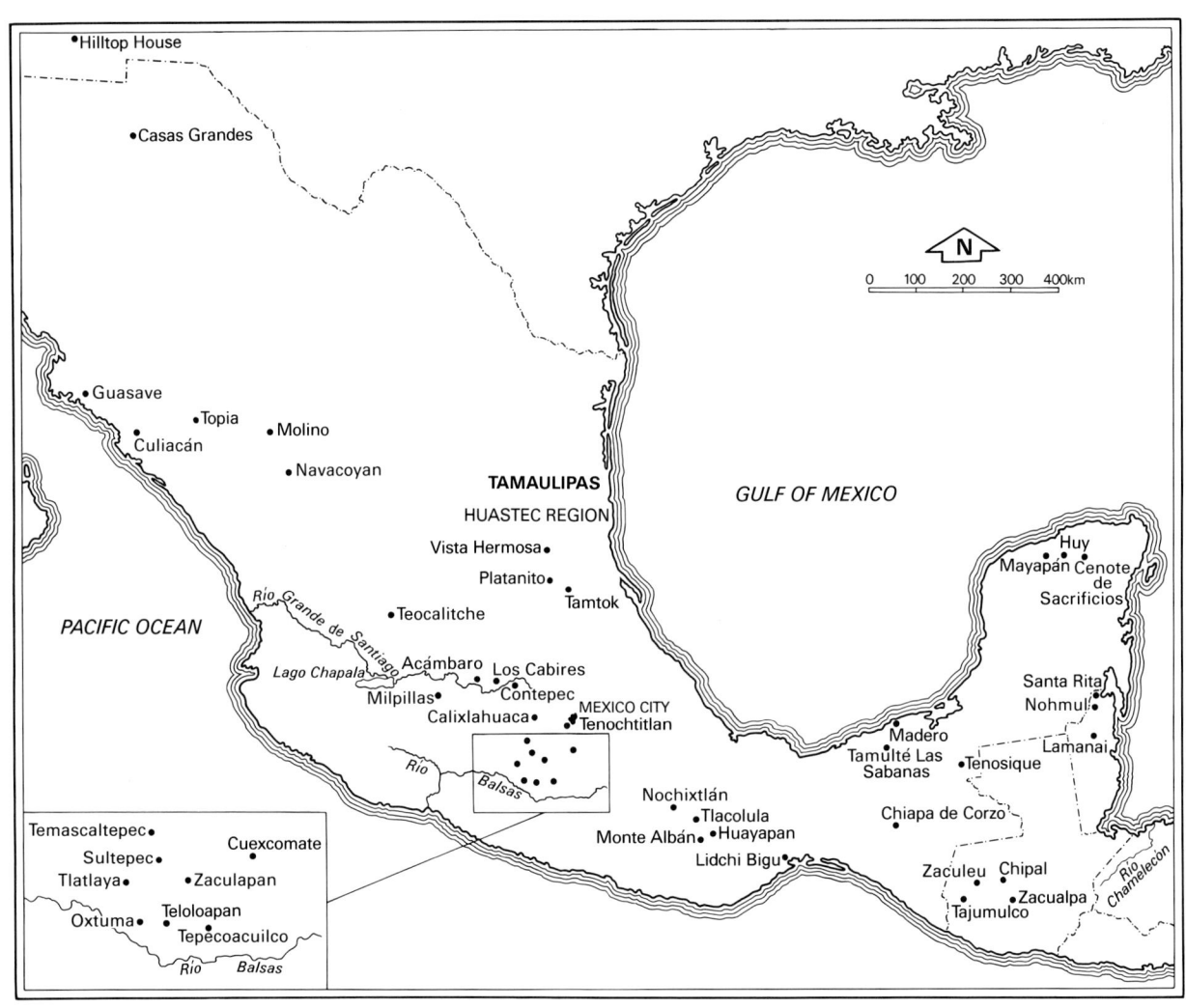

7.1

Mesoamerican sites where metal artifacts have been found and/or where metal production took place. Most assemblages date to after approximately A.D. 1200.

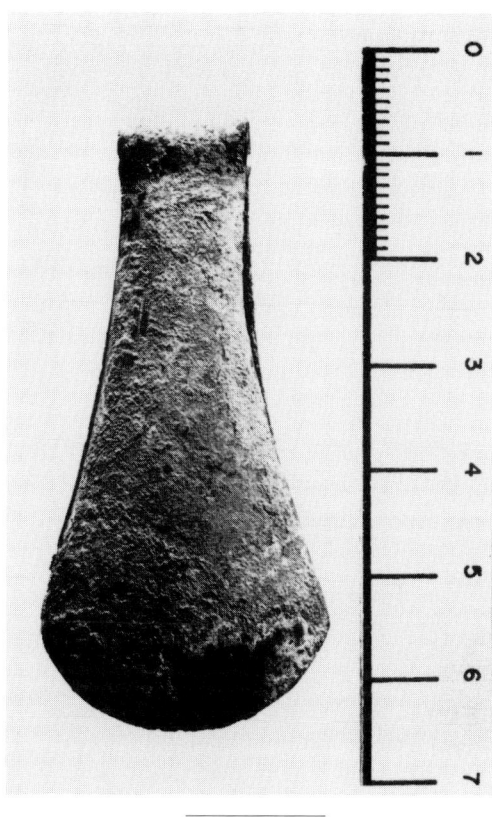

7.2

Copper-tin alloy shell design tweezers (type a) excavated at Lamanai, Belize; likely import from West Mexico. Late Postclassic Period.

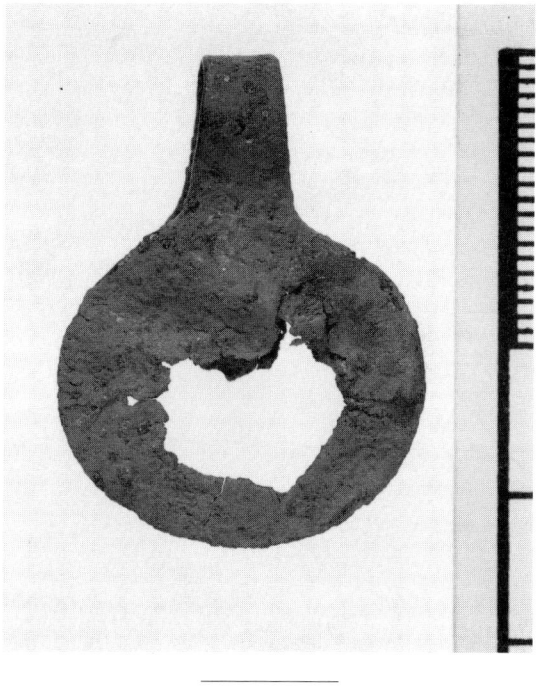

7.3

Copper-tin alloy shell design tweezers (type b) excavated at Lamanai, Belize; likely import from West Mexico. Late Postclassic Period.

state of Chiapas. They are also largely absent to the north, in the areas defined by the modern states of Chihuahua, Durango, Sinaloa, and Zacatecas.

Tin and/or tin bronze ingots are another product of West Mexican technology transported to other Mesoamerican regions. Artifacts containing tin recovered outside of the West Mexican metalworking zone but which are not West Mexican designs provide very strong evidence that these raw materials also circulated. Ethnohistoric sources provide corroborating data. In Diego de Landa's sixteenth-century account of native life in the Yucatán, he reports that ingots or blanks (sheets or plates) of a "hard metal" were brought to that region from Tabasco:

Tenían cierto azófar blanco con alguna poca mexcla de oro de que hacían hachuelas de fundición y unos cascavelazos con que bailaban. . . . Este azófar y otras planchas o láminas más duras las traían a rescatar los de Tabasco por las cosas (de Yucatán que eran). [They had a kind of white brass with a slight mixture of gold with which they cast axes and large bells that they used in dances. This brass and other plates or sheet that were harder were brought by people from Tabasco to trade for things from Yucatán.] (Landa 1978: 118)

The plates or sheets that Landa observed probably were copper-tin ingots. Brass (*azófar*) is a synonym for *latón,* an alloy of copper and zinc. Copper-zinc alloys were unknown in prehispanic Mesoamerica, although the term *latón* appears in Spanish texts describing indigenous metalwork. *Latón* or *azófar* were terms apparently used by the Spaniards in a generic sense to mean "copper alloy." ("White brass" describes a yellowish copper alloy.) The statement by Landa indicates that copper alloys were imported from metal-producing zones and used as stock material. Artifact compositions suggest that such alloys were of copper and tin. In addition, Berlin (1956: 146) maintains that the materials described by Landa came from the west, citing a statement made by Bernal Díaz del Castillo (1939 vol. 1: 142) that the *caciques,* or chiefs, from Tabasco informed Hernán Cortez in 1532 that their gold and jewels were brought to them from the region of the sunset.

Artisans outside of West Mexico thus used metal from copper-tin ingots or tin ingots to fashion bronze objects that were local designs. The ingots were imported from metal-producing regions in the west or possibly from areas in, or adjacent to, the Zacatecas tin province. Imported ingots may also have been used to make certain copper-tin bronze tools, especially in metalworking regions such as Oaxaca, although we cannot know without many more analytical studies. Other artifacts containing tin may have been made from metal produced by melting down tin bronze objects and reusing the metal, as I will argue subsequently. Objects fashioned from recycled metal usually can be distinguished on the basis of atypical concentrations of tin and the presence and level of other elements.

We do not yet know which cassiterite deposits in the Zacatecas tin province were mined for the tin exported as ingots to West Mexico and to other Mesoamerican regions. As chapter 2 indicates, a few cassiterite deposits do exist on the peripheries of the West Mexican region, in what constitutes an extension of the tin province (see figure 2.3). These probably were the primary ore sources for West Mexican smiths. We suspect that a number of deposits were exploited based on the nonlinear indium-to-tin ratios characteristic of the artifacts (an issue I also treat in chapter 2). The question can be explored systematically by locating and sampling cassiterite deposits. Until such studies are carried out, I will rely on other information to locate deposits that may have been exploited by these ancient metalworkers.

The Zacatecas tin province is not well known archaeologically, nor does much documentary evidence exist for the area. However, some information is available concerning the very few deposits that lie on the edge of the zone, near primary West Mexican metalworking areas. A tin deposit in Guerrero, north of Teloloapan, appears on the *Mapa Metalogénico de México.* Two objects made from pure tin and a cassiterite nodule have been reported from the Teloloapan area (Caley and Easby 1964), suggesting that this deposit may have been mined in the prehispanic era. In Michoacán, the *Mapa Metalogénico* shows a large deposit at Los Cabires, near Maravatío, although there is no information suggesting it was mined. What seems to be the same deposit is shown located at Contepec in Michoacán on the UNAM map for tin (see chapter 2). The same map shows three deposits characterized as secondary on the Jalisco-Aguascalientes border. Yet another, at Teocalitche (Jalisco), is also shown by Panczner (1987). We do not know whether any of these deposits served as sources for tin metal in the prehispanic era, however.

The most likely candidate for prehispanic exploitation is a cassiterite deposit located in the state of Mexico. Tribute lists assembled in 1536 report that 3 *cargas* of tin were tendered every 80 days to the Spaniards from Sultepec, a mining zone in the mountainous southern extreme of the state of Mexico (Quezada R. 1972). The *Relación de Sultepec* (Paso y Troncoso 1905–1906, vol. 7), written in 1582, also mentions tin mines there. The UNAM map for tin shows a deposit in the same area, at Zaculapan, whose commercial value is unknown; geological reports in the archives of the large Mexican mining company, Industrial Minera México, also refer to it. They relate that these cassiterite deposits are located near the town of Tlatlaya, which lies some 30 kilometers to the south of Sultepec (figure 7.1).

The provenience of copper-arsenic alloy artifacts (both ritual objects and tools) found outside West Mexico is more difficult to determine than for those made from tin bronze, because the parent materials, arsenopyrite and chalcopyrite, occur in many regions of Mexico. West Mexico, however, is the only area where we know objects made from arsenic bronze were manufactured on a large scale. Some copper-arsenic alloy bells appear in the Huastec region. Since they are types also known from Guerrero, we cannot be sure if they represent imports, were produced locally, or both (Hosler and Stresser-Péan 1992).

Apart from transporting artifacts and ingots to adjacent regions, West Mexican traders and/or metalworkers also conveyed information concerning ore processing. Period 2 West Mexican metallurgy sparked metallurgical developments in the Huastec region during the century prior to the Spanish invasion (Hosler and Stresser-Péan 1992). It also stimulated the development of a metalworking tradition characterized by the use of copper-lead alloys in or very near the Valley of Mexico and directly linked to the Period 2 technology in West Mexico. By at least A.D. 1438, metalworkers in or near the valley were employing this alloy to cast bells, some of which are identical to the earlier, large cylindrical type 10b West Mexican wirework design. These bells sporadically appear at sites elsewhere in Mesoamerica, where their composition readily distinguishes them from their West Mexican counterparts. William Root (in Lothrop 1952), who carried out many chemical analyses of Mesoamerican metal artifacts, argued on the basis of their distribution that these copper-lead bells must have been manufactured in the Valley of Mexico. Other authors also report copper-lead wirework bells from that area (Arsandaux and Rivet 1923; Grinberg and Franco V. 1987). Deposits of galena, the sulfide ore of lead, are widespread in Mesoamerica and appear in the west. (They occur in the states of Mexico, Morelos, Michoacán, Veracruz, and in other regions.) However, there are no copper-lead alloy artifacts from Milpillas or other West Mexican sites and none are found in the RMG collection. Copper-lead alloy objects, therefore, were not a significant element in West Mexican metallurgy and must have been fashioned outside of the metalworking zone.

The data presented here demonstrate that certain aspects of Period 2 metallurgy spread to other regions of Mesoamerica. Most of the evidence derives from my laboratory studies of large, Postclassic (A.D. 1200 to A.D. 1521) metal assemblages from outside of West Mexico—Lamanai in Belize, Cuexcomate and Capilco in Morelos, and Platanito and Vista Hermosa in the Huastec region—but it also includes occasional reports in the literature that may concern only one or two objects from a single site. In Morelos and Belize, West Mexican objects begin to appear around A.D. 1200. The Huastec evidence is later. There, by approximately A.D. 1440, metalworkers produced copper-arsenic-tin alloys and fashioned certain objects from them and from copper. Artifacts appearing at other sites rarely can be dated, except rather generally to the Late Postclassic Period.

Western Morelos: Cuexcomate and Capilco

Cuexcomate and Capilco were two towns in western Morelos (see figure 7.1) that flourished during the Postclassic Period. (I discussed them in chapter 5; because their location in Morelos places them slightly outside of the West Mexican metal-producing zone, I treat them here in greater detail.) Capilco was first inhabited around A.D. 1200; initial evidence for occupation at Cuexcomate dates to approximately 50 years later (Smith and Doershuk 1991). After A.D. 1438, both were incorporated into the Aztec empire. Production of cotton cloth and bark paper were major industries at both centers. Some of these products figured as tribute to the Aztec capital of Tenochtitlan; others made their way into regional markets. Michael Smith, who excavated the sites (Smith and Doershuk 1991), has defined three chronological periods on the basis of ceramic associations: Temazcalli (A.D. 1200 to A.D. 1350), Early Cuauhnahuac (A.D. 1350 to A.D. 1430), and Late Cuauhnahuac (A.D. 1430 to A.D. 1550).

Forty-five metal objects, mostly hand tools (needles, awls, and narrow-bladed chisels) were recovered from Capilco and Cuexcomate, primarily from house floors and midden deposits. All fit within the West Mexican tradition with respect to their manufacturing techniques. The tools were shaped from an original cast blank by cold work; bells were lost-wax cast. Table 7.1 shows the artifact types excavated and the number from each type analyzed for chemical composition.

Table 7.1 Metal Objects Excavated at Cuexcomate and Capilco

Types	Number Excavated	Number Analyzed
Awls/chisels	14	14
Needles	16	11
Bells	5	5
Tweezers	2	2
Wire	5	5
Sheet metal	2	2
Ornament	1	1
Total	45	40

The high proportion of tools relative to status objects recovered from these sites probably results from the fact that no adult burials were excavated. Artisans may have used the narrow-bladed chisels for wood carving and paper processing. The unusual presence of all three West Mexican needle types here suggests that intensive cloth production probably was carried out at Cuexcomate, since design differences likely correspond to differences in use. Smith and Heath-Smith (1994) have discussed the likelihood of intensive cloth production at Cuexcomate and Capilco.

We do not know where the Cuexcomate and Capilco objects were fashioned, but artifact chemistries provide important clues. The laboratory data show that from the Early Cuauhnahuac period on, three distinct artifact groups can be distinguished on the basis of metal composition. Artifacts or the artifact metal came from the West Mexican metalworking zone, from in or near the Valley of Mexico, and from an unknown source where metals were recycled. Smith recovered no crucibles, ingots, slag, or ore that would suggest that smelting or alloying operations were carried out at the sites themselves. It is likely, however, that Cuexcomate and Capilco artisans performed minor repairs on metal objects. For example, the probability is good that a loop eye needle reshaped into an ornament (figure 7.4) was reworked *in situ*. Two enigmatic ceramic tubes that could have served as blowpipes were also encountered in the excavations.

Table 7.2 presents the chemical compositions of 40 objects from Cuexcomate and Capilco as determined by quantitative analytical methods. The results are organized by period and compositional group or likely source. All but two objects are copper alloys, with the alloying element present in high enough concentrations to alter the working properties of the metal.

Group 1 (West Mexico) contains objects whose chemistries are identical to those of their counterparts in the RMG assemblage: they are made from alloys of copper-arsenic, copper-tin, and copper-arsenic-tin. Group 2 (Valley of Mexico) also contains artifacts made from alloys of copper-tin and copper-arsenic-tin, but unlike objects from West Mexico these also contain lead in concentrations above approximately 0.30%. The third group (recycled) consists of artifacts in which key elements, such as tin, are present in concentrations too high to have been introduced as a trace element with the ore, but too low to constitute an intentional alloy. The metal for these objects could have been smelted from an unusual mixed ore derived from a polymetallic deposit, although we know of no such ores in Mexico. The analytic results point more strongly to recycled metal made by melting down bronze artifacts so that the metal could be reused. Table 7.3 summarizes the data in table 7.2, showing the number of objects and object types in each group by period.

The artifacts in group 1, virtually identical to West Mexican types in composition and design, probably were fashioned in Guerrero or Michoacán. Of these, bells are the only items for which some information exists that may

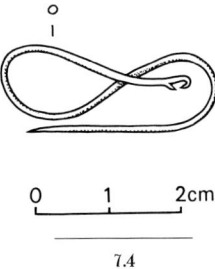

7.4

Needle reworked into an ornament, excavated at Cuexcomate, Morelos.

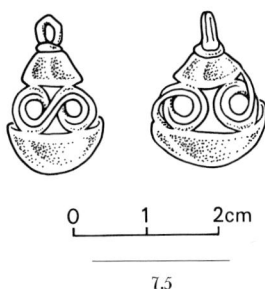

7.5

Copper-tin alloy bell excavated at Cuexcomate, Morelos. Similar designs are found in Guerrero.

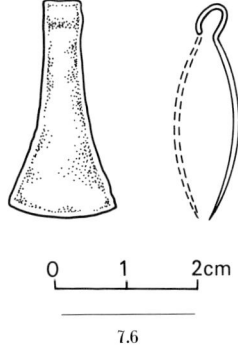

7.6

Copper-tin alloy West Mexican shell design tweezer excavated at Cuexcomate, Morelos.

Table 7.2 Quantitative Chemical Analyses of Artifacts from Cuexcomate and Capilco

		Composition (weight percent)					
Artifact Type	ID No.	Ag	As	Fe	In	Pb	Sn
Early Cuauhnahuac (A.D. 1350–1430)							
Group 1: West Mexico							
Bell	c5	0.02	—	—	—	—	6.61
Bell	c34	0.06	0.43	0.04	0.04	0.007	12.4
Needle	c2	0.07	0.84	—	—	—	1.02
Needle	c19	0.14	0.42	—	—	—	2.55
Needle	c24	0.05	0.49	—	—	0.13	1.80
Chisel	c8	0.07	0.81	—	—	0.09	5.12
Awl	c22	0.07	1.17	0.02	—	0.02	1.83
Wire	c11	0.06	0.36	—	—	—	1.18
Sheet	c39	0.12	0.68	0.01	0.013	—	6.47
Group 2: Valley of Mexico							
Chisel	c40	0.09	1.23	0.02	0.01	0.66	4.35
Group 3: Recycled							
Needle	c1	0.07	0.73	—	—	—	0.36
Needle	c21	0.14	1.95	—	—	—	0.45
Awl	c38	0.24	0.5	—	—	0.33	0.89
Wire	c23	0.04	0.76	—	—	—	0.41
Late Cuauhnahuac (A.D. 1430–1550)							
Group 1: West Mexico							
Bell	c10	0.23	0.20	—	—	—	7.39
Bell	c27	0.11	0.85	0.07	0.005	—	—
Tweezer	c3	0.01	—	—	—	—	10.6
Tweezer	c15	0.01	—	—	—	—	10.9
Needle	c6	0.10	1.69	—	—	—	3.14
Needle	c32	0.05	—	—	0.01	—	7.51
Needle	c41	0.12	1.89	0.001	—	—	—
Needle	c48	0.06	0.24	—	0.004	—	—
Awl	c17	0.09	1.00	—	—	—	6.99
Awl	c26	0.05	0.11	0.02	0.007	0.005	10.9

continued

Table 7.2 (continued)

Artifact Type	ID No.	Composition (weight percent)					
		Ag	As	Fe	In	Pb	Sn
Awl	c35	0.14	0.15	0.01	—	0.04	1.01
Wire	c30	0.17	—	—	—	—	—
Sheet	c31a	0.03	0.29	0.01	0.05	—	11.8
Group 2: Valley of Mexico							
Bell	c18	0.19	6.51	—	—	6.770	0.38
Awl	c25	0.12	1.14	—	—	0.35	2.2
Awl	c43	0.09	0.9	—	0.01	0.75	6.28
Group 3: Recycled							
Needle	c16	0.06	1.80	—	—	—	0.40
Needle	c20	0.08	0.37	—	—	0.550	0.51
Chisel	c7	0.04	2.69	—	—	—	1.26
Chisel	c33	0.09	1.26	—	0.004	1.780	1.01
Chisel	c42	0.83	0.53	—	—	0.920	0.37
Chisel	c47	0.07	3.53	—	0.020	—	2.68
Awl	c28	0.22	1.32	—	—	0.570	0.83
Wire	c44	0.07	0.30	—	—	0.560	0.34
Wire	c45	0.05	0.50	0.006	0.010	—	0.67
Ornament	c29	0.08	0.45	—	—	0.007	0.59

Note: "Chisel" refers to bladed awl. A dash indicates element not detected. Analyses by atomic absorption spectrometry

point to more specific manufacturing regions. These are of two types; one type is known from Guerrero (figure 7.5),[2] while the other appears both in Guerrero and at Milpillas in Michoacán. Three group 1 bells are made of tin bronze alloys, a fourth from an alloy of copper and arsenic. Group 1 also contains shell design tweezers. The tweezers (figure 7.6), like many of their West Mexican counterparts, are made from copper-tin alloys, with tin present at slightly above 10.5%. Tin concentration is high enough in both bells and tweezers to alter the color of the artifact metal and to increase fluidity during casting. As table 7.2 indicates, the tools in this group—small chisels, awls (unipointed, bipointed, and with narrow blades), needles, and wire—are made from copper-arsenic-tin alloys, with the alloying element generally present in lower concentrations than in bells and tweezers, allowing work hardening but avoiding brittleness. Group 1 (West Mexico) artifacts contain tin in concentrations above 1%. Lead, if present, appears below 0.20%; arsenic

Table 7.3 Possible Origins of Cuexcomate and Capilco Artifacts, by Period

	West Mexico (Group 1)		Valley of Mexico (Group 2)		Recycled (Group 3)		
	Early C	Late C	Early C	Late C	Early C	Late C	Total
Bells	2	2	—	1	—	—	5
Tweezers	—	2	—	—	—	—	2
Awls/chisels	2	3	1	2	1	5	14
Needles	3	4	—	—	2	2	11
Wire	1	1	—	—	1	2	5
Sheet	1	1	—	—	—	—	2
Ornament	—	—	—	—	—	1	1
Total	9	13	1	3	4	10	40

Note: Early C = Early Cuauhnahuac (A.D. 1350–1430).
Late C = Late Cuauhnahuac (A.D. 1430–1550).

can be present at higher levels but is always lower than the tin.

The closest source for tin is the deposit near Teloloapan. However, the indium-to-tin ratios for the Cuexcomate/Capilco objects suggest that more than one cassiterite source was utilized for the tin metal present in these objects. Artifacts from these sites, like their RMG counterparts, exhibit a nonlinear relationship between the concentration of indium and tin, as shown in table 7.4, suggesting that the tin derived from several sources.

Artifacts in group 2 (Valley of Mexico) contain lead in concentrations above 0.30% and tin or arsenic in concentrations above about 1%. Lead never appears at these levels in objects from West Mexico. Either the metal or the artifacts probably come from the Valley of Mexico, where copper-lead alloy bells were common, although we do not know how the lead was introduced into the metal. One of the artifacts in this group (no. 18) is a plain-walled bell with a vertical zigzag design, containing both lead and arsenic at levels above 6%. This alloy is highly unusual. Other objects made from this alloy have been found in the Valley of Mexico: Root (in Lothrop 1952) analyzed two wirework bells that contain both elements in concentrations above 2%. A bell from Tamtok, Tamaulipas, in eastern Mexico, also is made from this same alloy type and probably was imported from the Valley of Mexico (Hosler and Stresser-Péan 1992).

I consider the metal for the artifacts in group 3 recycled. These artifacts—awls, needles, and wire—contain alloying elements, particularly lead, tin, and arsenic, in anomalous concentrations. In some objects (for example, no. 20), they are present in concentrations too low to affect the working properties of the metal and to constitute a deliberate alloy (less than 0.75%), but too high simply to represent trace elements introduced in association with the ore. Tin concentrations are most illustrative; some artifacts contain tin in concentrations between 0.34% and 0.59% (nos. 44 and 29). Lead also appears in several artifacts in concentrations around 0.50%. The most plausible explanation for these low levels of such elements within a single object is that the metal is recycled, resulting from melting down other artifacts. Those group 3 objects that contain lead were probably made from metal that ultimately derived from objects made in or near the Valley of Mexico.

Some artifacts made from copper-arsenic-tin alloys appear in the recycled group due to their relative concentrations of tin and arsenic; the tin content, for example, is always lower than the arsenic content (for instance, no. 47). In general, smiths in West Mexico managed the properties of this ternary alloy very carefully, adding ore or metal so that the concentration of tin was always higher than that of arsenic, the other important alloying element. Copper-arsenic-tin ores do not exist in Mexico; this ternary alloy was made by deliberately adding cassiterite or molten tin to a copper-arsenic binary. These latter alloys were made by co-smelting chalcopyrite and arsenopyrite (an issue discussed in chapter 2). The relation between the relative concentrations of these two elements found in West Mexican artifacts does not hold for many of the group 3 artifacts (e.g., nos. 7, 33, and 47) and the presence of tin at such low levels with respect to arsenic is perplexing. The compositional evidence strongly suggests that these objects were fashioned from a stock copper alloy formed from remelted objects that contained arsenic and tin.

Table 7.2 shows that apart from ritual and status objects (tweezers and bells), which never were made from recycled metal, these three sources of metal were used for all artifact classes.[3] These data suggest, however tenuously, that factors regulating the production of status objects were different from those for tools, some of which were made from recycled metal. The major utilitarian

Table 7.4 Ratio of Indium to Tin in Artifacts from Cuexcomate and Capilco

ID No.	In (wgt %)	Sn (wgt %)	Ratio In:Sn
26	0.007	10.90	0.001
31a	0.050	11.83	0.004
32	0.010	7.51	0.001
34	0.040	12.40	0.003
39	0.013	6.47	0.002
40	0.010	4.35	0.002
43	0.010	6.28	0.002

artifact categories, needles and awls, contain some items whose composition is typical of West Mexico and others made either from recycled metal or from alloys produced in the basin of Mexico. No objects made from recycled metal appear in the RMG collection, and they are likewise absent at Milpillas (see chapter 5). Recycling may have been practiced in areas such as Cuexcomate and Capilco where the raw materials, ore or native metals, were difficult to procure. Nearly one-third of the objects studied were made from such metal. It is not surprising that recycling was also carried out at Lamanai, as I subsequently show.

The Cuexcomate and Capilco assemblages offer no surprises with respect to the kinds of metal objects people used. They employed the same ritual and status objects as in West Mexico: gold-colored bells (copper-tin alloys), a silvery-looking bell (a copper-arsenic-lead alloy), and gold-looking tweezers (high-tin bronze). Craftspeople likewise employed the same assemblage of small hand tools—needles and awls (copper-tin and copper-arsenic alloys)—as in the west, and these items probably were imported from there. What is striking about this assemblage is that the artifacts or the metal come from two distinct geographical regions—the Valley of Mexico and West Mexico (Guerrero and Michoacán)—and that recycled material constitutes a third source of metal for Cuexcomate and Capilco artifacts.

What is more, these sources did not change through time, even when the two towns were incorporated into the Aztec empire. We know that alloying developed earlier in West Mexico than in the Valley of Mexico, where the technology was derivative, but we do not know precisely how either West Mexican objects or those from the Valley of Mexico made their way to these centers. Based on the metallurgical evidence alone, Cuexcomate and Capilco were looking to West Mexican metalworking regions for about half their artifacts and using recycled metal for most of the others. The only significant change that occurs through time is in the number of recycled artifacts containing lead: these increase disproportionately after this region was incorporated into the Aztec empire, and seem to reflect increased contact with that region.

Lamanai, Belize

The Maya site of Lamanai is a large ceremonial center on the New River in Belize (Pendergast 1981). Lamanai was continuously occupied from the Preclassic (circa 200 B.C.) through the Historic Period (A.D. 1641). No appreciable deposits of metallic minerals exist in this vast lowland area. The closest sources for copper lie to the south in the Guatemalan highlands. Metal objects were first used at Lamanai around A.D. 1150 (Middle Postclassic), and their use continued through the Late Postclassic Period (A.D. 1300 to A.D. 1544) into the Historic Period (A.D. 1544 to A.D. 1641+). Excavations at Lamanai have yielded approximately one hundred metal objects. They include bells, tweezers, needles, axes, buttons, pins, fishhooks, and metal ingots, and were recovered from a variety of contexts including structures and burials (Pendergast 1981).[4] Table 7.5 shows the numbers of artifacts found by period and the number sampled for laboratory studies. Bells are by far the most numerous objects excavated.

Sixty-three artifacts recovered at Lamanai are ritual or status display items: tweezers, buttons, bells, finger rings, and pin bells or rattles. Twenty-two are tools. The function of the remainder could not be determined. Information from artifact chemistries (table 7.6) coupled with observations of artifact design features show that two distinct metallurgical traditions were represented at Lamanai during the Middle and Late Postclassic periods: one West Mexican, the other southeastern Mesoamerican. By the Historic Period, the situation becomes far more complex, with strong evidence for production at the site of Lamanai itself.

Table 7.6 presents Lamanai artifact chemistries. The artifacts are grouped by period and probable manufacturing region. Not all of the objects analyzed qualitatively were examined quantitatively. However, in most cases the correspondence between the two determinations is sufficiently well established to group most artifacts based on qualitative data alone. Table 7.7 summarizes those data, showing the number of objects belonging to each functional category by period and compositional group.

During the Middle and Late Postclassic periods (see table 7.6) some metal objects used by people at Lamanai represent a local southeastern Mesoamerican metalworking tradition. That tradition is characterized by lost-wax cast copper status ornaments; sometimes the castings are from copper-gold alloys. In the Lamanai case, the objects include filigree finger rings, elaborate filigree buttons, and plain-walled bells. At Lamanai, all were made using a very pure copper. These objects either were fashioned *in situ* from imported raw materials or were imported from metalworking centers elsewhere in the region. The buttons (there are two varieties) and finger rings are common

Table 7.5 Artifact Types and Numbers Sampled from Lamanai, by Period

Type	Middle Postclassic (A.D. 1150–1300)		Late Postclassic (A.D. 1300–1544)		Historic (A.D. 1544–1641+)		Total	
	Number Sampled	Total in Assemblage	Number Sampled	Total in Assemblage	Number Sampled	Total in Assemblage	Number Sampled	Total in Assemblage
Bell	5	9	2	2	6	28	13	39
Tweezer	1	1	2	2	—	—	3	3
Needle	—	—	—	—	3	5	3	5
Awl	—	—	1	1	—	—	1	1
Sheet metal	—	—	1	1	—	—	1	1
Axe	1	1	—	—	5	12	6	13
Button	6	13	—	—	—	—	6	13
Finger ring	1	2	1	1	1	3	3	6
Fishhook	—	—	—	—	1	3	1	3
Fragment	1	1	—	—	1	11	2	12
Ingot	—	—	—	—	4	4	4	4
Pin bell	2	2	—	—	—	—	2	2
Total	17	29	7	7	21	66	45	102

in this southeastern area and also in lower Central America. Buttons are also found in Oaxaca. The bells are types that have been recovered from the Cenote de Sacrificios, at Chichén Itzá. Three of the six Lamanai bells from the Middle Postclassic Period are of type 11b, which appears during Period 1 in West Mexico (see chapter 3). Similar types, for example 11a, are found still earlier in lower Central America and Colombia (see chapter 4; also see Sharer 1985; Strong 1935). Figure 7.7 illustrates one of the filigree buttons; the photomicrograph of the cross section (figure 7.8) shows a cast structure with exceptionally large grains, resulting from extremely slow cooling in the mold.

During this same interval, people at Lamanai also used metal objects imported from West Mexico (see table 7.6). They are made from bronze alloys and shaped primarily by cold work. Tweezers, needles, sheet metal, and awls of copper-tin and copper-arsenic bronze appear in this group. Pendergast and colleagues found two of the four classic West Mexican tin bronze shell design tweezer types at Lamanai (see figures 7.2 and 7.3), and these artifacts are identical in design and composition to tweezers in the RMG assemblage. One (figure 7.2), excavated from a Middle Postclassic context, is a typical West Mexican form of type a; it contains tin in a concentration (8.99%) consistent with other West Mexican bronze tweezers. The likelihood is extremely high that this tweezer was imported from Michoacán or northern Guerrero. Other Middle Postclassic artifacts, an axe and a bell, are made from copper-arsenic alloys with arsenic present

Table 7.6 Lamanai: Chemical Analyses of Artifacts, by Period and Probable Manufacturing Region

Type	ID No.	Ag	As	Au	Bi	Co	Cr	Fe	Ga	In	Mg	Mn	Ni	Pb	Sb	Sn	Zn
Middle Postclassic Objects (A.D. 1150–1300)																	
Grouping by probable provenience on the basis of qualitative analysis (emission spectroscopy)																	
Group 1: West Mexico																	
Axe	557/2	v+	m–	—	v	—	—	v	—	—	v	?	m	v	v	v	—
Bell	69/6	v	m–	—	?	—	—	m–	—	—	v	m	?	v	—	?	?
Tweezer	557/3	m	v	—	m+	—	—	m	—	v	v	—	?	v	?	m++	—
Group 2: Southeastern Mesoamerica																	
Bell	61/13	m	v–	—	?	—	—	m	—	—	v	m–	?	v	?	?	—
Bell	69/7	v	—	—	—	—	—	v	—	—	v–	v–	—	v+	—	—	?
Bell	69/8b	v+	—	—	—	—	—	v	—	—	v–	v–	?	v	—	—	—
Bell	69/8c	v+	—	—	—	—	—	v	—	—	v–	v	?	v	—	—	—
Button	69/9b	v	?	—	—	—	—	v–	—	—	v–	?	?	v	?	?	—
Button	69/9c	v+	?	—	—	—	—	v	—	—	v	m–	?	v+	—	?	—
Button	69/9f	m	—	—	—	—	—	v	—	—	v	—	—	?	—	—	—
Button	90/8a	v+	—	—	—	—	?	v	?	—	v	m–	?	?	—	—	v–
Button	90/8c	v+	—	—	—	—	?	v	?	—	v	m–	?	?	—	—	v–
Button	90/8f	m	—	—	—	—	—	m	—	—	v	v+	—	?	—	—	—
Finger ring	68/3	—	v	—	—	—	—	m	—	—	v	m	?	?	—	?	—
Fragment	118/11	v–	?	—	—	—	v–	v	—	—	v–	m	?	v+	—	?	v–
Pin bell	91/2a	v	—	—	—	—	—	v	—	—	v–	v	?	v–	—	—	?
Pin bell	91/2b	v	—	—	—	—	—	v	—	—	v–	v	?	v–	—	—	?
Grouping by probable provenience on the basis of quantitative analysis (atomic absorption)																	
Group 1: West Mexico																	
Tweezer	557/3	0.11	0.01	0.006	na	na	na	na	na	na	na	na	na	0.04	0.1	8.99	na
Bell	69/6	0.06	1.33	na	na	na	na	na	na	n	na	na	na	0.002	na	na	na
Group 2: Southeastern Mesoamerica																	
Bell	69/7	0.1	0.02	na	na	na	na	na	na	na	na	na	na	0.01	na	na	na
Button	69/9b	0.04	na	na	na	na	na	na	na	na	na	na	na	na	na	na	na
Button	90/8a	0.03	0.004	na	na	na	na	na	na	na	na	na	na	na	na	na	na
Finger ring	68/3	0.08	0.01	na	na	na	na	na	na	na	na	na	na	0.01	na	0	na
Pin bell	91/2a	0.03	na	na	na	na	na	na	na	na	na	na	na	na	na	na	na

continued

Table 7.6 (continued)

Type	ID No.	Ag	As	Au	Bi	Co	Cr	Fe	Ga	In	Mg	Mn	Ni	Pb	Sb	Sn	Zn
Late Postclassic Objects (A.D. 1300–1544)																	
Grouping by probable provenience on the basis of qualitative analysis (emission spectroscopy)																	
Group 1: West Mexico																	
Awl	922/2	m	v	—	?	?	?	v	—	—	v–	v	?	m	v–	m++	?
Bell	774/20	v+	?	—	—	—	—	v	—	—	v–	v+	?	?	?	m–	?
Sheet	614/2	—	v+	—	m	—	—	m–	—	v	m	v+	v	m	m–	M	—
Tweezer	614/1	m	m–	—	v+	—	—	v+	—	v	v	v+	v+	m	v+	M	—
Tweezer	905/1	v	?	—	?	?	—	?	—	v–	?	?	v–	m–	?	m+	—
Bell	774/23	v+	v–	—	v–	—	—	v	—	v–	v–	v+	?	v+	?	m+	?
Group 2: Southeastern Mesoamerica																	
Finger ring	774/3	m	v	—	?	—	—	v–	—	—	v–	v–	?	v+	v	v+	?
Grouping by probable provenience on the basis of quantitative analysis (atomic absorption)																	
Group 1: West Mexico																	
Bell	774/23	0.11	0.3	na	na	na	na	na	na	na	na	na	na	0.06	0.08	1.89	na
Tweezer	614/1	0.15	0.39	0.007	na	na	na	na	na	na	na	na	na	0.005	0.13	4.3	na
Tweezer	905/1	0.04	0.09	na	na	na	na	na	na	na	na	na	na	0.05	0.09	4.63	na
Historic Objects (A.D. 1544–1641+)																	
Grouping by probable provenience on the basis of qualitative analysis (emission spectroscopy)																	
Group 1: West Mexico																	
Bell	834/3	v	v	—	?	—	?	m–	—	—	v–	v	v–	v–	v–	v+	?
Bell	867/2	v+	v	?	v	—	—	v	—	v	?	?	v–	v	v	m+	?
Bell	878/7	v+	v	—	v–	—	?	v	—	v–	?	v	v	v+	?	m+	?
Needle	858/21	v	v	?	v	?	?	v	—	—	?	?	v	v	m	v–	m	?
Group 2: Southeastern Mesoamerica																	
Bell	834/4	v–	v–	—	?	—	—	v	—	—	v	?	v	v–	?	v–	?
Group 3: Recycled																	
Axe	855/1	v+	v	v	v	?	—	v	—	?	v–	?	v	m	v–	m	?
Axe	856/1	v+	v	v	?	?	—	v	—	—	v–	?	v–	m	?	m	v–
Axe	856/6	m	v	v+	?	—	—	?	—	?	?	?	v–	m–	?	m	?
Axe	871/1	v+	v	v+	v–	—	—	v	—	?	?	?	v	m	?	m+	v

continued

Table 7.6 (continued)

Type	ID No.	Ag	As	Au	Bi	Co	Cr	Fe	Ga	In	Mg	Mn	Ni	Pb	Sb	Sn	Zn
Axe	908/2	m	v–	?	?	?	—	?	—	—	v–	v	?	m	?	m	?
Bell	885/23	m	v+	?	v	v	—	v–	—	—	v–	?	v	m+	v	m–	?
Fishhook	834/1	m	v	—	m	—	—	v–	—	—	v–	v–	v–	m+	v–	m+	?
Finger ring	822/1	v	v	v	?	—	—	v	—	—	?	m	v	m+	v	v+	v
Fragment	878/3	m–	v–	?	?	?	?	v+	—	—	?	v	v–	v	?	m	?
Ingot	858/11	v	v–	?	?	—	—	v–	—	—	v–	v–	v	m	?	m	?
Ingot	881/1	v	v	?	v+	?	—	—	—	—	v–	—	v–	m	v	m	—
Ingot	894/1	m–	v	?	?	—	—	v–	—	—	v–	—	v	m–	?	m	?
Ingot	908/1	m	v–	v	?	?	—	?	—	—	v–	?	?	m	?	m	?
Needle	916/26	m–	v–	—	v+	—	—	v	—	?	v–	v–	v	m+	?	m+	—
Group 4: Uncertain																	
Bell	823/10	v	v–	?	?	?	—	m–	—	—	v–	v	v	m	v–	v	?
Needle	856/3	v+	v–	—	v	?	—	?	—	—	v–	—	v	m	?	m	?

Grouping by probable provenience on the basis of quantitative analysis (atomic absorption)

Group 1: West Mexico

Type	ID No.	Ag	As	Au	Bi	Co	Cr	Fe	Ga	In	Mg	Mn	Ni	Pb	Sb	Sn	Zn
Bell	834/3	0.07	1.21	na	na	na	na	na	na	na	na	na	na	0.008	0.09	0.06	na
Bell	867/2	0.06	0.6	0.01	na	na	na	na	na	na	na	na	na	0.01	0.15	2.59	na
Needle	858/21	0.07	1.06	0.01	na	na	na	na	na	na	na	na	na	0.12	0.09	0.16	na
Group 3: Recycled																	
Axe	855/1	0.14	0.16	0.013	na	na	na	na	na	na	na	na	na	0.14	0.09	0.38	na
Axe	856/1	0.21	0.31	0.73	na	na	na	na	na	na	na	na	na	0.15	0.07	0.73	na
Axe	856/6	0.25	0.19	0.86	na	na	na	na	na	na	na	na	na	0.04	0.07	0.28	na
Axe	871/1	0.25	0.32	1.44	na	na	na	na	na	na	na	na	na	0.33	0.01	1.14	na
Finger ring	822/1	0.08	0.96	0.02	na	na	na	na	na	na	na	na	na	0.76	0.21	0	na
Fishhook	834/1	0.24	0.7	0.02	na	na	na	na	na	na	na	na	na	0.43	0.11	0.69	na
Fragment	878/3	0.09	0.52	0.01	na	na	na	na	na	na	na	na	na	0.004	0.09	0.72	na
Ingot	858/11	0.1	0.06	0.011	na	na	na	na	na	na	na	na	na	0.07	0.05	0.65	na
Ingot	881/1	0.18	0.38	0.05	na	na	na	na	na	na	na	na	na	0.08	0.11	0.51	na
Ingot	894/1	0.13	0.38	0.1	na	na	na	na	na	na	na	na	na	0.05	0.07	0.04	na
Ingot	908/1	0.16	0.51	0.43	na	na	na	na	na	na	na	na	na	0.07	0.08	0.69	na

continued

Table 7.6 (continued)

Type	ID No.	Ag	As	Au	Bi	Co	Cr	Fe	Ga	In	Mg	Mn	Ni	Pb	Sb	Sn	Zn
Group 4: Uncertain																	
Bell	823/10	0.07	0.61	0.02	na	na	na	na	na	na	na	na	na	0.23	0.09	na	na
Needle	856/3	0.09	0.18	0.011	na	na	na	na	na	na	na	na	na	0.8	0.08	0.14	na

Note: Figures for quantitative analyses represent weight percent; na = element not analyzed. In the qualitative analyses, a dash indicates element not detected; ? = questionable; v–, v, v+ = visible; m–, m, m+, m++ = minor element; M = major element.

Table 7.7 Lamanai: Number of Objects Belonging to Each Functional Category, by Period and Manufacturing Region

Type	West Mexico (Group 1)			Southeastern Mesoamerica (Group 2)			Recycled (Group 3)			Uncertain (Group 4)		
	MPC	LPC	HIST	MPC	LPC	HIST	MPC	LPC	HIST	MPC	LPC	HIST
Awl		1							5			
Axe	1											
Bell	1	1	3	4	1	1			1			1
Button				6								
Finger ring				1	1				1			
Fishhook									1			
Fragment				1					1			
Ingot									4			
Pin bell				2								
Needle		1							1			1
Sheet metal		1										
Tweezer	1	2										

Note: MPC = Middle Postclassic (A.D. 1150–1300); LPC = Late Postclassic (A.D. 1300–1544); HIST = Historic (A.D. 1544–1641+).

7.7

Copper filigree button excavated at Lamanai, Belize. Middle Postclassic Period. Lost-wax cast.

in high enough concentrations to affect the properties of the metal; the bell contains 1.33% arsenic, and the qualitative analyses of the axe indicate that it contains arsenic in comparable concentrations. The bell is a very general type, but it does possess a double suspension ring, characteristic of some West Mexican designs. Similar bells, ascribed to Michoacán, are found in the collections of the Museo Nacional de Antropología e Historia in Mexico City. Although we can be less sure of the provenience of these copper-arsenic alloy objects, the evidence to date indicates that they were fashioned in the west.

All seven metal objects that date to the Late Postclassic Period were sampled; five are made from tin bronze alloys, and four of those five were cold-worked to shape: two tweezers, a square of sheet metal, and an awl. Both tweezers are West Mexican types. The fifth object is a bell, probably of central or north-central Mexican design. Although fewer metal objects appear in this period than previously, it is remarkable that such a high proportion consists of nonlocal types made from nonlocal materials. No evidence exists during this time for on-site manufacture.

During the subsequent Historic Period, substantial changes take place in the raw materials used to make Lamanai metal objects. For the first time the evidence strongly indicates that artisans were crafting objects at Lamanai itself, and they seem to have been doing so using recycled metal. Pendergast recovered four ingots dating to this period. One is illustrated in figure 7.9. All four were analyzed, and their highly anomalous compositions point to stock metal derived from melted-down artifacts. Compositional data (see table 7.6) also suggest that similar stock metal was used to fashion some Historic Period artifacts, particularly axes. Two of the four ingots contain arsenic, gold, and tin. The ingots, the axes, and other Historic Period objects exhibit an array of minor elements unlikely to have been contributed by any single ore source: gold, silver, arsenic, lead, antimony, and tin. Gold, while unusual in West Mexican objects, would have been introduced readily to Lamanai crucibles, because *tumbaga* or gilt copper objects were present in this area.

During the Historic Period, people at Lamanai continued to use metal objects representing the West Mexican bronze-working tradition, in addition to objects characteristic of the southeastern Maya lost-wax casting technology (see table 7.6). Smiths used the two bronze alloys for bells and a needle. At least one of the bells (figure 7.10) is a type appearing in the collections of the Museo Nacional de Antropología e Historia in Mexico City, and is ascribed to West Mexico although it does not appear in the RMG collections. The needle, whose provenience is less secure because it is made from a low-arsenic copper-arsenic alloy, is a loop eye design.

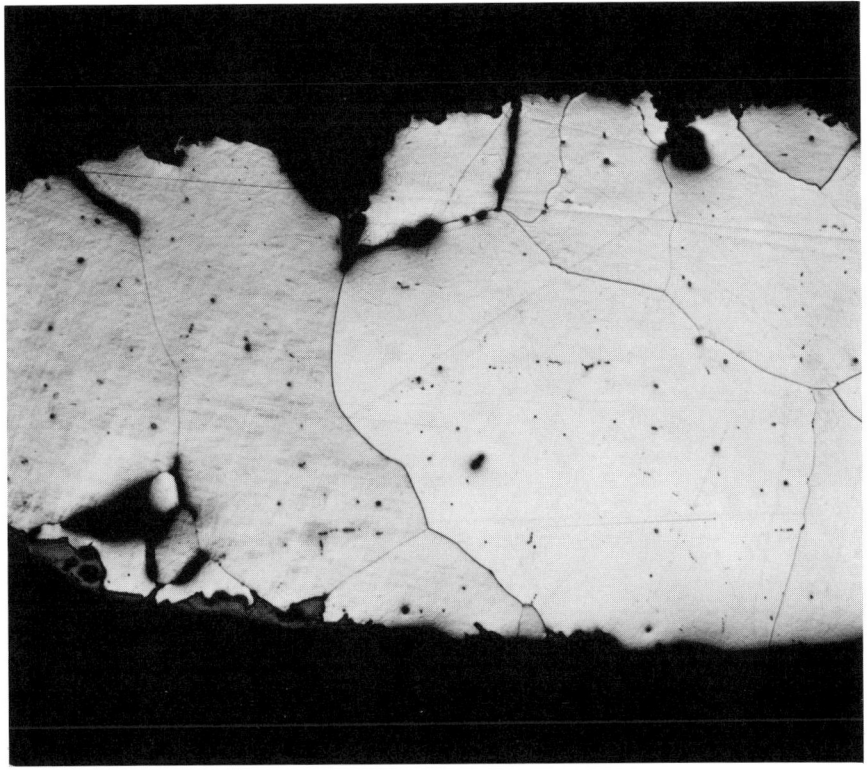

7.8

Longitudinal section from Lamanai filigree button illustrated in figure 7.7. Note extremely large grains resulting from very slow cooling in the mold. A hot mold and slow rate of cooling assisted in the retention of fine surface detail in the casting. Sample etched in potassium dichromate (mag.: 50).

The Lamanai data, taken as a whole, are striking in demonstrating that West Mexican tin bronze objects were moved so far and so early (by about A.D. 1200) from their manufacturing regions. Artifacts representing a local southeastern copper-based casting tradition, exemplified by the button in figure 7.7, also appear at Lamanai by A.D. 1150, although we do not know if these objects actually were made at Lamanai. By the Historic Period, unambiguous evidence exists for local manufacture in the form of cast ingots that seem to be made from recycled metal. Pendergast argues that local metalworking at Lamanai preceded the arrival of the Spaniards (Pendergast 1991).

7.9

Ingot from Lamanai, top and side views. Historic Period (A.D.1544–1641+).

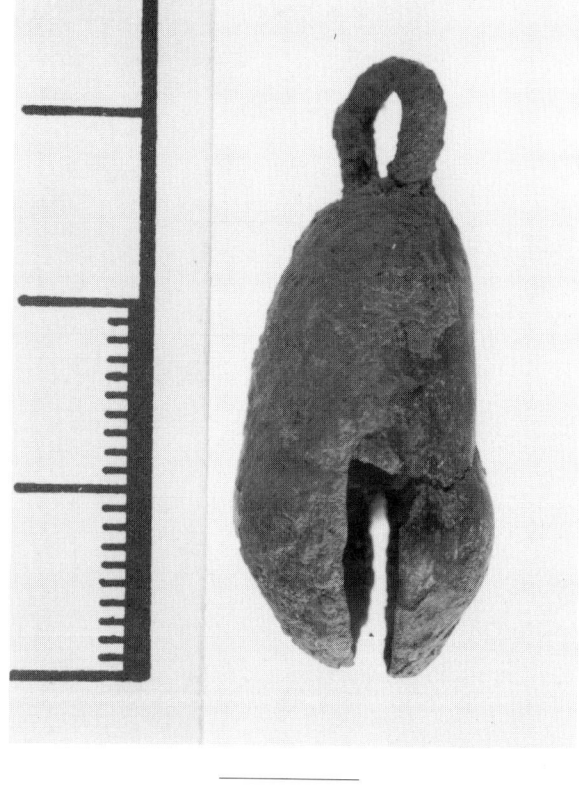

7.10

Bell from Lamanai. Historic Period (A.D.1544–1641+).

The Huastec Region: Vista Hermosa and Platanito

The Huastec region of eastern Mexico exhibits intriguing evidence for metal production in the very Late Postclassic Period. Here 118 metal artifacts (101 of which are bells), two samples of intermediate processing metallurgical material, and an ingot have been recovered from the large Huastec centers of Platanito and Vista Hermosa (Hosler and Stresser-Péan 1992).[5] Both were occupied during the 100-year period prior to the Spanish invasion in 1519. Platanito had been extensively looted at the time it was surveyed and only nine of the 87 artifacts recovered there come from excavated contexts. Others were purchased. The Vista Hermosa assemblage comes from test pits,

7.11

Ingot from Vista Hermosa.

7.12

Partially processed metallurgical material from Vista Hermosa. Late Postclassic Period.

burials, and the backdirt of looted burials (Hosler and Stresser-Péan 1992).

The most convincing evidence for *in situ* metallurgy is at Vista Hermosa, where archaeologists found processing material. It consists of two samples of partially processed metal and one ingot (figures 7.11, 7.12). Chemical analytical studies show that all three samples are copper-arsenic-tin alloys. These items make clear that the alloy was produced locally (Hosler and Stresser-Péan 1992). Cassiterite and other ores were easily accessible since the Huastec region lies along the eastern edge of the Zacatecas tin province (see figure 7.1); thus such alloys could have been produced in the area (Hosler and Stresser-Péan 1992).[6] Huastec metallurgy was virtually unknown until these large assemblages were recovered.

The bells from these sites can be divided into the five distinct designs illustrated in figure 7.13. Evidence that Huastec smiths themselves fashioned these objects is far less secure than evidence for Huastec smelting operations. Types 3 and 4 are found in Guerrero and may have been imported. Only one bell type (type 5) is unique to the region; type 1 appears in the RMG collection, and type 2 is known from other West Mexican assemblages. Other artifacts include bell clappers, beads, and metal sheet.

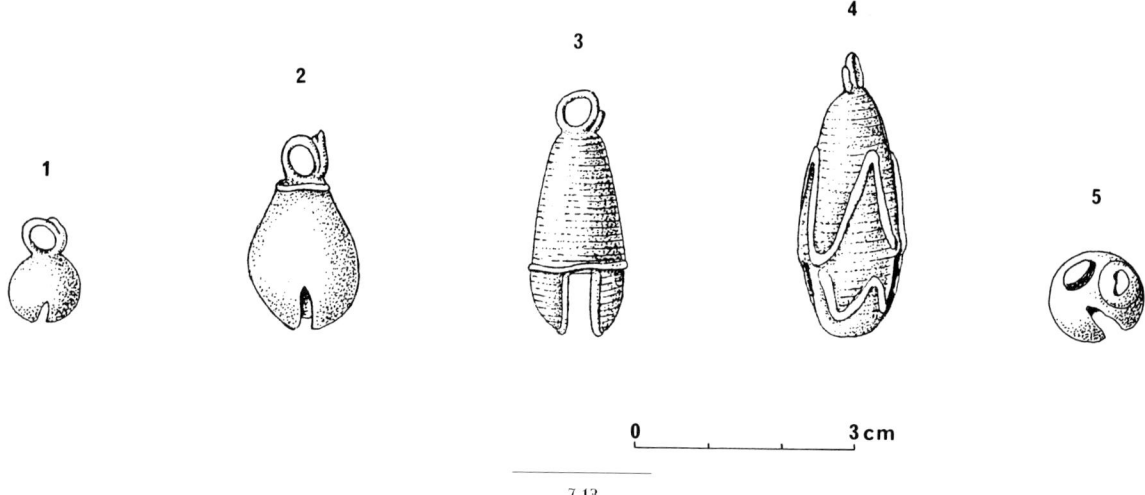

7.13 Five bell designs found in the Huastec region. Late Postclassic Period. (From Hosler and Stresser-Péan 1992.)

Tools are uncommon, although a few examples of axes and needles were found.

Forty-five bells analyzed quantitatively were found to be made of copper and alloys of copper-tin, copper-arsenic, and copper-arsenic-tin (Hosler and Stresser-Péan 1992). The copper bells are plain-walled (types 1 and 2), and the wirework bells are made from copper alloys (types 3 and 4). The copper bells were fashioned locally. One group, from a burial at Vista Hermosa, contains bells that are small, fragile, and frequently miscast. They correspond to type 1a in the RMG collection. The other plain-walled design is also made from copper but was purchased at Platanito from individuals who had informally excavated the site. These bells also are sometimes miscast; that, coupled with the fact that they occasionally retain their casting sprues, provides solid evidence for local production.

The two wirework designs are made from alloys of copper-tin, copper-arsenic, and copper-arsenic-tin.[7] One fragmentary arsenic bronze bell, which could not be classified, contained a charcoal nucleus and was covered with charcoal, suggesting *in situ* casting. However, since both wirework bell designs appear in Guerrero, some or all of them may have been imported from there. Unfortunately no Guerrero examples have been analyzed for chemical composition. Nonetheless, at least one of these wirework bells, a type 3 specimen, may have been cast in the Huastec region. The composition of that bell, an alloy of copper, arsenic, and tin,[8] is essentially the same as that of the pieces of intermediate smelted material.

Apart from producing the ternary alloy and copper-arsenic alloy bells, the possibility is good that Huastec smiths also produced copper-tin alloy metal. Eight bell clappers were recovered from the same Vista Hermosa burial that contained the ingot; one was analyzed and proved to be a copper-tin bronze. In addition, the type 5 bells, unknown in other regions of Mesoamerica, are made from tin bronze alloys.

Thus these data from Platanito and Vista Hermosa indisputably support Huastec production of copper objects, copper-arsenic-tin alloy metal, and at least one copper-arsenic alloy bell. The probability is high that copper-tin alloys were also produced locally and that at least one bell design (5) was cast locally from this bronze. Some of these raw materials may also have been exported as ingots from the Huastec region to other areas of Mesoamerica. At the same time, the fact that the bronze objects found in these Huastec assemblages are *not* the RMG artifact types commonly found in other regions of Mesoamerica (type 10b wirework bells) strengthens the argument that the type 10b bell design was manufactured only in West Mexico. What we still do not know is whether the wirework bell types found at Platanito and Vista Hermosa were fashioned by Huastec smiths; these bells (and technical know-how concerning alloying) may have been introduced from Guerrero through Aztec market systems.

The Huastec assemblages are striking for their large numbers of bells. Ethnographic studies show that Huastec people still wear metal bells in traditional dances. Around the calf they fasten tiny, high-pitched bells that sound when the dancer stamps on the ground. Larger, lower-pitched bells are fastened at the lower part of the back, and the dancer makes them sound with a lateral motion of the hips. Until recently, local people sometimes collected these bells from archaeological sites.[9] Although Huastec peoples seem to have been as interested in metallic sound as their West Mexican neighbors, the bells, whether produced locally or imorted, do not contain arsenic or tin in high enough concentrations to alter bell color. In the bells studied and reported here, tin concentration never exceeds 4%.

OTHER SITES AND REGIONS

Artifacts belonging to some of the compositional groups identified at the sites just discussed have been found in southern and southeastern Mesoamerica. They include objects made from tin bronze and copper-lead alloys, or objects that contain tin in concentrations suggesting they were manufactured from recycled tin bronze artifacts. There are never large numbers of them; often only one or two are found at any one site. The most common artifact class is wirework bells—either the cylindrical type 10b wirework bell made from tin bronze, or wirework bells cast from copper-lead alloys typical of the Valley of Mexico. They sometimes appear together at the same sites. Archaeologists also have recovered tin and arsenic bronze tools with design characteristics similar to West Mexican varieties. Very occasionally, metalsmiths employed the tin bronze alloys for designs that are clearly local, indicating that ingots as well as finished artifacts were transported from West Mexican metal-producing zones. Figure 7.1 locates sites where such objects have been reported.

Southern Mesoamerica: The Lowlands. Most artifacts from southern Mesoamerica have been found at sites in the lowland coastal zones. The Cenote de Sacrificios in the Yucatán Peninsula (see figure 7.1), an ancient well outside the site of Chichén Itzá, has yielded one of the largest assemblages of metal artifacts in Mesoamerica (Coggins 1984; Lothrop 1952). These objects were thrown into the well as offerings. The most common Cenote artifact type is bells. Some are copper-gold alloys fashioned in lower Central America and Oaxaca (Root, in Lothrop 1952). However, among the 82 Cenote bells made principally from copper, eight contain tin in concentrations above 0.75%; three of these are typical West Mexican alloys. One bell was probably imported directly

7.14

Cenote de Sacrificios: jaguar-faced Maya bell. (From Lothrop 1952, figure 84i.)

from West Mexico; it is a classic West Mexican type 10b wirework design. Another is a Maya jaguar-faced bell (figure 7.14). The design is typical of bells found in this region (Lothrop 1952, figures 83a and 84i). The bell must have been cast locally from imported materials.

Other Cenote bells contain a suite of elements, in addition to tin, that are present in concentrations around 1%, such as gold, arsenic, and silver (Root, in Lothrop 1952). Such alloys resemble the stock metal used in the Lamanai ingots and, like them, may have been made by melting down tin bronze and other objects. A few bells were alloys of copper, arsenic, and lead (like the bells from Cuexcomate and Tamtok),[10] with both alloying elements present at levels above 2%. These were probably imports from Central Mexico. Ten others are made from the more typical Central Mexican copper-lead alloy; in these, lead occurs in concentrations between 4% and 21.76% (Root, in Lothrop 1952).

Copper-tin bronze wirework bells imported from West Mexico also have been excavated at other sites in this southeastern region. At Tamulté Las Sabanas, Tabasco, investigators (Berlin 1956; Root 1962) recovered two wirework bells: one of tin bronze, and the other made from a copper-lead alloy. Wirework bells are also reported (Berlin 1956: 146) from Madero and Tenosique, also in Tabasco, but have not been analyzed. At Mayapán, in Yucatán, objects belonging to the southeastern casting tradition—rings with anthropomorphic or zoomorphic heads—contain tin and lead in concentrations between 0.1% and 1% (Root 1962). They have been made from reused metal. One piece of sheet metal, a copper-tin alloy, may be imported. Two type b shell design tweezers from Belize—one from Nohmul and another from Santa Rita (Gann 1918)—have not been analyzed (Bray 1977; Gann and Gann 1939), but their counterparts from West Mexico and from Lamanai are made from tin bronze alloys, a requirement of the design when the metal is made very thin. One small tin bronze bell has been reported from Huy, Yucatán, and another from the Quemistlan "bell cave," near the Río Chamelecón in Honduras (Root, in Lothrop 1952);[11] the Honduran example is from a cave in which a large cache of bells was recovered. Some of these were analyzed; their chemistries indicate they were made from reused metal. Three, none of which are West Mexican types, contain tin at levels too high to have constituted a trace element in chalcopyrite ore but too low to represent an intentional alloy.

Southern Mesoamerica: The Highlands. Few tin bronze objects have been recovered from the southern highlands. The most significant assemblage is from Chiapa de Corzo, Chiapas (see figure 7.1). There, West Mexican wirework bells (type 10b) made from tin bronze have been recovered from the same depositional context as wirework bells made from the Central Mexican copper-lead alloy. In both cases the alloying element, tin or lead, is present in concentrations higher than 2% (Root 1969). Tools made from tin bronze—awls and axes—were also excavated at Chiapa de Corzo (Root 1969). At Chipal in the Guatemalan highlands, archaeologists recovered a Maya-style ring and an object described as a bead (Butler 1959), and analyses show them to be made from tin bronze with tin and copper present in "quantity," along with traces of other elements. Also, a fragment of metal sheet from Chipal contained 0.40% tin (Root, in Lothrop 1952). However, in metal assemblages analyzed at other Late Postclassic archaeological sites in the southern highlands—at Zaculeu (Woodbury and Trik 1953), Zacualpa (Wauchope 1948), and Tajumulco (Dutton and Hobbs 1943), for example—no objects have been recovered made from copper-tin or copper-lead alloys.

Oaxaca. Tin bronze artifacts have been found in Oaxaca (see figure 7.1), although relatively few chemical analyses have been performed on Oaxacan assemblages. The site of Lidchi Bigu near Juchitán has yielded small tools made from tin bronze alloys (Delgado 1965; Lowe 1959; Root 1969). A tin bronze tweezer (type a), a copper-lead wirework bell, and a copper-tin axe from Oaxaca were analyzed by Arsandaux and Rivet (1923) but lack specific provenience and temporal association. Root's studies (in Lothrop 1952) showed that tin bronze was used for a wirework bell from Tlacolula, a piece of sheet metal from Nochixtlán, and a tin bronze filigree mask from Oaxaca. More recently, 30 axes from Huayapán, Oaxaca, have undergone qualitative analysis; six proved to be tin bronzes, and the remainder are made from copper (Macías 1990).

Various West Mexican artifacts have been reported from Oaxaca but have not been analyzed, including the wirework bells Caso excavated at Monte Albán (Caso 1965). These same bell designs appear in the Frissel Museum in Oaxaca. Macroscopic examination of the Frissel objects indicate that at least some artifacts, particularly shell design tweezers, are made from copper-tin bronze.[12] No copper-arsenic alloy artifacts have been identified from Oaxaca thus far.

The Northern Regions. Few chemical analyses have been performed on metal artifacts found to the north of the metalworking zone, in Sinaloa, Durango, Zacatecas, and farther north in the southwest United States. However, no tin or arsenic bronze alloys appear among Root's analyses (in Lothrop 1952) of objects from Guasave and Culiacán in Sinaloa, and from Casas Grandes in Chihuahua.[13] Although we have no analytic data, no typically West Mexican artifact designs appear in assemblages from Navacoyan and Topia in Durango. West Mexican artifacts such as wirework bells occasionally have been reported from this area, for example at Molino (Durango) where Kelley (1986) describes Tarascan-style bells,[14] and from Casas Grandes and Hilltop House. One wirework bell is reported from each of these two sites. In general, the metal objects found thus far in the north seem not to be types imported from the west but probably are products of a local metalworking tradition.

Summary. Certain objects and materials characteristic of Period 2 metallurgy were exported to other Mesoamerican regions, including tin bronze artifacts and ingots made from tin or tin bronze. The objects tend to be elite and ritual items such as type 10b wirework bells and shell design tweezers. They are identical in design and composition to their West Mexican counterparts. Tools were sometimes moved as well. In some cases, smiths outside of West Mexico used tin or tin bronze metal to craft local designs. In addition, at sites such as Cuexcomate, Lamanai, the Cenote, and the Quemistlan "bell cave," objects were made locally from metal whose composition is so highly anomalous—containing tin, lead, arsenic, and other elements—that it very likely originated from recycled artifacts, and some of these had to have been West Mexican tin bronzes. Copper-arsenic alloy objects found outside the West Mexican metalworking zone raise more complex problems, because the parent materials are so widespread. For now, the West Mexican region seems to be the only area where significant numbers of objects were made from these alloys. In addition to making the raw materials and artifacts available outside of the metalworking zone, the west also stimulated the development of a local bronze metallurgy in the Huastec area a few decades prior to the Spanish invasion.

I have considered three large metal assemblages here: Cuexcomate and Capilco in Morelos, Lamanai in Belize, and Platanito and Vista Hermosa in the Huastec region. In Morelos and Belize, metal artifacts appear that are identical in design and chemistry to types found in the RMG assemblage. These were transported by West Mexican metalworkers or intermediaries to those areas. There is no evidence for local bronze-working technologies. At Platanito and Vista Hermosa the evidence is equivocal. The presence of copper-arsenic-tin alloy intermediate processing material *and* wirework bells made from copper-tin, copper-arsenic-tin, and copper-arsenic alloys suggests that these bells, which *do not* appear in the RMG collection but do appear at sites in Guerrero, may have been fashioned in the Huastec area. However, they could also have been trade items from Guerrero, and until we have analytical data from their Guerrero counterparts, it is impossible to say where they were manufactured. The Huastec data do indicate, however, that at least one copper-arsenic alloy bell was cast locally, and others probably were as well. What is significant is that a local Huastec metallurgy was taking shape during the century prior to the Spanish invasion along the eastern edge of the Zacatecas tin province and that this technology, like the metallurgy in the west, focused chiefly on bells.

Apart from these examples, West Mexican artifacts also appear in Tabasco, Chiapas, the Yucatán Peninsula, Oaxaca, and elsewhere. Bronze tools found in known metalworking regions such as Oaxaca were either imported from West Mexico or crafted locally from imported raw materials. The copper-lead alloy wirework bells found outside West Mexico and recovered at some of the sites mentioned here were apparently manufactured near to the Basin of Mexico. These were also exported to other regions of Mesoamerica and are found at some of the same sites where tin bronze objects appear.

The best evidence that raw materials (tin or tin bronze ingots) were moved to these regions outside West Mexico are those few indisputably Maya or southeast Mesoamerican-style objects made from tin bronze alloys. The plates or sheets that Diego de Landa mentions describe one form in which these raw materials circulated. We may even have recovered fragments of such plates or sheet; at least one piece of tin bronze sheet occurs in Oaxaca, and one worked fragment was also recovered at Lamanai. These "ingots" in theory could have been produced anywhere in the Zacatecas tin province, including the Huastec area far from the metalworking zone. In contrast to tin, lead sources are abundant in Mesoamerica, but the evidence so far indicates that copper-lead alloy objects, specifically wirework bells, were cast in or near the Valley of Mexico. High concentrations of lead appear only in bells, and the highest concentrations (more than 4%) only in bells of the wirework type.

Discussion

We can now consider how these elements of West Mexican metallurgy were moved to the settlements and regions discussed here. Documentary sources and other evidence suggest that tin bronze artifacts, and possibly ingots, were distributed through both Aztec and Tarascan tribute and marketing systems and perhaps also through groups such as the Otomí and Matlazinca who may have operated as middlemen.

Who controlled exploitation of the tin sources? The cassiterite deposit in Sierra de Tlatlaya, at Sultepec (see figure 7.1), provides the best possibility for a source close to major manufacturing regions, and the documentary evidence cited earlier indicates that it was a mining center in the prehispanic era. At the time of the Spanish invasion, Berdan et al. (in press) place Sultepec in the Aztec strategic province of Temascaltepec. Temascaltepec was ruled by local nobility loosely overseen by the Aztec. The area as a whole contained a number of polities, several of which the Aztec had conquered repeatedly. Sultepec itself paid imperial tribute to Tenochtitlan. The region was populated by several linguistic groups, including speakers of Nahuatl, the major language in the region, Matlazinca, and Mazahua. Several towns in that area, one of which was Sultepec, were at war with the Tarascan empire (Gorenstein 1985).[15] Given these complex political and social arrangements, if this particular region served as a source for cassiterite or tin, the metal was probably distributed through middlemen to the producers, and the Tarascans were the most significant of these. Matlazinca groups in the Valley of Toluca may have been involved in these transactions because of their proximity to the Sultepec cassiterite deposits.[16] Whether or not the Matlazinca were involved in production on any significant scale, the chances are good that they did engage in some aspect of distribution. Independent evidence exists for trade across the Aztec-Tarascan frontier (Berdan et al. in press; Gorenstein 1985).

A second possible source for tin is the deposit at Teloloapan shown on the UNAM map. This area, like that around Sultepec, lay on the Tarascan frontier, a zone of active conflict between Aztec and Tarascan peoples following the Aztec conquest in 1442. It was also a region of shifting economic and political alliances, populated by speakers of Nahuatl, Chontal, and Malme (Malme may actually be the same language as or similar to Matlazinca) all of whom had been conquered by a succession of Aztec rulers. The Aztec maintained a garrison at Oxtuma, a large settlement in this zone. Based on the presence of Tarascan sherds at Teloloapan, Michael Smith (1990: 159) suggests that this region was also linked to the Tarascan empire. We know that this was a zone where metal objects were circulating and may have been produced, because at the time of the Spanish invasion copper axes were listed as tribute items to the Mexica from Tepecoacuilco, the tribute province in which Teloloapan is located (Berdan et al. in press; M. E. Smith 1990). Also, metal artifacts attributed to this region are housed in the Museo Nacional de Antropología e Historia in Mexico City.

Thus, neither of the two tin sources closest to the Tarascan state was under direct Tarascan control before or after the Aztec imperial expansion in 1440. Tarascan smiths either obtained tin from more distant sources or acquired it through intermediaries. Even in regions dominated by the Tarascans, in the border areas, more than one ethnic group lived in a single settlement (Gorenstein 1985). At Acámbaro, for example, Otomí, Chichimec, and Tarascan peoples lived, and all three languages were spoken. Pollard suggests that Otomí speakers may have mediated Aztec and Tarascan political and economic relations.[17] We know that Otomí groups were not only incorporated into the Tarascan domain but, according to Pollard (1993), were trusted by the Tarascans. The Matlazinca may also have played a key role in managing these relations, especially with respect to Tarascan access to Sultepec tin deposits. Some evidence suggests that Tarascans themselves were present at Sultepec; the documents indicate that both Tarascan and Nahuatl speakers lived there (Gorenstein 1985).

Assuming that cassiterite was mined at Sultepec or Teloloapan, tin was then moved from these deposits, probably through regional markets operating within the multiethnic communities along the Aztec-Tarascan border. Its destination was those areas, under Tarascan control, where metal objects were manufactured. The Tarascan tin bronze objects I have identified in other regions of Mesoamerica were probably then transported through intermediaries who participated in these same networks, since little evidence exists that Tarascan merchants operated outside of Tarascan borders. However, some tin bronze artifacts were also fashioned in regions of Guerrero under Aztec control, and these goods moved directly into Aztec market systems. The Codex Mendoza, for example, mentions that 40 "cascabeles grandes de latón" [large, *latón* bells] constituted tribute items to Tenochtitlan from the province of Quiauteopan, in Guerrero. *Latón* in this context must mean copper-tin bronze. The objects illustrated in the Codex Mendoza are large, cylindrical, type 10b wirework bells, which are common in the regions under Tarascan control and also are found at various sites in Guerrero. In this case they comprised items of imperial tribute to Tenochtitlan, and they may have been subsequently passed on to Pochteca traders who moved them to the south and to the east. They also may have circulated through the regional market systems that handled ceramics, obsidian, and other Aztec goods (M. E. Smith 1990).

We do not know how metallurgy was introduced to the Huastec region from West Mexican metalworking zones. The most likely possibility is that the technology was transmitted through the merchants responsible for distributing Aztec trade goods rather than through direct contact with the key Period 2 metal-producing areas of West Mexico, such as highland Michoacán. The archaeological evidence suggests that during this period contact between the Huastec region and these West Mexican zones was limited. However, we do know that relations existed between the Huastec peoples and the Aztec settlements in the Basin of Mexico. Huastec pottery has been reported at Aztec sites there, as well as to the south in the neighboring state of Morelos. We also know that Platanito was located in an Aztec tribute province. Furthermore, tin bronze bells were tribute items to Tenochtitlan from Guerrero, and some of those bells were probably fashioned in Guerrero. Given that two of the primary Huastec bell designs (types 3 and 4) also appear in Guerrero, and the late dates for Huastec metallurgy, the likelihood is high that the technology was introduced to the Huastec area through Aztec marketing networks.

The wide geographical distribution of tin bronze artifacts suggests that these artifacts, some of which were fashioned in regions of West Mexico not controlled by the Aztec, were nonetheless circulating in Aztec networks during the Middle and Late Postclassic periods. The fact that copper-lead alloy bells, made in or near the Basin of Mexico, associate with West Mexican tin bronze bells—at sites such as Chiapa de Corzo, Tamulté, and the Cenote—provides solid evidence for this assertion. However, the presence of objects found in these distant regions and that date prior to the Aztec conquests cannot be explained at the present time.

From A.D. 1200 to the time of the Spanish invasion, metal objects and materials were moved from West Mexico to other regions of Mesoamerica. In the southeastern zone—the Yucatán, Belize, Honduras, Guatemala—they appear at about the same time that local metalworkers elaborated their lost-wax casting technology, a technique that moved northward from lower Central America following its introduction by sea to West Mexico during Period 1. The West Mexican contribution to these assemblages was objects made from copper-tin bronze. Most were status objects—bells and tweezers—for which use of the alloy was a requirement of the design.

Although the data are still partial, the regional metallurgies explored here generally seem to have been shaped by the same principles and choices as those that gave rise to the metallurgy that developed in West Mexico. Metal was a ritual and elite material, and it was its sound, and to a lesser extent its color, that also interested these non–West Mexican peoples. The closing chapter will examine the social propositions that underlay these choices, and that give this West Mexican technology its distinctive and original character.

8

The Sounds and Colors of Power

Había un Dios Principal que estaba en el cielo y lo había criado todo, y que ha de haber jucio final; y que el mundo tuvo principio, y que hizo Dios un hombre y una mujer de barro, y q[ue] se fueron a bañar y se deshicieron en el agua; y que los volvió a hacer, de ceniza y [de] ciertos metales, y los envió al río a bañar, y que no se deshicieron; y q[ue], de aquéllos, empezó el mundo. [There was a principal god who lived in heaven and had created everything, and there was to be a final judgment; and the world began and god created a man and a woman of clay and they went to bathe and they came apart in the water; and the god made them over again from ash and certain metals and sent them to the river to bathe and they did not come apart and it was from those two that the world began.] (Acuña 1987: 36)

This account of human origins appears in the *Relación Geográfica* from Ajuchitlán, to answer a question the Spaniards posed concerning the beliefs of the native peoples.[1] The document reports that the deity created the original human beings from "ceniza y ciertos metales" [ash and particular metals]. The narrative suggests that metal was sacred and animate: an optimal material for the first human couple. I argue here that the metallurgical technology that coalesced in West Mexico was shaped by technical decisions based on this premise, manifest in the properties of metal that smiths chose to elaborate and in the ways metal objects were used. Most were worn or used in ritual: bells, hair ornaments, sheet metal disks and diadems, and large ornamental tweezers. These metal objects identified the individuals who wore them with supernatural forces through their form and through two key material properties: sound and color. Through sound and through color these objects created a sacred domain of experience in which priests and other religious functionaries could enact in ritual basic societal propositions, contributing structure and meaning to the lives of these ancient peoples.

Sacred power inhered in many materials and substances known in ancient Mesoamerica. Clay was formed into ritual censers; cloth was woven into religious regalia. Jade, amethyst, and obsidian were chipped, abraded, or ground into small objects representing divine and sacred beings; artisans used basalt and other rock for monumental sculptures representing humans emerging from the mouths of jaguars or snakes, and events such as the death of Coyolxauhqui (figure 8.1), the goddess of the moon and sister of the great Aztec god of war Huitzilopochtli. Nonetheless, the proportion of metal objects made to serve ritual, symbolic, and status ends is unusually high when compared, for example, to the metallurgies of Ecuador, Peru, and Bolivia, where bronze alloys also were widely available and used for utilitarian ends. In West Mexico and elsewhere in Mesoamerica, metal, including the characteristically utilitarian bronze alloys, was endowed with exceptional and sacred qualities.

Both of the properties that shaped the technology, sound and color, were developed in bells. Bell sounds and metallic colors themselves are linked linguistically at least in Nahuatl, a language widely spoken in Central Mexico and in some areas of the metalworking zone. As I have shown, bells were fashioned elsewhere in the ancient Americas, but only in West Mexico did they become central, orienting elements of the technology. Metallic color, while vivid in bells, was most dramatically apparent in the shimmering, reflective, brilliantly golden and silvery colors of sheet metal, crafted into diadems, disks, and other objects. Tarascan nobility, religious functionaries, and other elites, as well as rulers and elites elsewhere in West and Central Mexico, wore these objects, usually in ritual, affirming their own sacred power and affiliation with the supernatural.

Metalsmiths devoted quite a small but technically accomplished portion of their activities to toolmaking, as

8.1

The Coyolxauhqui stone found at the Templo Mayor, Mexico City. Sculpture was placed at the landing of the stairway leading to the temple of Huitzilopochtli. (From Townsend 1992, plate 98.)

I have established, taking advantage of the hardenability and toughness of copper and the bronze alloys. West Mexican tools exhibit extensive technical knowledge and skill, but the sheer volume of brilliantly colored golden and silvery ritual objects, especially bells made from bronze and other alloys, reveals that these artisans' real interests lay in the varied pitches and myriad colors the new material provided. I explore these technical preferences in the following discussion, drawing on evidence from a variety of sources. The most direct information comes from sixteenth-century Mesoamerican native and Spanish documents, from terms related to metal and sounding instruments in Nahuatl and Tarascan, and from ethnographic studies of indigenous Mexican peoples. Among the sixteenth-century documents, the Florentine Codex from Central Mexico (Aztec) is particularly helpful, since the Aztecs occupied areas contiguous with the metalworking zone, and during Period 2 controlled parts of it. Information from South America is sometimes useful as well: various scholars have argued that indigenous American societies share certain ideological precepts (see, for example, Willey 1962; Lathrap 1982). If so, those precepts or fragments of them still may be manifest materially or in other forms among indigenous American societies.

Color

Golden and silvery metallic colors were associated with deities in many regions of the Americas, including Mesoamerica. These metals were identified respectively with the sun and moon by Tarascans, Aztecs, and probably other Mesoamerican groups. In Nahuatl, the word for gold, *cuztic teocuitlatl*, literally means "yellow divine excretions" and is taken to mean excretions of a solar deity (Campbell 1985).[2] According to the Florentine Codex, that deity was Tonatiuh. The name Tonatiuh comes from the verbal root *tona*, which means to shimmer, to shine, or to give off heat; when the Florentine Codex describes the visual effect of gold, it uses *tona*, translating it to mean "to give off rays" (Sahagún 1950–1982 book 11: 234). *Tona* also means for the sun to shine; thus gold shines, shimmers, and gives off rays, or heat, like the sun or like Tonatiuh. Metal conducts heat and it is also highly reflective, and here both qualities are explicitly recognized. The Florentine Codex is unambiguous about the origins of gold: the name "derives from [the fact that] sometimes, in some places, there appears in the dawn something like a little bit of diarrhea.... It was very yellow, very wonderful, resting like an ember, like molten gold" (Sahagún 1950–1982 book 11: 233).

Similarly, the Nahuatl word for silver, *iztac teocuitlatl*, signifies "white divine excretions" (Campbell 1985), and could mean excretions of the moon.[3] The terms make clear that these metals were divine substances, produced, smelted, or emitted by the deities.

The Tarascans impute similar origins to gold and silver. In the *Relación de Michoacán*, Hiripan, one of three brothers responsible for preconquest Tarascan expansion, wanted to recover gold, silver, and other items from conquered villages.

"Hermanos, ¿qué haremos? que la gente de los pueblos se llevan huyendo los plumajes y joyas, con lo que fueron señores en los pueblos que conquistamos. . . . Id a retenellos, que se vengan los dioses a sus pueblos." [What shall we do, brothers? The people of the villages run away and carry off the plumes and jewels that made them nobles in the villages we have conquered. Go get them so the gods [the gold and silver objects] come back to their villages.]

The *Relación* continues,

Viendo aquel oro amarillo y la plata blanca dijo Hiripan, "Mirá, hermanos, que esto amarillo debe ser estiércol del sol que echa de si; y aquel metal blanco estiércol de la luna, que echa de si. . . ." [Seeing that yellow gold and white silver, Hiripan said, "Look brothers, this yellow metal must be manure which the sun casts off, and that white metal must be manure cast off by the moon. ,. ."] (Tudela 1977: 152)

These two metals, gold and silver, were divine emissions and divine property; they also validated the authority of the nobility. The root *tiripeti* means gold in Tarascan, and *Tiripeti* was also the name given to gods that were individual manifestations of the sun (Brand 1951). In fact, Tarascan state treasurers stored gold diadems and disks in chests to honor the sun, and disks of silver in honor of Xaratanga the goddess of the moon (Tudela 1977: 257). Tarascan kings were interred with gold shields at their back and golden bells on their ankles (Tudela 1977: 219); in battle they were protected by a shield of silver (Tudela 1977: 192). Some of these objects were made from the pure metals. Others were apparently made from copper-silver and, probably, gold-silver alloys, judging from sixteenth-century accounts of the variable quality, or assay, of the gold and silver objects that the Spaniards demanded and received from the Tarascan ruler.

In South America, metalsmiths produced gold and silver colors by using those metals or alloying them with copper, producing golden and silvery surfaces using sophisticated depletion gilding techniques. West Mexican smiths devised ingenious new methods to produce these same colors in objects whose design characteristics disallowed the use of pure gold or silver. They developed high-arsenic copper-arsenic alloys (as well as the more common copper-silver alloys) for a silver color; and for gold, high-tin copper-tin bronzes. West Mexican smiths also fashioned some objects from pure gold and silver, when design requirements permitted it.

In the central Andes, golden and silvery colors were likewise identified with the solar and lunar deities, and by the Late Horizon (A.D. 1476 to A.D. 1532) objects made from these metals were worn and used exclusively by the Inka. The myth reported by Calancha from the central coast of Peru relates that the *caciques* (lords) had descended from a golden egg and their wives from a silver egg. The egg responsible for commoners was made from copper.[4] Andean smiths nonetheless achieved these metallic colors in ritual objects with alloys of copper-silver and copper-silver-gold, although they also sometimes used the pure metals.

Thus, throughout the ancient Americas, one way metalsmiths created golden and silvery colors was by alloying. We do not know how the process itself may have been interpreted by Mesoamerican peoples. The indigenous copper-based technology has disappeared and the ideology that supported it is fragmentary. Perhaps the only published ethnographic information that may help explain the meaning of alloying gold, silver, and copper comes from the Desana, a Tukanoan people in the northwest Amazon. Given broadly shared attitudes toward metal in the Americas, these data could apply to Mesoamerica as well; insights will also come through studies of indigenous terms for metals processing.

The Desana live in a region where metal was used extensively in the prehispanic era. The Desana wore metal ear ornaments, nose rings, and pendants at one time, which were probably imported from the highlands. A Desana myth that speaks of two metals, according to Reichel-Dolmatoff (1981), who has interpreted it, alludes to a "sun which fertilizes a brilliant New Moon which . . . then passes through a sequence of yellowish, reddish, and copper-colored phases which are compared to . . . the process of embryonic development." He argues that Desana concepts of generation and growth are "a model of metallurgical combinations," that is, of alloying (Reichel-Dolmatoff 1981: 21).

Metallurgical combinations, or alloys, thus represent transformations; in fact, they do transform through surface enrichment procedures in which the coppery color of the initial alloy is changed to silver or to gold. Lechtman (1993) suggests that the Desana account of alloying may perpetuate a deep-rooted highland and coastal Andean tradition of alloying copper with silver and gold to produce *tumbaga*. Ritual objects made from these alloys and the pure metals eventually communicated the continuity of the Inka royal lineage, and therefore of the state.

We have no evidence for whether West Mexican metalsmiths conceived alloying in this way; even from the Aztec only tangential data are available. Those data are nonetheless worth noting. Xipe Totec, the Aztec god of goldsmiths, also represented fecundity and renewal. In fact his name, Xipe, could derive from *xipintli*, foreskin, or *xipehua*, which means to scrape or to peel, or, as a nominal (*tepul quaxipeuhcatl*), the head of the penis (Campbell 1985: 411). It is also related to *xiptli* or layer. His name means "he who wears the skin."[5] His insignia was the rattlestick, a pointed rattling implement often made from metal, used in rites and connected to human and agricultural fertility. Metal or metalsmithing (putting on layers of gold, or possibly removing layers of copper, through depletion gilding) and reproduction and agricultural fertility (layers of new vegetation) thus may be explicitly linked.

The creator in the *Relación Geográfica* from Ajuchitlán made people from ash and metal. However, the deity apparently did not fashion the first human couple from pure metal. The god created the first pair from an alloy (*ciertos metales*). Artifact chemical analyses show that the alloying elements, silver, tin, and arsenic, were often added to copper in concentrations high enough so that ritual and status objects displayed a range of silvery and golden "divine" colors—or layers, in the case of sheet metal made from copper-silver alloys. If alloying—or metalsmithing—in Mesoamerica stood for generation, fertility, and growth, then in creating people, especially from alloys, the act of creation for the deity was the act of sexual union and fertilization. In that act, the creator not only symbolically engaged in sexual union, but fashioned two people who themselves could reproduce. The account goes on to tell us that after the creation "it was from those [two] that the world began."

In most cases, as I have shown, the properties of the West Mexican alloys were required to optimize design at the same time as they altered color: they allowed thinner, harder, and finer tools and brilliantly golden and silvery status objects. Perhaps what mattered most was alloying, mixing, and the metaphor for fertilization; the designs became gradually thinner and more streamlined as smiths experimented with and came to understand alloy properties.

Tweezers are one of the more intriguing of the golden and silvery sheet metal objects in the corpus we have been considering. Large, elegant, golden and silvery tweezers became symbols of priestly office, at least in the Tarascan state. Their design usually required an alloy because of the extreme thinness of the metal hinge and blades. The alloying element, most often tin, is present in concentrations between approximately 8 and 12 percent. Some tweezers are made from high-silver copper-silver alloys, and others probably from copper-silver-gold ternaries.[6] The tweezers' shimmering golden and silvery colors visually evoked the power of the solar and lunar deities. The computer simulation studies have also established that even the large multispiraled tweezers were designed so that they could function as depilatory tools; the objects had to be mechanically capable of performing the task they symbolized.

Ethnographic and historical evidence concerning when and how tweezers were used amplify the technical information identifying fabrication methods and materials. However, we do not know why a tool used for facial depilation became a symbol of office among Tarascan elites. *Catzicutaqua* and *matirehperaqua*, Tarascan terms for tweezers, also mean to pinch and to squeeze (Anonymous 1991 vol. 1: 668). The following verbs also relate to plucking, pulling out, and shaving hair:

vanduhcuni, muruhcuni: Arrancar o pelar los pelos de las manos [to pull out or shave hair on the hands].

vandumpzscani, murumpzscani: Arrancarles las [barbas] [to pluck or pull out someone's beard hairs].

vandumpzquareni, murumpzquaren: Pelarse las [barbas] a si mismo [to shave one's own beard hairs].

vanduqua: Tenazuelas de quitar cejas [tweezers to remove eyebrow hairs].

vanduxucuni: Arrancar las alas o pelos [debajo] de el [brazo] [to pluck or pull off wings or hair from under the arms]. (Anonymous 1991 vol. 2: 672–673)

This extensive lexicon suggests that removing body and facial hair apparently was a subject of some interest to Tarascan peoples. Tweezers were, and are, widely used for facial depilation throughout the indigenous Americas. These implements, made from metal, shell, and occasionally wood, also appear in archaeological contexts. In coastal Ecuador caches of hundreds have been recovered from burials; González (1979) has established a chronology for archaeological (metal) tweezers from northwest Argentina. Villagers in remote areas of highland Michoacán report that men regularly tweeze beard hairs with commercial or homemade tweezers.[7] In highland Peru, depilation of beard hairs also continues and is quite casual; men in public tweeze while informally conversing.[8] Numerous ethnographic descriptions refer to facial depilation elsewhere in Peru (see, for example, Farabee 1922: 58, 83).

The presence or absence of body and facial hair, its length, and its treatment are universally important, and mark gender, age, status, and other socially significant attributes. A Tarascan man plucked and shaved his body hair and facial hair (and from the sound of the verb *vandumpzscani*, he plucked other peoples' beard hairs as well). In fact, the Spaniards and the state executioner (see figure 3.16) are the only bearded individuals depicted in the entire *Relación de Michoacán* (Tudela 1977: 11). We have no information regarding how Tarascans and other West Mexican groups thought about body hair, but evidence from South America may again prove illuminating. Terence Turner (1980) reports that the Tukano of the northwest Amazon equated facial and body hair with dirt, and that dirt is considered dangerous to the health of the individual. "Health" signifies full membership in the social world, and "illness" is the presence within the social domain of natural and animal forces. Cleaning removes dirt, and thus evidence of illness, from the body. Turner argues that the same principle is embodied in removing facial and body hair, which transforms the skin from a natural envelope of the physical body into a kind of social filter. An account of a raid of the Shipibo, who live on the eastern side of the Peruvian Andes, against the neighboring Cashibo relates that the Shipibo shaved the beards and cut the hair of their Cashibo captives. Roe (1982) comments that Shipibo consider facial hair monkeylike and ugly. They turn the Cashibo into civilized people by removing the symbol of their unconscious sexuality, thereby controlling it (although Shipibo may also cut their hair to humiliate them).[9] The idea that body hair and body dirt are unconsciously equated has also been treated by Edmund Leach (1958), among others. Leach contends that body hair is universally identified with excrement, sexuality, and powerful supernatural forces. Although his argument has been widely debated (virility rather than excrement can be associated with facial and head hair), the lexical evidence and the physical objects (tweezers) make it clear that facial and other body hair was systematically plucked or shaved by the Tarascan peoples; whether this practice was aimed at making humans more "civilized" remains to be investigated. Paul Friedrich in a linguistic analysis of Tarascan does point out that "a critical line between human and animal" is one of nine Tarascan semantic fields (Friedrich 1984: 60); while such a critical line is present in all human cultures, its semantic representation is significant.

If, for the Tarascans, facial hair was tantamount to being animal-like, the use of large silvery and golden tweezers as emblems of priestly office becomes more explicable. The tweezers proclaim membership and high rank in the social world by making dramatically visible the tool that removes evidence of the nonsocial: facial hair. The tweezers' size and elegant design, and their golden and silvery colors created by alloying, identify the priests who wear them with the supernatural. By wearing tweezers, the priest embodies the power and sanctity accorded to those deities, and simultaneously affirms his own moral probity and adherence to the social code.

The other golden and silvery sheet metal objects made from tin bronze and copper-silver alloys and examined in this work also announced status and supernatural affiliations. The laboratory evidence has shown that a primary technical objective of West Mexican smiths was to produce these highly reflective surfaces. What more specifically can we say about their meaning? The Aztec paradise, termed a "cult of brilliance" (Burkhart 1992; Hill 1987), was populated by glittering and glowing objects and beings. Nahuatl devotional literature describes this sacred domain as a shimmering garden. The garden is called into being through song, by manipulating garden imagery in ritual contexts. Thus sound, or in this case song, creates luminous colors. Burkhart describes the garden:

In this symbolic garden, one came into direct contact with the creative, life-giving forces of the universe and with the timeless world of deities and ancestors. The garden is a shimmering place filled with divine fire; the light of the sun reflects from the petals of flowers and the iridescent feathers of birds; human beings—the souls of the dead or the ritually transformed living—are themselves flowers, birds, and shimmering gems. . . . This garden is not a place of reward for the righteous, existing on some transcendent plane of reality separate from the material world. It is a metaphor for life on earth, a metaphor that ritual transforms into reality by asserting that, in fact, this is the way the world is. (Burkhart 1992: 89)

Iridescent shining, shimmering, glimmering entities—objects such as feathers, bells, flutes, and flowers—fill this world and are metaphorically used to describe the beings inhabiting it.

There is no doubt that shimmering effects of metallic colors reflecting off the iridescent entities helped create this sacred garden: Nahuatl devotional texts frequently describe glowing and shining golden blossoms, golden petals, leaves, birds, feathers, and raindrops. I provide several examples below, drawn from two Nahuatl texts, *Psalmodia Christiana* and *Cantares Mexicanos*:[10]

A golden quetzal-colored dew formed drops. The golden flowers are shedding petals; they are raining down. (Sahagún 1993: 373)

Quetzal rattle-bell flowers, red solandras are outspreading like the early light of dawn; they glow like gold. (Sahagún 1993: 373)

The cacao flowers, the popcorn flowers, the red basket flowers lie waving with quetzal feather dew, lie glistening like gold. (Burkhart 1992: 97)[11]

We do not know whether a similar sacred domain also existed in Tarascan conceptions of the divine; we will not know until the Tarascan language has received the same attention accorded to Nahuatl.[12] Such a domain may indeed exist, especially in light of the vast quantities of golden and silvery sheet metal ornaments metalworkers made, and the sacred qualities ascribed to golden and silvery colors.

Sound

The power accorded to bell sounds also determined the course of West Mexican metallurgy. Bells and other rattling instruments figure prominently in ritual and ceremony throughout the indigenous Americas and, at the broadest level, the sounds and the instruments express similar concepts. Here I will explore what these instruments and sounds may have meant in the metalworking zone, drawing on the sources mentioned earlier. Most documentary information comes from the central highlands (Aztec) but can be generalized to areas of the metalworking zone controlled by the Aztec. Since Tarascan ideology shares basic principles with all Mesoamerican societies (including autosacrifice, flaying, the ball game, and heart sacrifice; Pollard 1993), we can also assume that ritual was broadly similar, as were ritual accoutrements: bells and other percussion (rattling) instruments. Not all Central Mexican deities (Quetzalcoatl and Tlaloc, for instance) were worshiped in the Tarascan region, but assorted Tarascan deities nonetheless represent aspects of them. Furthermore, the musical cultures of the Tarascan, Aztec, and Mixtec were fundamentally similar (Stanford 1966), so that the instruments and the meanings ascribed to their sounds should also resemble one another.

Bell sounds and the sounds of composite bell instruments played a major role in at least three sacred contexts. One was in ritual celebrating human and agricultural fertility and regeneration. Another was in warfare, where bell sounds could protect. The third was in the sacred paradise, created through song and sound. One of those sounds—the sound of bells—was associated with the shimmering, colorful, singing birds and with human voices that represented deities and their human transformations.

Religious functionaries in Mesoamerica used rattles and wore bells (made from ceramic) long before the introduction of metal. Instruments such as hand-held rattles and ankle rattles are very ancient. For example, Formative Period figurines belonging to the Tlatilco tradition are sometimes shown holding rattles and wearing rattles on the calf or ankle (figures 8.2 and 8.3). Figure 8.4, from the murals at Bonampak (A.D. 300 to A.D. 900), shows Classic Maya figures holding a rattle in each hand.

8.2
Tlatilco period figurine wearing ankle rattles. (From Coe 1965, plate 150.)

8.3

Tlatilco period figurine holding rattle. (From Coe 1965, plate 127.)

8.4

Maya figures holding rattles illustrated in the Bonampak murals. (From Martí and Kurath 1964, plate 20.)

Rattlesticks apparently came into use slightly later; a Quetzalcoatl figurine from Teotihuacán (Brundage 1985) holding a rattlestick is the earliest example. When metal appeared, it offered a novel and particularly resonant material for some of the same rattling instruments artisans had made previously from other materials.

The Sounds of Bells: Fertility and Regeneration. The deities most frequently associated with bells and other composite percussion instruments containing bells represent fertility, life, and regeneration. Tlaloc, Xipe Totec, and Quetzalcoatl are primary.

In this region, rain, water, storms, thunder, lightning, rattlesnakes, and new vegetation repeatedly appear as symbols of fecundity and new life. Bell sounds replicate the sounds of thunder, rain, and the rattle of the rattlesnake, and we probably can add to that the roar of the jaguar, since Hunt (1977) links thunder to it. The jaguar is the sacred progenitor feline associated with water and caves. While all bells reproduce these sounds, two West Mexican designs (and many variations on them) constitute visual metaphors for rain and for lightning as well: the goggle-eyed Tlaloc bells (see figure 3.5) and wirework bells with vertical zigzag designs (see figure 5.2). The zigzag design is a visual convention for lightning. These are the only two bell designs in the entire RMG corpus with recognizable iconographies.

Apart from bells, the percussion (rattling) instruments most frequently illustrated or described from Central Mexico and surrounding territories are the hand-held rattle (known in Central and South America as the *maraca*), the rattlestick, and the mist rattleboard. Several archaeological examples of hand-held metal rattles exist, from both Peru and West Mexico. One from West Mexico comes from Apatzingán, Michoacán (Kelly 1947). The only rattlestick-like instruments I know of come from Peru and consist of wooden sticks outfitted with three metal bands; bells are suspended from these bands. I know of no archaeological examples of mist rattleboards.

Terms for some of these instruments, at least in Nahuatl, connote fertility. The Aztec rattlestick (figure 8.5) or *chicauaztli* (Stevenson 1968), a long pointed stick containing rattles or bells, was widely employed in agricultural rites. *Chicaua,* the verbal root of the noun *chicauaztli,* means to strengthen and fortify (Karttunen 1992: 46). Seler argues that the "word and symbol obviously refer to the strengthening of the reproductive function, to fertilizing" (Sahagún 1950–1982 book 1: 40, note 118). The Codex Borgia illustrates a rattlestick above the first human pair, who appear to be covered with a common quilt, explicitly associating the instrument with human sexuality (Neumann 1974). Neumann thinks that the *chicauaztli* serves as a symbol for sexual union. The term for mist rattleboard, *ayauhchicauaztli,* also contains *chicaua,* the same verbal root.[13] Seler describes the instrument as a board with holes; bells were strung in them, and the bells sounded as the person carrying it walked (Sahagún 1950–1982 book 1: 17, note 53). Seler says

8.5

The rattlestick or *chicauaztli*. (From Broda 1971, figure 5.)

that the sounds reproduced the thunder of Tlaloc, who, according to Pasztory (1974), is a rain deity in one aspect and a deity of the earth, caves, and the underworld in another.

Tlaloc is identified with thunder, lightning, rattlesnakes, caves, and the underworld. The god's physical representations contain visual metaphors for these concepts (figure 8.6). Specifically, illustrations of Tlaloc usually show him (a) holding an object, referred to as a lightning snake (the visual image of the lightning flash

Instruments containing bells or rattles were sounded in rites for Tlaloc to attract rain and thunder. The Florentine Codex describes the behavior of the fire priest in Tlaloc's temple when the rains break out. The images graphically associate snakes (probably rattlesnakes), their rattling sounds, and rain:

And when the rain broke out, then he forthwith arose; he seized his incense ladle. . . . The incense ladle rattled. It was in the form of a serpent. And the serpent's head also rattled. . . . Then he offered incense; to the four directions he raised [the incense ladle]. Much did it rattle; [the incense] spilled out. . . . Thus he attended to the matter; thus he called upon the Tlalocs; thus he prayed for rain. (Sahagún 1950–1982 book 2: 151)

Etzalqualiztli, the sixth month of the Aztec ritual calendar, marked the end of the dry season. Priests carried and shook the *ayauhchicauaztli,* the mist rattleboard, in rites honoring Tlaloc. Four priests marched in procession,

and leading them went a man, an old priest. He went bearing on his shoulders the mist rattleboard, also called the sorcerer's staff. It was wide, very wide, excessively wide; and it was long, very long. And it rattled; he went along rattling it. (Sahagún 1950–1982 book 2: 81)

In a description of a procession to the temple of Tlaloc, the priest

put on his mist jacket, his rain mask or his Tlaloc mask. . . . Thus they went to the Temple of Tlaloc. . . . And when this was done, then he scattered *yauhtli* [an herb]. And when he had scattered it, then they gave him the mist rattleboard. He rattled, he shook it; he raised it in dedication [to the god]. (Sahagún 1950–1982 book 2: 87)

The Florentine Codex describes other deities related to Tlaloc wearing bells, rattling hand-held rattles, and carrying the *ayauhchicauaztli.* One of those was Napa Tecutli, who wore ankle bells and belonged among the

8.6

Tlaloc. (From Broda 1971, figure 1.)

descending), in his hands, and (b) wearing a face mask with a jaguar mouth and two eyes made up of two coiled serpents. According to Hunt, the eyes repeat the visual code of the image of rain lightning, which is a snake, and the jaguar mouth symbolizes thunder (Hunt 1977). Tlaloc's associations with jaguars fit with the more recent work linking him with the caves and the underworld. Sometimes Tlaloc is represented as a snake uncoiling. The water:snake association is again reflected in language: in Nahuatl, the reflexive verbal form *mo-mana* means for a snake to be coiled up or for water to be dammed up; the nominal *atl mo-mana* means puddle (Campbell 1985: 171, 172).

tlalocs. Mist rattleboards were also sounded at rites for Opochtli, an aspect of Tlaloc linked to fishermen and water (Sahagún 1950–1982 book 1: 37). Chalchiuhtli Ycue, the goddess of the waters for whom rites were also performed during the sixth month, also carried a mist rattleboard, and when the fire-priests came to receive her "with the mist rattleboard they went speaking; the elders of the *calpulli*, her singers, sang for her" (Sahagún 1950–1982 book 1: 22).

The sounds of ankle bells (as well as bells in the mist rattleboard and the *chicauaztli*) played an important part in these ceremonies. Describing the impersonator of Uixtociuatl, the elder sister of the rain gods, Sahagún relates:

And on her ankles she had placed bells, golden bells, or rattles. On the calf of her legs she had bound ocelot skins on which were the bells. And when she walked, much did she rustle, clatter, tinkle, continuously tinkle. (Sahagún 1950–1982 book 2: 92)

The final verse of the song of Tlaloc directly associates rattles and rain:

To all places go
To all places reach
To Poyauhtlan
With mist-bringing rattles [*ayauhchicauztica*]
To Tlalocan taken. (Sahagún 1950–1982 book 2: 225)

In one of Quetzalcoatl's (the feathered serpent's) myriad aspects as wind and storm deity, he represents fertility, wind or breath, and life (Brundage 1982; Hunt 1977). Quetzalcoatl becomes the wind storm that occurs just before the rainy season begins. Sahagún calls him the "roadsweeper" of the rain gods. According to Townsend (1992: 114), the best portrayal of Quetzalcoatl is an Aztec sculpture depicting a coiled serpent rising from the earth, with a Tlaloc mask on the bottom. Quetzalcoatl's twin, Xolotl, appears as a deified form of lightning (Hunt 1977: 126). As snake, Quetzalcoatl is often depicted with

8.7
Tezcatzoncatl wearing ankle bells. (From Sahagún 1950–1982 book 1, plate 21.)

prominent rattles; the rattles are sometimes shown as bells.[14] According to Sahagún, Quetzalcoatl in human form wears ankle bells. Quetzalcoatl occasionally is depicted holding the *chicauaztli*. Mayahuel, the goddess of *pulque* with 400 breasts whom Brundage (1982: 96) calls an avatar of Quetzalcoatl, holds the *chicauaztli* as well. Tezcatzoncatl, one of the rabbit gods, is also illustrated with ankle bells (figure 8.7); Hunt thinks that these rabbit gods and gods of *pulque* are also linked to fertility (Hunt 1977: 86–87).

Xipe Totec (figure 8.8), patron of the metalsmiths and god of new vegetation, is another deity invariably represented with percussion and rattling instruments. Xipe's origins are confusing, but one good possibility is that he came from the Yope-Tlapanec area of Guerrero, a zone within the metalworking area. As I mentioned earlier, Xipe's name may derive from Nahuatl terms related to foreskin, to scrape or peel, to the head of the penis (Campbell 1985), and to layer. His insignia is the *chicauaztli* (see figure 8.5). Heyden (1986: 38) and others have noted that the rattlestick, which initiates thunder to attract rain, is an analog to the phallus. The rattlestick sometimes is represented as a serpent. Xipe is the god who looses (or peels) the skin, or puts on a layer of skin; the earth that loses its skin after the harvest and gains a new one with new vegetation; or perhaps the snake that loses its skin but grows a new one. Xipe is also the god of metalworkers, and metals are something that comes from the earth and the gods: they are divine emissions, powerful substances that are lost, shed, and excreted but that may also provide a new layer, such as the silvery surface layer of copper-silver alloys.

Xipe is the principal deity of the Aztec second month, Tlacaxipeualiztli (the flaying of people), when sacrificial victims were flayed. Flaying ceremonies are related to agriculture, fertility, and prosperity (Broda 1970).[15] The objects associated with Itztli, a deity related to Xipe, make this obvious: he is depicted grasping a digging or planting stick, wearing a flayed skin, and holding a basket of corn (Heyden 1986: 40). Xipe's song, collected by Sahagún, is an "invocation for rain" and focuses on the growth of the maize plant (Broda 1970: 259). It is therefore not surprising that mist rattleboards were sounded during Tlacaxipeualiztli, the ceremonies dedicated to Xipe:

8.8

Xipe Totec. Note the rattlestick he is holding. (From Broda 1970, figure 2.)

And when it was this [time], the rattleboards were sown there at Yopico. And only they, the old men of the *calpulli* at Yopico, sat singing, sat rattling their rattleboards until the day was done. (Sahagún 1950–1982 book 2: 57)

The phrase "the rattleboards are sown" refers to a dance with bells that commoners danced (Broda 1970: 229), but sowing rattleboards also unambiguously links Xipe to rain.

The *chicauaztli*, Xipe's insignia, appears in Sahagún as an attribute of six gods related to fertility (Broda 1970:

241): Xipe, Opochtli, Yauhqueme (identified with Tlaloc), Chalchiuhtli Ycue (goddess of the waters), Xilonen (goddess of young maize), and Tzapotlan Tenan (who was related to Xipe and was the patroness of skin diseases). Tzapotlan Tenan has also to do with agricultural fertility, as does Xipe. Her impersonator also carried the mist rattleboard, and priests rattled rattles for her (Sahagún 1950–1982 book 1: 17).

In the Tarascan realm, as in Aztec territories, agricultural and rain-making rites must also have been accompanied by bell sounds, in view of common ideologies and similar musical cultures. Apart from the vast numbers of bells I refer to in this work, specific evidence also exists that Tarascan musicians and religious functionaries used two other instruments: hand-held rattles and some other sort of rattle. Hand-held rattles, shaken by dancers, are illustrated in the *Relación de Michoacán* (Tudela 1977: 160), and another rattle is also described among metal objects exported from Michoacán by the Spaniards:

Further, I received in native rattles of the same metal [copper-silver alloys] 4 arrobas and 33 pounds. These rattles are put together like the breast leather of a saddle with heavy cords. (Warren 1985: 119)

No direct correspondence exists between these Aztec deities and divinities in the Tarascan pantheon. Nonetheless certain Tarascan deities are associated with rain and with agricultural and human fertility; the bells that were so pervasive in the region must have played a part in rites dedicated to them. The deity controlling rain and fertility is Cuerauaperi, who also controlled birth and death; she was the mother of all gods and venerated throughout the Tarascan realm. Pollard (1993) thinks that her worship was widespread by the Late Preclassic or Formative Period. Flaying rites, analogous to those carried out for Xipe, were dedicated to her, and she was also associated with snake imagery. Curicaueri was the sun's messenger, the sky god, a warrior, the god of the hunt, and the patron god of the Tarascan royal dynasty. Xaratanga, the moon deity, was the wife of Curicaueri and daughter of the earth goddess; she is depicted as snake, half-moon coyote, and vulture. Xaratanga was associated with childbirth and fertility. Although we know something about ceremonies held for them, we simply do not have the detail that is available for the Aztec.

It is tempting, on the basis of transcriptions appearing in the *Diccionario Grande* (Anonymous 1991), to link the Tarascan terms for rain, rattling sounds, rattlesnakes, and metallic sounds. The Tarascan term for to rain hard is *charancharamahcuni hanini;* the root, *chara*, means to make noise, to purr, to rattle, and red. *Chara* could also be the root for rattlesnake (*chariqueri*), the snake's rattle (*chariraqua*), and copper (*tiamu charapeti*) (Anonymous 1991 vol. 2: 152–154). However, the roots involved here are in fact distinct. This becomes obvious in examining more modern materials in which roots are cited in an orthography that represents the phonological distinctions in the language. Three roots are involved: *chara* (to thunder or to burst), *charha* (in *charhanda* or red earth and *tiamu charhapiti* or copper), and *shari* (in *sharhirakua* or rattlesnake and *sharhisharhikasi* or reddish object). These roots cannot be related to each other, so that copper, rattlesnake, and rain do not appear to be directly associated. However, it is interesting to note that, despite phonological differences, certain concepts are to some extent related: *shari* is a root for reddish objects, including rattlesnakes, and *charha* is one for red earth, including copper. In addition, a relationship *does* exist between the terms meaning metal and metal bell:[16]

tiamu: Hierro, campana, etc. [iron, bell, etc.].
tiamu charapeti: Hierro, cobre [iron, copper]. (Anonymous 1991 vol. 2: 597)

The Sounds of Bells: Warfare. Bell and other rattling sounds not only attract rain but also protect in warfare. Among Huitzilopochtli's insignia were ankle bells. In his cosmic battle with his 400 brothers, Sahagún (1950–1982 book 3: 3) tells us, the latter arrayed themselves as if for war, binding little bells to the calves of their legs. Nonetheless Huitzilopochtli quickly dispatched them with his invincible fire serpent. The event is commemorated in myth, translated by León-Portilla and discussed by Matos (1987):

In vain they tried to do something against him,
In vain they turned and faced him,
To the sounds of their bells,
And they slapped their shields.
They could do nothing,
They could achieve nothing,
They could defend themselves with nothing.

Huitzilopochtli then arrays himself with their regalia, which may explain why he is shown wearing bells:

He took from them their finery, their adornments,
Their destiny, put them on, appropriated them,
Incorporated them into his destiny,
Made of them his own insignia. (Matos M. 1987: 53–54)

The reference to slapping their shields coincides with a statement by Durán (1967 vol. 2: 167), who reports that Huastec warriors battling the Aztecs wore bells that made "a strange sound" [un ruido extraño] attached to their shields or at their shoulders. Shields with bells attached to them are illustrated in the Florentine Codex.

Bell and rattle sounds are also associated with warfare in the Tarascan region. For example, the *Relación de Michoacán* (Tudela 1977: 190) illustrates the chief warrior wearing an ankle band with bells attached as he and his men attack a village. It also reports that when Curatame, the son of the Tarascan leader Tariacuri, dressed for war, he hung many snake rattles from his temples (Tudela 1977: 132).

In some cases no apparent relation exists between a particular deity and fertility, rain making, or protective sounds. For example, Titlacauan, a god related to Tezcatlipoca, is also associated with bell sounds. His impersonator

went placing his bells on both sides, on his legs. All gold were the bells, called *oyoalli*. These [he wore] because they went jingling, because they went ringing; so did they resound. (Sahagún 1950–1982 book 2: 69)

However, Brundage links one facet of Tezcatlipoca with shamanism and in fact thinks that Tezcatlipoca is most pervasively modeled on the American Indian shaman (Brundage 1988: 82). As I subsequently show, rattles and bells are the primary instrument of the American shaman. Tezcatlipoca's bells could be explained by that association. Whether or not this explanation holds in Tezcatlipoca's case, there is nonetheless a strong relation between bell sounds and rain or fertility, and between bell sounds and protection; other categories of ritual associated with these sounds remain to be identified.

The Sounds of Bells: The Sacred Garden. The third arena in which bell sounds shaped experience of the sacred is in the Aztec divine garden that Burkhart describes. The garden is created through song, represented by bell sounds, bird songs, and the sound of the human voice singing, but the garden also shimmers and glows with iridescent colors. I draw again on examples from the *Cantares Mexicanos* and the *Psalmodia Christiana*. The first sentence of the following quotation illustrates the metaphorical associations of bird songs, bell sounds, and singing. The key verb, *icauaca*, means also to warble, murmur, or clamor (Campbell 1985: 114). The Aztec sacred is extraordinarily sensual: a garden where smells, colors, and sounds are intensely and simultaneously experienced.

The spirit swans are echoing me as I sing, shrilling [warbling] like bells from the Place of Good Song. As jewel mats, shot with jade and emerald sunray, the Green Place flower songs are radiating green. A flower incense, flaming all around, spreads sky aroma, filled with sunshot mist, as I, the singer in this gentle rain of flowers sing before the Ever Present, Ever Near. (Bierhorst 1985: 141)

Birds, songs, flowers, trees, and golden colors are all associated with bell sounds in the Aztec paradise:

Hear it! He's shrilling, warbling on the branches of the flower tree. He's shaking! It's the golden flower-bell, the rattle hummingbird, the swan, Lord Monencauhtzin. Like a gorgeous troupial fan he spreads his wings and soars beside the flower drum. (Bierhorst 1985: 165–167)

We crave the reed-thrush plume who sings like a precious bell, who sings for the Only Spirit. (Bierhorst 1985: 277)

My songs are shrilling like gold bells. (Bierhorst 1985: 259)

Lord, your songs come ringing as bracelet bells. (Bierhorst 1985: 273)

White feather flowers blossom where Ixtlilcuechahuac as a plume, a rattle bird, is shrilling, singing. (Bierhorst 1985: 357)

The links between bell sounds and human voices are not limited to singing but also include speech:

We are Huaxtecs, hey! And they come jingle-shouting. (Bierhorst 1985: 361)

Speech is also associated with birds, song, and bell sounds:

May your speech resound! May there be chattering, may your songs resonate like bells! (Burkhart 1992: 95)

Other lexical evidence also shows that rattles, bells, and rattling sounds are linked to song and related to speech. A single word means a clear sound like a bell and metallic sounds:

tzilictic: Something which has a clear sound, like a bell or something similar.
tzilini: For metal to sound or ring.
tzilinia: To ring a bell or something similar. (Campbell 1985: 380)

In fact, the root *nahuatl*, meaning clear sound or order, also appears in compound nominals, one of which (*nauatillalia*) means town statutes; used in derived verb forms such as *nauati* (to speak loudly, or for a bell to have a good sound, or something similar), it can mean "something that sounds good like a bell or a man who speaks well" (Campbell 1985: 201). These definitions link good bell sounds, order, and laws or statutes; they suggest that sound transforms, it creates. But sound not only creates order: it creates color. It is through song and sound that the sacred garden comes into being, as the following stanza from *Cantares Mexicanos* shows:

As colors I devise them. I strew them as flowers in the Place of Good Song. As jewel mats, shot with jade and emerald sunray, the Green Place flower songs are radiating green. A flower incense, flaming all around, spreads sky aroma, filled with sunshot mist, as I, the singer, in this gentle rain of flowers sing before the Ever Present, the Ever Near. (Bierhorst 1985: 141)

The idea that sounds can create color appears in other metaphors also. However, they show that sound and color come into being through precisely the same process, and that process can be a metallurgical one:

I drill my songs as though they were jades. I smelt them as gold. I mount these songs of mine as though they were jades. (Bierhorst 1985: 207)

Smelting, that is, making a solid material liquid through heat, gives birth to sound (and song) and also gives birth to the golden metallic colors. Sound and golden metallic colors come into being in the same way, and the material results are sound-and-color objects: the high-tin and high-arsenic bronze bells whose fabrication technology I have described in this work. Sound and color are also associated in other verbal forms:

cahuantimani: To jingle. From *cahuani* [to catch fire], thus, crackling sounds or flaming colors. (Bierhorst 1985: 137; Karttunen 1992: 21)

comontoc: To ring. From *comoni* [for the fire to light and burn], thus, crackling sounds or flaming colors. (Bierhorst 1985: 169; Campbell 1985: 78)

tlazocoiolmilintimani: To flare forth like precious rattle bells. From *tlazoa* [precious], *coyolli* [jingle bell], *milini* [to shine]. (Campbell 1985: 359; Karttunen 1992: 43, 147; Sahagún 1993: 279)

In many regions, bells were worn by rulers and members of the upper classes. We do not know what their bells signified; they may have served to symbolically protect them. For example, the *Relación de Michoacán* shows the Tarascan ruler wearing bells around his ankles (Tudela 1977: 251) and reports that at death he was buried with bells of gold. Bells also figured as items of elite dress among Aztec upper classes and rulers. Sahagún (1950–1982 book 8: 28) reports that when the Aztec ruler prepared to dance, he bedecked himself with gold bells and shook golden gourd rattles. When Axayacatl, who ruled from 1469 to 1481, died, ankle bells were one of the tribute items (Durán 1967 vol. 2: 297). Aztec court dignitaries wore jewels and gold bells bound to their calves (Durán 1967 vol. 2: 364). Ceremonial attire for the king of Texcoco, a powerful state in the Basin of Mexico and ally of the Aztec, included golden bells worn at the instep (Durán 1967 vol. 2: 301). And in Oaxaca, members of the upper classes wore bells (Dahlgren de Jordan 1979).

The sounds of rattles and bells thus promote human and agricultural fertility, they create order, and they protect. They also create color, a shimmering sacred paradise full of lustrous beings. The importance of bells is not only apparent in their variety and abundance, and in the ways they are used in specific rites, but is encoded in language. The material, metal, and a kind of object made from it, bells, were inextricably associated. Metal and bells were cultural synonyms.

Let us now turn briefly to ethnographic data to explore whether any of these meanings may still obtain.

The Meaning of Rattling: Ethnographic Data. The primary context in which rattles, bells, and other rattling instruments are used presently or in the recent past in the Americas is in shamanistic activity (Izikowitz 1935). Although the Tarascan and Aztec polities were states, and shamanism is not generally associated with the state, elements of shamanistic practices certainly were incorporated into religious thought and activity. Brundage's remarks about Tezcatlipoca are a case in point. Less complex political entities in which shamans may have played a more central role were typical of Period 1, and these persisted, in some cases, through the Spanish invasion in peripheral areas. Some examples I cite here come from those peripheral areas; the evidence they provide, while incomplete, is consistent with the previous discussion. I strongly suspect that the meaning of bell sounds in the metalworking zone during Period 2—especially for protection and fertility—derives from the meanings ascribed to them in simpler societies. Some pieces of those ancient belief systems still seem to be in place and link rattle and bell sounds with agricultural fertility and regeneration, and to the sacred flowery world of the Aztec paradise.

The most common instrument of the American shaman is the hand-held gourd rattle, but shamans also use bells, bows, bone rasps, and drums (Bahr et al. 1974; Furst 1974; Gossen 1974; Lumholtz 1973; Métraux 1949). The bottle gourd, which frequently serves as a rattle, was the first plant domesticated in the New World but was never used for food. Sullivan (1988) thinks that the plant was disseminated for religious reasons, and calls the gourd rattle the single most sacred instrument in the indigenous Americas. The examples I give show that the shaman uses the gourd rattle (hand-held rattle) to communicate with supernatural forces; the sounds and the object itself have supernatural qualities. The rattle and its sounds induce ecstasy and accompany ecstatic states in deities and humans alike.

Descriptions from northern and western Mexico show how gourd and other rattles and bells were used in ceremony. Cabeza de Vaca, a member of Pánfilo Narváez's 1527 expedition from Spain to Florida, reported them in his description of a village in what was then northern Mexico (Rio Concho, Texas):

At sunset we reached a village of a hundred huts. All the people who lived in them were awaiting us at the village outskirts with terrific yelling and violent slapping of their hands against their thighs. They had with them their precious perforated gourd rattles which they produce only at such important occasions as the dance or a medical ceremony and which no one but the owner dares touch. They say there is virtue in them and that, since they do not grow in that area, they obviously come from heaven. (Covey 1961: 101–102)

Karl Lumholtz, the Swedish botanist who wrote extensively about the indigenous groups he encountered in his travels in the early 1900s through northern and western Mexico, described a rite he observed in Nayarit among the Huichol. In this rite, the supernatural wears rattles, and they are associated with protection, curing, and trance. The shaman ingested *peyote*, a hallucinogen that the Huichol conceive as a god, then sang a song describing how the sacred *peyote* god walked with his rattles and with his staff of authority. The god came to cure, to guard the people, and to grant "beautiful" intoxication (Lumholtz 1973 vol. 1: 368). The shaman's assistants wore deer hoof rattles or bells in the dance accompanying the rite.

Lumholtz's observations of a Huichol shrine underscore the notion that a rattlesnake and its rattle are associated with supernatural protection. He visited a house that he was told housed a god. The god was folded up in a bundle of cloth containing the winged parts of arrows—which the Huichol considered the arrow's vital part or its heart—small, soft, woven back shields, and a rattlesnake's rattle. According to Lumholtz's informant, the serpent belonged to the god; the god was a warrior who always carried his rattling bell with him (Lumholtz 1973 vol. 2: 56). This deity sounds very much like Curatame, the Tarascan warrior who attached rattlesnakes' rattles to his temples. The term "rattling bell" probably comes from *cascabel,* which in Spanish means both a metallic bell and the rattle of a rattlesnake; however, a single Huichol term means both *sonajas de los indios,* the term used to describe native people's metal bells, and *víbora cascabel,* which means rattlesnake.[17]

Agricultural fertility is linked to snakes, Tlaloc, rain, and lightning in dances described by Alain Ichon (1990) from the Sierra de Puebla, an area east of the metalworking zone. Hand-held rattles and a violin are the only instruments accompanying these dances. The dances reflect aspects of the belief system I have described for those areas of the metalworking zone that supported state-level societies. The central prehispanic element in the dances is a serpent, who is ritually killed or sacrificed. One dancer, known as the mother of thunder, holds a wooden serpent that represents lightning, rain, and corn;

the serpent has to die to be reborn as corn. Her husband is guardian of the lightning-serpent, a concept that Ichon thinks incorporates Tlaloc. Ichon maintains that this dance is a fertility rite representing the arrival of rain, the death of the serpent, and the resurrection of the grain of corn. Both Ichon (1990) and Baudez (1992) relate this dance to the Quiché snake dance, which has sexual connotations.[18]

Fragmentary evidence for a sacred world filled with flowers and created through sound comes from the Huichol, Yaqui, and Mayo peoples in northwestern Mexico. These groups share a common religious system that Spicer (1964) terms central Uto-Aztecan; Nahuatl is also a language in the Uto-Aztecan family. Analyses of the songs accompanying the Yaqui deer dance stress that the connection to the supernatural world is created through song and flowers (Spicer 1964). Participants play various rattling instruments: they shake hand-held rattles and wear ankle rattles as well as the *cinturón-sonaja,* the bell waistband described in chapter 3. Varela has described the movements of one of the dancers:

Con increíble destreza estampa los pies en el suelo produciendo un complicado contrapunto rítmico entre el arpa y los capullos-sonaja de sus piernas: salta, arrastra la planta o la orilla izquierda o derecha del pie sobre el suelo, camina hacia delante y hacia atrás, y cada movimiento repercute en la cadera, de modo que indirectamente hace sonar los cascabeles del cinturón que lleva puesto. [He stamps his feet on the ground with incredible agility, producing a complicated rhythmical counterpoint between the harp and the rattles on his legs; he jumps, drags the left or right edge of his foot on the ground, walks forward and backward, each movement reverberating at his hips so that he indirectly sounds the bells on the belt he is wearing.] (Varela R. 1986: 47)

In the Huichol deer dance, participants also wear and hold rattles, bells, and gourds. The individual representing the deer wears "a leather belt from which hang deer claws. Around his legs are strings of ténabari, dried cocoons filled with gravel, and he carries a large gourd rattle in each hand." The other dancers wear *cinturón-sonajas* and hold gourd rattles in their right hands, "and from their leather belts hang brass or copper bells. They wear the same kind of necklaces and strings of cocoons as the deer, but their rattles are of wood, hollowed out above the handle to admit some metal disks" (Toor 1947: 331–332).

The dance sodalities of Mayo groups in Sonora, another Uto-Aztecan people, perform at religious observances and rites of passage. The sodalities have special relationships to sacred animals (for example snakes, deer, and others), which serve as their insignia and through which the sodalities derive personal magical power. The Mayo group enact a deer hunt (Crumrine 1977: 100–101) culminating in a dance with another sodality in which rattle sounds, bell sounds, and the sounds of an instrument that seems suspiciously like an offshoot of the Aztec *ayauhchicauaztli* figure prominently. The costume of one of these sodalities includes "a white blanket secured at the waist . . . [a] belt from which small round bells hang, and long ankle strings of cocoon rattles." They also carry a rattle "and a wooden frame in which are mounted metal discs" (Crumrine 1977: 97).

Rattles and their sounds also have supernatural powers in Amazonian South America. Deities go into trance accompanied by these sounds. In Avá-Chiripá (an indigenous group in eastern Paraguay) myth, Kuarahy, the divine twin and the sun, used shamanistic techniques to communicate with his father, the creator god Nanderú Guazú. "Fashioning himself a rattle (*mbaraká*), he danced himself into ecstasy until his father [took] him away with him" (Sullivan 1988: 428). The rattle of Thunder, a powerful supernatural being of the Tapirapé people in central Brazil, causes the shaman who touches it to die

(Wagley 1940: 258). Rattles are also sacred. Staden reports that the Tupinambá of southern Brazil

> changed their rattles into gods, [then] each man takes his rattle away, calling it his beloved son, and building a hut apart in which to place it, setting food before it, and praying to it for whatever he desires. These rattles are their gods. (Stevenson 1968: 35)

Rattles, rattling sounds, snakes, sexuality, protection, and creativity are also linked. According to the Warao people in Venezuela, the rattle symbolizes the joining of male and female elements in the universe by the fastening of the rattle handle onto the head of the rattle. This act makes the instrument effective and creative (Wilbert 1974: 91). Civrieux reports that the Kari'ña, also in Venezuela, view the rattlesnake as the wisest and most powerful species and that the shaman identifies with it:

> El cascabel es considerada como la especie de mayor sabiduría-poder. Es el único crótalo que hace uso del sonido para llamar, alzando su cascabel o maraca. . . . La maraca del *puidei* humano es una réplica del cascabel. Cuando el *puidei* levanta y sacude su maraca la convierte en instrumento de intimidación, y ahuyenta los espíritus peligrosos. [The rattlesnake is considered the all-wise all-powerful species. It is the only snake that uses sound to call, lifting its rattle or maraca. . . . The maraca of the human shaman is a replica of the rattlesnake's rattle. When the shaman lifts and shakes the rattle he transforms it into an instrument of intimidation and chases off the dangerous spirits.] (Civrieux 1974: 21)

Referring to snakes, Civrieux comments that "ellos son los Grandes Maestros que, 'en el Principio,' poseían el secreto del baile. Bailando saben apoderse de los enemigos y de las hembras." [They are the great masters who in the beginning possessed the secret of dance. By dancing they know how to overpower their enemies and to possess women.] (Civrieux 1974: 21)

He then describes a dance in which the shaman shakes his maraca to defeat his invisible enemies and another, called the rattlesnakes' dance, in which one of the participants finally claims victory and then engages in a courtship dance followed by symbolic copulation.

These data provide a sense of the persistent use and meaning of rattles and rattling instruments in the Americas, especially in Mexico. The ethnographic data from Mexico show general similarities to what we have seen in the prehispanic era, especially with respect to agricultural fertility and the protective function of rattling sounds. Fragments also appear of a sacred flowery world, called into being by these sounds.

Sound, Metal, and Creation

Returning again to the *Relación Geográfica* from Ajuchitlán, this recounts that metals were the materials from which the creator fashioned the first human beings. Like the creators in the *Popol Vuh*—the origin myth of the Quiché Maya of highland Guatemala—the deity performed repeated experiments to find the material whose physical properties were suitable for the first couple. The first attempt was from *barro*, a term for clay, but when the creator sent the couple to bathe, they came apart in the water. The deity then tried a different material for the human design. The next and successful creation was from ash and several metals, and "this time when they went to bathe they did not come apart in the water and it was from those [two] that the world began."

"Ash and various metals" refers to bell-making technology. West Mexican metalworkers used these materials for lost-wax casting: Sahagún (1950–1982 book 9: 73) tells us that ash was incorporated into the clay molds used in such castings. The creator chose several metals for the first human couple, casting them from an alloy. By suggesting that the fabrication process was lost-wax casting, the *Relación* intimates that the first bells were human and that the first humans were bells; furthermore, that they were cast and created—appropriately enough, mixed and fertilized—from metal alloys, most probably from the alloys of copper-tin and copper-arsenic bronze. The newly created beings possessed various essential properties: they were indestructible, they manifested their divinity and fertility visually in their golden and silvery colors, and they were animate and could communicate with their creators through sound: through the powerful and clear sounds of metal bells, rattles, and rattlesticks.

Similarly, the *Popol Vuh* makes it clear that in Mesoamerican cosmologies, creation demands reciprocity, and that the deities sought to craft beings that could reciprocate through sound—specifically through speech—to praise their creators. According to the *Popol Vuh*, the deities first created animals, instructing the birds, the jaguars, the deer, and the pumas they made:

"Name now our names, praise us. We are your mother, we are your father. . . . Speak, pray to us, keep our days [worship us]," they were told. But it didn't turn out that they spoke like people: they just squawked, they just chattered, they just howled. It wasn't apparent what language they spoke; each one gave a different cry. . . . "It hasn't turned out well, they haven't spoken," they [the creators] said among themselves. "It hasn't turned out that our names have been named. Since we are their mason and sculptor, this will not do," the Bearers and Begetters said among themselves. (Tedlock 1985: 78)

The deities experimented anew:

"So now let's try to make a giver of praise, giver of respect, provider, nurturer," they said.

So then comes the building and working with earth and mud. They made a body, but it didn't look good to them. It was just separating, just crumbling, just loosening, just softening, just disintegrating, and just dissolving. Its head wouldn't turn, either. Its face was just lopsided, its face was just twisted. It couldn't look around. It talked at first, but senselessly. It was quickly dissolving in the water.

"It won't last," the mason and sculptor said then. "It seems to be dwindling away, so let it just dwindle. . . ."

"There is yet to find, yet to discover how we are to model a person, construct a person again, a provider, nurturer, so that we are called upon and we are recognized." (Tedlock 1985: 79–80)

The deities then carved men and women out of wood, but destroyed them because they were unable to speak. Eventually they created human beings from corn and from water; the water was transformed into blood. These humans were capable of speech, they were able to communicate with the deities and to praise them. However, ultimately even words and speech did not suffice. Certain tribes lacked fire. When they pleaded for it, Tohil, the creator, assented but demanded human sacrifice in return:

Don't they [the tribes] want to be suckled on their sides and under their arms? Isn't it their heart's desire to embrace me? I who am Tohil? But if there is no desire, then I'll not give them their fire. . . . When the time comes, not right now, they'll be suckled on their sides, under their arms. (Tedlock 1985: 174)

The *Popol Vuh* then explains:

And this is what Tohil meant by being "suckled": that all the tribes be cut open before him, and that their hearts be removed "through their sides, under their arms." (Tedlock 1985: 175)

The creators demanded that they be suckled, nourished, by the hearts and the blood of the beings they created. The archaeological and documentary records testify that this and other forms of human sacrifice were widespread in Mesoamerica, but that they were carried out with particular institutional zeal by the Aztec in Central Mexico.

In the version of creation presented in the *Relación* from Ajuchitlán, the first beings were lost-wax cast, and the *Relación* intimates that they were bells. If this couple, like their counterparts from the highlands of Guatemala, were also required to nourish their creators—and nothing in Mesoamerican cosmology suggests that creation was ever undertaken as an act devoid of reciprocal responsibilities—here it was through sound, in this case through the strengthening, fertilizing, singing voices of rattles and bells.

The Social Context

I have argued that the premise that metal was divine, indestructible, and powerful gave rise to the choices that shaped West Mexican metallurgy. The technical interest in bells arose from the creative power of their sounds; in metallic colors, from their associations to solar and lunar deities and the glittering, shimmery, sacred garden these colors evoke. Convictions such as these probably influence the course of most technologies. Multiple other factors do so as well, and the relative weight of each in any particular social context probably depends on historical circumstance. As an approach to understanding the forces that gave this technology its distinctive orientation, I will describe the significant contemporaneous circumstances and suggest the ways in which they may have impinged on the technical choices I have identified here.

Before doing so I want to emphasize that this discussion is speculative. On one hand, we lack comprehensive studies of other preindustrial or prehistoric technologies that would allow cross-cultural generalizations. Ideally, we could compare West Mexican metallurgy to another preindustrial technology also introduced from outside. On the other hand, we also lack data from within this large and diverse metalworking zone to explore it as a single case. Chronologies are incomplete. Information concerning subsistence practices, population densities, population growth rates, resources, settlement patterns, and craft specialization, among other topics, is generally sparse except in a few well-studied locales. Inter- and intraregional relationships have yet to be fully defined. With respect to metallurgy, we lack technical data from processing sites and sites where metal objects were crafted, as noted elsewhere. Few studies have even plotted the distribution of metal objects within sites. Little information exists about the organization of metal production. Nonetheless, with such caveats in mind, I will consider the particulars of the historical situation in which West Mexican metallurgy emerged, and comment on how they may have contributed to the technical choices I have described in this work.

The most significant event that occurred about the time that metallurgy was introduced to this area was the decline of Teotihuacán. Teotihuacán's demise affected polities not only in the metalworking zone but in many other Mesoamerican areas. The collapse of this powerful empire created an economic and ideological vacuum. Teotihuacán's presence in the West Mexican metalworking zone has been well documented. It is visible in many regions and in various media, including pottery at sites in Colima (Kelly 1949), Michoacán (Kelly 1947; Pollard 1993), Jalisco (Weigand 1992), Nayarit (Weaver 1981), central Guerrero, and the lower Río Balsas (Cabrera C. 1976), and ceremonial centers exhibiting Teotihuacán-style architecture in Jalisco (Weigand 1985), Michoacán (Pollard 1993), and elsewhere.

The nature and impact of the contacts between Teotihuacán and the West Mexican metalworking zone varied considerably, and we have few means of measuring them. Archaeologists have argued that one effect of Teotihuacán's interactions with the west had been "to increase the process of social differentiation already taking place [there], and stimulate the emergence of territorial discrete and competing polities" (Pollard 1993). Weigand thinks that, just as the Teuchitlán tradition in the lake region of Jalisco may have coalesced as a local response to Teotihuacán expansionism, so Teotihuacán's decline may have triggered Teuchitlán's demise. Regardless of the particulars, there is no doubt that Teotihuacán served as an overwhelmingly powerful symbol of sacred power, and that its collapse caused major ideological upheavals. Some of those must have been felt in the metalworking zone.

In fact, Teotihuacán's decline could explain the timing of metallurgy's appearance. We know that Teotihuacán was interested in the west for mineral resources, marine shells, and possibly salt, cotton, cacao, and other perishables. The most important of these to this discussion is *Spondylus,* a ritual item. Teotihuacán imported great quantities of *Spondylus* and other Pacific shells, and Millon (1981) thinks that long-distance shell trade must have been a major concern of that state. Teotihuacán's collapse interrupted this trade throughout Mesoamerica (Starbuck 1975). I have suggested that *Spondylus* was one item the Ecuadorian seagoing traders sought in West Mexico (see chapter 4). *Spondylus* grows along the Pacific coast as far north as the Gulf of California, and Teotihuacán used the shell extensively (Marcos 1978; Millon 1981; Starbuck 1975). Later, Aztec tribute records indicate that *Spondylus* was supplied to the central highlands from the Pacific coast, and these suppliers may have been the same as those who provided the shell earlier. Worked *Spondylus* beads have been recovered at Amapa, for example (Mountjoy 1992). Teotihuacán's decline must have seriously affected west coast *Spondylus* suppliers, since the city had been the major market for the shell. These coastal peoples' search for other groups interested in this ritual item may have precipitated interactions with Ecuadorian traders (see chapter 4), who then acquired and distributed west coast *Spondylus* products to Peru and Ecuador through the maritime trading organization. In exchange, they introduced to West Mexico metal objects and knowledge of certain facets of production technology.

We cannot be certain how Mexican polities responded to the political and ideological disruptions brought about by the collapse of Teotihuacán, but whatever the responses, there is no doubt that metal objects provided a novel, visually and aurally powerful means of communicating with (and recreating) the sacred through bell sounds, a new and unusual material with which to mark hierarchy and social status, and a new item to trade. We know that by the beginning of Period 1 metallurgy, around A.D. 650 to A.D. 800, large centers were emerging, for example at Teuchitlán, at Amapa, and in other areas. Emerging polities vie for prestige and power (Earle 1989; Helms 1979), and the social disruptions following the collapse of Teotihuacán must have exacerbated those tendencies.[19] Helms's (1979) argument that elites, especially chiefs, in societies at this scale need to convince others that they can control both the worldly and the supernatural aspects of life is germane to this situation. Chiefs control their followers through myths and beliefs about the source of exotic goods. These are associated and identified with divine beings. Elites seize control of existing principles of legitimacy, both supernatural and natural (Earle 1989), as well as the objects that represent them. Metal must have served as an ideal material for these ends because of its exotic origins and the sacred and animate qualities attributed to it.

If we knew who controlled metal production, we could also say more about why this technology developed as it did. During Period 1, production seems to have taken place in many different regions, and it was during this period that the technology took on its distinctive nonutilitarian character. Some Period 2 evidence is available from the Tarascan realm and the Aztec empire. Production of ritual items, particularly golden and silvery bells, tweezers, and sheet metal ornaments, seems to have been controlled by elites; we have to assume that Period 2 metallurgy was greatly influenced by their ritual and status requirements. We also know that even during Period 2, neither control nor production was centralized, except perhaps in Michoacán during the last years of the Tarascan empire (see Pollard 1987). The archaeological and the laboratory evidence have consistently pointed to multiple production centers: there is slag at Amapa, metalworkers' tools at Tomatlán, ingots in the Lago Chapala region of Jalisco, slag from the Churumuco mine in Michoacán, and slag along the middle Río Balsas. The laboratory data show that standard designs and standard alloys were likewise absent. However, Tarascan sources do indicate that the state directly controlled some mines and smelting operations (Pollard 1987). At Tenochtitlan, documentary evidence indicates that elites controlled production of wealth and status items (and clearly ritual items), including metal (Brumfiel 1987). Laboratory evidence to corroborate the documentary sources is thus far unavailable.

The fact that this technology came from the outside, and via a maritime route, perhaps may be the single most important factor determining its overall trajectory. One has only to imagine traders arriving from the south in canoes, bringing with them objects made from an entirely unknown material, a material about which rumors may have been circulating but which no one had seen. It was a material that could produce tones never heard before and colors never seen before, and was associated with ancient beliefs and stories about the kingdoms to the south. Metal was as exotic and magical as those imaginary kingdoms, and it is no wonder that it was considered divine, indestructible, and sacred: an optimal material for the first human beings.

If, ultimately, the trajectory of this technology resulted from its introduction from an exotic and distant source, the final and least answerable question concerns why metallurgy did *not* develop autochthonously in Mesoamerica. We cannot answer that question without case studies identifying those circumstances that predispose the development of certain technologies and mitigate against the development of others. What we do know is that when this exotic material appeared, production and distribution systems for other technologies and their products (stone, bone, cloth, ceramic, etc.) had been in place for hundreds of years. Although certain properties of metal are redundant with, and sometimes superior to, the properties of these other materials, metal axes did not replace stone axes, and metal projectile points and knives did not replace projectile points and knives made from obsidian. Rather, artisans used metal for those properties that cannot be easily replicated in other media and that express its sacred power: its reflectivity, the ability to develop color through alloying, and the varied range of pitches it produced.

The metallurgy of West Mexico does not constitute a radical departure or an entirely new perspective on metal in the Americas. Rather, this technology was a local expression of broader, pan-American themes shaped by the circumstances surrounding its introduction. The ideological vacuum created by the collapse of Teotihuacán predisposed certain technical choices, but the exotic origins of the new material did so as well. Also, the sacred meanings of sound and certain metallic colors were deeply rooted in the experience of these and other ancient American peoples; one of the West Mexican artisans' most distinct and imaginative contributions was to transform even the utilitarian bronze alloys into materials that communicated religious and social power in objects whose *design* (including materials and fabrication methods) and *meaning* required those particular materials. However, predispositions toward a particular technical outcome do not necessarily mean that such an outcome was predetermined; they simply provide a means of explaining that can stand until researchers provide other cases for comparison. What has become plain in the course of this discussion is that the metallurgy that emerged from within this particular historical and social context arose from West Mexican artisans' encounter and experiment with the immutable physical and mechanical attributes of materials, filtered through social and economic needs, aesthetic sensibilities, and ideologies, making this metallurgy, as any other, "a fully human experience" (C. S. Smith 1977).

APPENDIX 1

Technical Studies: Data and Methods

The laboratory studies provided fundamental and reproducible data concerning the technical characteristics of the metallurgies of the metalworking zone, other regions of Mesoamerica, and Ecuador. The studies had three objectives: to identify artifact fabrication techniques, artifact chemical composition, and metal/alloy properties, such as microhardness. These determinations also made it possible to assess the ways the objects were used or could have been used.

Laboratory studies were performed on objects in the RMG (Regional Museum of Guadalajara, Jalisco) in Mexico and the MAG (Museo Antropológico de Guayaquil) in Ecuador. Additional Mexican data were gained through laboratory analyses of artifacts from six Mexican archaeological sites. Macroscopic studies also were carried out on objects in the Museo Nacional de Antropología e Historia in Mexico City, and on the collections mentioned subsequently in Mexico and the United States. I also carried out laboratory studies of artifacts from five Ecuadorian archaeological sites. The methods used to determine fabrication techniques, chemistries, and artifact function or use are outlined below.

Measurements

Fabrication Techniques. Fabrication methods for all objects were identified using standard metallographic techniques for the interpretation of microstructure. To examine a metal artifact using these techniques, one or more intact samples (as distinct from filings or drillings) have to be removed from the artifact. The sample is mounted, ground flat, and polished to produce a plane and scratch-free surface; it is then etched with a chemical reagent to reveal the metallic microstructure. The microstructure of a metal artifact records the history of the procedures used to fashion the object. Shaping methods such as casting, hot and cold working, and other common techniques produce highly characteristic microstructures that have been experimentally reproduced, studied, and catalogued. By examining and interpreting the microstructure of a metal artifact it is possible to describe, step by step, the fabrication history of that object. Often artifact microstructure can also indicate whether or not the object was used. Use produces characteristic deformation of the metal crystals or grains.

Metallographic examinations can also provide an indication of the bulk chemistry of an object. If the metal is reasonably pure (unalloyed), it can frequently be identified through metallographic examination both by its color and by its characteristic response to certain specific etching reagents. It is almost always possible to determine from its microstructure whether an object is made from an alloy, and to identify the primary alloying element (although the alloy concentration can only be determined within general ranges), since particular compositions produce specific kinds of structures.

Chemical Composition. Several terms or concepts related to metals, alloys, and their chemical composition need clarification. Metals frequently used in nonindustrial metallurgies and especially important in the prehistoric Americas include copper, gold, and silver. These metals occur in their native state as naturally occurring metal deposits. Native metals can be melted and cast or worked directly, and require no further processing once they have been mined. For the most part, however, copper and silver occur as ores, where they are present in the metallic minerals that have been deposited in association with a host rock. These metals are "won," or extracted, from their ores by smelting, a process activated by heat that both chemically and physically separates the metal from the nonmetallic portions of the mineral as well as from the rocky matrix. A specific metal or alloy may often be produced in more than one way, each method requiring different raw materials and techniques but yielding the same, or a comparable, final product.

Alloys are mixtures of two or more metals. Some alloys are natural. For example, deposits of gold in Colombia often contain as much as 20% silver, and such deposits are effectively natural alloys of gold and silver (electrum). In other cases the metals that eventually form the alloy co-occur in the metallic minerals of their ores. This is not infrequent in highland Andean ores of copper, for example, which contain both copper and arsenic in mineral form. When such ores are smelted, the direct product is an alloy of copper and arsenic known as arsenical copper or arsenic bronze (the term "arsenic bronze" is discussed in chapter 2, note 8). Other alloys require the intentional mixing of two metals by melting them together after each has been won from its ore. Tin bronze is usually made in this way, by melting metallic tin and metallic copper after each metal has been smelted from its own ore. Common alloys in prehistory, whether natural or intentional, include copper-silver, copper-gold, the two bronze alloys (copper-arsenic and copper-tin), and others. Although iron and steel (an alloy of iron and carbon) were used in the Old World during the prehistoric era, they were not developed in the Americas.

There are many techniques used to determine chemical composition of metal artifacts. The choice of technique depends upon the question posed and the extent to which the object can be sampled for chemical analysis. Techniques include qualitative methods, which provide information about the elements present in the artifact and their relative concentration levels, and quantitative methods, which yield precise determinations of the concentration of each element, usually given as weight percent. The qualitative determinations of chemical composition of all but a few artifacts included in this study were carried out using emission spectrographic techniques. The results of those studies are cited here; the raw data can be found in Hosler (1986). The primary quantitative method used was atomic absorption spectrometry, which provided concentrations of major, minor, and trace elements. Neutron activation analyses were performed for one group of objects in which the concentration of an important trace element, indium, occurred at levels below the detection limits of the atomic absorption method. The electron microbeam probe, which identifies chemical compositions of very small areas on a polished metal cross section, was also used in one case.

Function or Use. Several sources of evidence, including laboratory data, have been considered in determining how these metal objects were used. One was to evaluate the objects' design and mechanical properties (see chapter 1), which were measured in various ways. Some, such as hardness, were determined directly by microhardness tests made on the sampled cross sections using the Vickers diamond indentation method. In other cases (such as a measure of springiness for the tweezers), mechanical properties were derived through standard calculations that use experimentally determined data about the properties of metals and alloys as a function of their fabrication techniques. In these cases, physical data obtained from the sampled cross sections were incorporated into the engineering formulas to arrive at the appropriate value for the property being measured. The basic data about both fabrication techniques and composition were obtained in the laboratory studies, and in one case provided the information to simulate artifact function using computer methods.

The Study Corpus

The RMG Collection. Study of this collection involved macroscopic examination (and classification) of approximately 3,200 prehispanic metal artifacts, from which approximately 400 were selected for extensive laboratory analyses. The artifacts are primarily from the West Mexican states of Michoacán, Jalisco, Colima, and Nayarit.

The RMG collection was systematically assembled over many years by Ingeniero Frederico Solórzano of Guadalajara, Jalisco, Mexico, with the express intent that it serve as a study corpus. Solórzano thus acquired every object made available to him, including fragments and damaged items, without selection bias. The collection became the property of the Mexican government and is housed at the Museo Regional de Guadalajara under the auspices of INAH (Instituto Nacional de Antropología e Historia). To confirm its representativeness, I compared the artifact types in the collection to those published in Pendergast's

(1962b) distributional study of Mesoamerican metal objects and to other collections of Mesoamerican and West Mexican metal objects: in the American Museum of Natural History, New York; the Museo Nacional de Antropología e Historia, Mexico City; the Heye Foundation Museum of the American Indian, New York; the Peabody Museum, Harvard University, Cambridge, Massachusetts; and the Frissel Museum, Mitla, Oaxaca. The RMG collection contains all of the major artifact types found in West Mexico and identified by Pendergast, but also contains various types that have not yet been described in the literature. Their relative frequencies are generally in keeping with the frequencies observed in other smaller collections and in assemblages that derive from controlled excavations.

The RMG artifact types and their relative frequencies are shown in table A1.1. The collection contains 12 artifact classes. By far the most frequent type is bells, which constitute 60% of the total; 20% of the collection consists of open rings. Visual inspection indicated that all objects are made from copper or copper alloys except the sheet ornaments, which are made of gold, silver, or alloys of those metals. The question of the authenticity of these artifacts must be addressed, since they lack specific archaeological provenience. The indisputable evidence of their authenticity lies in the presence of internal and external corrosion alterations that characterize copper and copper-alloy objects that have undergone corrosion at slow rates over long periods of time (hundreds to thousands of years). The characteristics of such mineralization include internal corrosion, almost always intergranular (corrosion proceeds along grain boundaries between adjacent grains), and external corrosion (mineral products generally form in layers that correspond to the alteration from metal to mineral and from one mineral type to another). In this sequence of layers, the mineral that almost always forms in direct contact with the metal is cuprite (cuprous oxide), which subsequently alters to the green copper minerals, such as malachite or atacamite. Research in conservation laboratories has shown that the rapid rates of corrosion induced by attempts to patinate faked objects produce neither intergranular corrosion nor the

Table A1.1 RMG Collection: Number of Artifacts by Type

Type	Number	% of Total
Bells	1934	60.5
Open rings	685	21.4
Axe-monies	186	5.8
Sheet ornaments	136	4.3
Needles	87	2.7
Axes	41	1.3
Tweezers	42	1.3
Awls	23	<1
Bell ornaments	22	<1
Pins	17	<1
Fishhooks	14	<1
Beads	9	<1
Other	?	<1
Total	circa 3196	

formation of a coherent cuprite layer. All of the collection objects exhibit the reddish-brown cuprite layer; all of those studied metallographically also exhibited intergranular corrosion.

The sample for laboratory studies to determine fabrication techniques, the metals and alloys used, and the use of the various object types was selected from among the eight most frequent artifact classes. Those types and the number of artifacts sampled from each are shown in table A1.2.

A few unique artifact types, or types containing only a few examples, were also sampled, to identify patterning in chemical composition, fabrication technique, and alloy properties, and to determine reliably how the objects were used. In all major artifact classes but one, the general goal of a 10% sample was achieved. The exceptional category, bells, contained far too many artifacts (1,934 items, of which a 6.5% sample was made) to have made

Table A1.2 RMG Artifact Types Sampled for Laboratory Studies

Type	Total in Collection	Number Sampled	% Sampled
Bells	1934	125	6.5
Open rings	685	81	11.8
Axe-monies	186	33	17.7
Sheet ornaments:			
gold and gold alloys	60	0	0.0
silver and silver alloys	76	28	36.8
Needles	87	30	34.5
Axes	41	41	100.0
Tweezers	42	39	92.9
Awls	23	15	65.2
Other	>62	8	circa 13.0
Total	circa 3196	400	

that goal feasible, given the extremely high cost of chemical analyses. At the same time, since fabrication technique for all the bells proved virtually the same—they were cast by the lost-wax method—it was unnecessary to verify that process through frequent and repetitive metallographic examination. In most cases, the sample goal of 10% was far exceeded, and in other cases, where technical questions posed by design, composition, fabrication techniques, and cultural function were especially complex (axes and tweezers), all or virtually all artifacts within the type were sampled. The analytical procedures used in this study and the total number of artifacts examined by these procedures are presented in table A1.3.

Artifacts from Mesoamerican Archaeological Sites. Laboratory studies were carried out on artifacts from the six archaeological sites listed in table A1.4. The differences in relative frequencies of particular types reflect depositional context. Excavations at Lamanai included several large structures and burials, whereas at Cuexcomate and Capilco no burials were excavated. The Cuexcomate assemblage predictably contains many tools and few sumptuary objects. At Milpillas, excavations were of burials. Most material from Platanito and Vista Hermosa also derived from burials. Apart from laboratory studies of these objects, I also carried out macroscopic studies of objects from Urichu (Michoacán), Amapa (Nayarit), and Tomatlán (Jalisco).

The MAG Collection. Data on the metallurgy of northern South America were obtained principally through macroscopic studies of the collections of the Museo Antropológico del Banco Central in Guayaquil, Ecuador. The collection contains some 7,900 objects (in the frequencies shown in table 4.8), of which 154 were selected for analysis. In general the artifacts selected for the sample from the MAG were types with counterparts in West Mexico. This collection cannot be considered representative in the same sense as the RMG corpus, since the objects were acquired with a specific preference for sumptuary items as opposed to tools.

In addition to the objects housed in Guayaquil, I also studied the metal collections of the Museo del Banco Central in Quito macroscopically, to acquire a sense of the range and abundance of distinct artifact classes.

Artifacts from Ecuadorian Archaeological Sites. Data from excavated collections, presented in table A1.5, suggest that the proportion of tools, especially needles and axes, was undoubtedly far higher than appears from the numbers represented by the MAG collection. Of the five Ecuadorian sites from which metal objects were excavated, Salango and Loma de los Cangrejitos may provide the most accurate reflection of the relative frequencies of artifact types. All objects for which there was sufficient weight of nonmineralized material were analyzed, except those made from gold; the preservation of metal objects at Salango, El Azúcar, Cerro Alto, and OGSE-Ma-172 was generally poor, and many of the objects from Salango were completely mineralized and could not be analyzed.

Other. To gain a sense of South American metal artifact types I also examined macroscopically the South American collections of the American Museum of Natural History in New York, the Metropolitan Museum of Art in New York, and the collections of copper objects housed in the Museo del Oro, Bogotá, Colombia.

Research Permissions

Study of the RMG material was performed under an official government permit issued by the Consejo de Arqueología, Instituto Nacional de Antropología e Historia in Mexico City, under the auspices of the Instituto de Investigaciones Antropológicas of the Universidad Nacional Autónoma de México. Studies of material from Milpillas, Cuexcomate and Capilco, and Vista Hermosa and Platanito were carried out under permits granted by the Consejo de Arqueología; the Lamanai objects were studied with the permission of the Commissioner of Archaeology of Belize.

Study of the MAG material was performed under a formal *Convenio* signed by the Museos del Banco Central del Ecuador and the Center for Materials Research in Archaeology and Ethnology at the Massachusetts Institute of Technology. The materials from Salango, El Azúcar, OGSE-Ma-172, Cerro Alto, and Loma de los Cangrejitos were included in this study but also received study permits issued by the Instituto Nacional de Patrimonio Cultural del Ecuador.

Table A1.3 Analytical Method and Number of Analyses Performed

Analytical Technique	Number of Artifacts Analyzed
Metallography	175
Microhardness tests	95
Qualitative chemical analyses (emission spectrography)*	374
Semiquantitative chemical analyses (emission spectrography)*	25
Quantitative chemical analyses (atomic absorption)**	214
Quantitative chemical analyses (neutron activation)	102***

* The results of these analyses are found in Hosler (1986).
** Twenty of these analyses are for axe-monies and are published in Hosler, Lechtman, and Holm (1990).
*** Of this number, 53 had also been analyzed by atomic absorption; these results are presented in appendix 2.

Table A1.4 Artifact Types from Excavated Mesoamerican Assemblages: Total in Assemblage and Number Analyzed, by Site

Type	Milpillas Total in Assemblage	Milpillas Number Analyzed*	Capilco and Cuexcomate Total in Assemblage	Capilco and Cuexcomate Number Analyzed*	Lamanai Total in Assemblage	Lamanai Number Analyzed*	Vista Hermosa and Platanito Total in Assemblage	Vista Hermosa and Platanito Number Analyzed*
Awls/chisels**	—	—	14	14	1	1	—	—
Axes	—	—	—	—	13	6	4	4
Bells	20	19	5	5	39	13	101	45
Bell clappers	—	—	—	—	—	—	8	1
Buttons	—	—	—	—	13	6	—	—
Finger rings	—	—	—	—	6	3	—	—
Fishhooks	—	—	—	—	3	1	—	—
Ingots	—	—	—	—	4	4	3	3
Needles	1	1	16	11	5	3	2	0
Open rings	11	1	—	—	—	—	—	—
Ornaments	—	—	1	1	—	—	—	—
Sheet	3	0	2	2	1	1	—	—
Tweezers	—	—	2	2	3	3	—	—
Wire	—	—	5	5	—	—	—	—
Miscellaneous	—	—	—	—	14	4	3	1
Total	35	21	45	40	102	45	121	54

* Number indicates qualitative analyses only.
** "Chisels" refers to awls with narrow but flaring blades.

Table A1.5 Artifact Types from Excavated Ecuadorian Assemblages: Total in Assemblage and Number Analyzed, by Site

Type	Salango Total in Assemblage	Salango Number Analyzed	El Azúcar Total in Assemblage	El Azúcar Number Analyzed	Loma de los Cangrejitos Total in Assemblage	Loma de los Cangrejitos Number Analyzed*	Cerro Alto and OGSE-Ma-172 Total in Assemblage	Cerro Alto and OGSE-Ma-172 Number Analyzed
Awls	—	—	—	—	—	—	—	—
Axes/chisels	—	—	—	—	6	3	—	—
Axe-monies	—	—	—	—	14	3	4	0
Beads	9	0	2	0	—	—	—	—
Bells	1	1	2	0	27	6	—	—
Earrings	6	—	—	—	—	—	—	—
Fishhooks	—	—	—	—	—	—	2	2
Knives	—	—	—	—	1	1	—	—
Needles	9	4	2	1	11	1	2	2
Noserings	2	0	—	—	—	—	—	—
Pendants	—	—	—	—	2	2	—	—
Open rings	5	3	2	2	3	0	—	—
Stars	—	—	2	2	—	—	1	1
Tools	10	1	1	1	20	5	1	1
Tweezers	2	1	1	1	16	5	1	1
Wire	—	—	—	—	—	—	1	0
Other	6	—	2	0	1	1	1	0
Total	50	10	14	7	101	27	13	7

* Number indicates qualitative analyses; quantitative data are presented in table 4.7.

APPENDIX 2

Quantitative Chemical Analyses of Artifacts in the RMG Collection

Table A2.1 Quantitative Chemical Analyses of Artifacts in the RMG Collection*

Artifact Type**	ID No.	Composition (weight percent)												
		Ag	As	Au	Bi	Cu	Fe	In	Mg	Ni	Pb	Sb	Sn	Zn
Copper														
Awl	114	0.03	na	na	na	na	0.01	na	0.0096	na	0.0006	na	na	na
	115	0.05	na	na	na	na	0.01	na	0.001	na	na	na	na	na
	117	0.05	na	na	na	na	0.06	na	na	na	na	na	na	na
	796	0.04	na	na	na	na	0.04	na	na	na	na	na	na	na
	799	0.17	na	na	na	na	0.07	na	0.0016	na	na	na	na	na
	872	0.63	0.22	—	—	na	0.0077	na	—	—	—	0.44	—	na
Alloy														
Awl	112	0.035	2.01	na	—	na	na	na	na	—	—	0.01	0.19	na
Copper														
Axe	354	0.05	na	na	na	na	0.02	na	0.001	na	0.0072	na	na	na
	357	0.1	na	na	na	na	0.01	na	0.001	na	na	na	na	na
	359	0.08	na	na	na	na	0.01	na	0.0001	na	0.0057	na	na	na
	371	0.23	na	na	na	na	0.01	na	0.001	0.0023	0.0068	na	na	na
	372	0.198	0.14	—	—	na	0.04	na	—	—	—	—	—	na
	373	0.22	—	—	—	na	0.25	na	—	0.03	—	—	—	na
	380	0.09	na	na	na	na	0.03	na	0.0044	na	0.02	na	na	na
	381	0.05	na	na	na	na	0.01	na	0.001	na	0.0071	na	na	na
	387	0.15	—	na	—	na	na	na	na	na	—	—	—	na
	388	0.22	—	—	—	na	—	na	—	—	—	0.02	—	na
	391	0.28	na	na	na	na	0.01	na	na	na	na	na	na	na
	393	0.04	na	na	na	na	0.04	na	na	na	na	na	na	na
	396	0.28	—	—	—	na	0.051	na	—	—	—	—	0.01	na
	397	0.05	na	na	na	na	0.01	na	0.001	na	na	na	na	na
	406	0.07	na	na	na	na	0.01	na	na	na	na	na	na	na
	853	0.06	na	na	na	na	0.01	na	na	na	na	na	na	na
	2396	0.05	na	na	na	na	0.04	na	0.001	na	0.01	na	na	na

continued

Table A2.1 (continued)

Artifact Type**	ID No.	Composition (weight percent)												
		Ag	As	Au	Bi	Cu	Fe	In	Mg	Ni	Pb	Sb	Sn	Zn
Alloy														
Axe	28	0.06	1.22	na	0.049	na	na	na	na	na	—	0.0097	0.0097	na
	351	0.07	0.71	na	—	na	na	na	na	na	0.1	0.04	1.33	na
	351^n	0.0622	0.716	0.0006	na	na	—	0.0021	na	na	na	0.0891	1.15	—
	367	0.12	0.09	na	na	na	na	0.06	na	na	na	na	2.48	na
	367^n	0.1099	0.007	—	na	na	0.0133	0.0052	na	na	na	0.0175	2.56	0.0017
	369	0.11	0.64	na	na	na	na	0.02	na	na	na	0.18	5.31	na
	370	0.61	4.84	—	—	na	0.042	na	—	0.021	0.116	0.42	—	na
	370^n	0.5662	5.675	—	na	na	—	—	na	na	na	0.4831	—	—
	374	0.04	0.08	na	na	na	na	na	na	na	na	na	7.92	na
	374^n	0.0402	0.113	0.0001	na	na	—	0.0269	na	na	na	0.0847	8.72	0.0008
	378	0.03	0.96	na	—	na	na	na	na	—	0.015	—	0.12	na
	379	0.03	0.06	na	—	na	na	—	na	na	0.01	—	0.77	na
	385	0.06	0.71	na	—	na	na	na	na	na	—	0.28	—	na
	386	0.27	1.31	na	—	na	na	na	na	na	0.11	0.1	3.10	na
	401	0.05	—	na	—	na	na	na	na	na	—	—	6.22	na
	402^n	0.1	0.12	—	na	na	0.0159	0.0018	na	na	na	0.006	1.26	0.0082
	403b	0.12	0.07	na	—	na	na	0.05	na	na	—	0.03	8.06	na
	2249	0.6	0.32	na	na	na	na	0.29	na	na	0.008	na	1.94	na
	2249^n	0.0564	0.615	0.0003	na	na	—	0.003	na	na	na	0.123	1.45	—
	2311	0.16	1.83	na	—	na	na	na	na	na	0.06	na	0.15	na
Copper														
Axe-money***	487	0.004	0.3	na	0.008	na	na	na	na	0.05	—	0.008	0.047	na
	501	0.036	—	na	—	na	na	na	na	0.027	—	0.43	0.027	na
	510	—	0.05	na	—	na	na	na	na	0.13	—	0.09	0.03	na
Alloy														
Axe-money***	264	0.04	0.44	na	—	na	na	na	na	—	—	0.04	0.024	na
	302^n	0.2312	3.381	—	na	na	—	0	na	na	na	0.0077	—	—
	449^n	0.1671	3.707	—	na	na	—	—	na	na	na	0.0332	—	—

continued

Table A2.1 (continued)

Artifact Type**	ID No.	Composition (weight percent)												
		Ag	As	Au	Bi	Cu	Fe	In	Mg	Ni	Pb	Sb	Sn	Zn
	463	0.46	6.35	na	—	na	na	na	na	0.05	—	0.28	0.023	na
	471[n]	0.2198	3.954	—	na	na	—	0.0001	na	na	na	0.171	—	—
	486	0.175	3.07	na	—	na	na	na	na	0.093	—	0.23	—	na
	489	0.047	0.81	na	—	na	na	na	na	0.103	0.0373	0.05	—	na
	496[n]	0.6918	5.333	0.0003	na	na	—	—	na	na	na	0.1449	—	0.003
Copper														
Bell	124b	0.08	—	na	—	na	na	na	na	—	—	0.42	na
	130a	0.19	—	—	0.0014	na	0.34	na	—	0.0043	0.008	0.01	—	na
	205[n]	0.0562	—	0.0025	na	na	—	0.0006	na	na	na	0.2999	—	—
	209	0.06	na	na	na	na	0.02	na	0.001	0.0014	0.0043	na	na	na
	213	0.06	0.14	na	—	na	na	na	na	—	—	—	0.96	na
	219b	0.054	—	na	—	na	na	na	na	0.014	—	0.09	—	na
	723	0.163	—	—	—	na	0.043	na	—	—	0.0109	0.01	—	na
	838	0.047	—	na	—	na	na	na	na	—	—	0.008	—	na
	1228	0.22	0.09	—	0.0026	na	0.041	na	0.001	0.024	0.0026	—	—	na
	1246[n]	0.0873	0.006	0.0009	na	na	0.0166	0.001	na	na	na	0.0023	0.53	—
	1437	0.011	—	na	0.023	na	na	na	na	0.023	—	—	—	na
	1446	0.03	0.35	na	—	na	na	na	na	na	—	—	0.09	na
	1539	0.11	—	—	0.0013	na	0.29	na	0.001	0.0013	0.0066	0.0159	—	na
	1546	0.019	—	na	—	na	na	na	na	—	—	—	—	na
	1608	0.148	0.3	na	—	na	na	na	na	—	—	0.16	0.022	na
	1825	0.032	0.11	na	—	na	na	na	na	0.024	—	—	na	na
	2080	0.053	—	—	—	na	0.16	na	0.0011	0.009	—	0.01	—	na
	2126	0.089	—	na	—	na	na	na	na	0.063	—	0.09	—	na
	2411	0.08	na	na	0.03	na	na	na	na	0.04	na	na	na	na
	2413	0.07	na	na	na	na	0.02	na	na	na	na	na	na	na
	2440b	0.11	0.34	na	—	na	na	na	na	na	0.078	0.068	0.58	na
	2538	0.02	0.38	na	—	na	na	na	na	na	—	—	0.16	na
	2550	0.127	0.19	na	—	na	na	na	na	na	—	0.009	0.16	na
	2720	0.104	0.25	na	—	na	na	na	na	—	0.019	0.07	—	na

continued

Table A2.1 (continued)

Artifact Type**	ID No.	Composition (weight percent)												
		Ag	As	Au	Bi	Cu	Fe	In	Mg	Ni	Pb	Sb	Sn	Zn
	2786	0.07	0.15	—	—	na	0.015	na	—	—	—	0.02	—	na
	2791	0.03	—	na	—	na	na	na	na	na	—	—	0.27	na
	Fa1l	0.093	—	na	0.031	na	na	na	na	0.02	—	—	0.17	na
Alloy														
Bell	128a	0.03	0.45	na	na	na	na	na	na	na	0.01	0.02	6.93	na
	128b[n]	0.0251	0.436	—	na	na	0.0277	0.007	na	na	na	0.0237	7.77	0.0015
	195[6]	0.2	0.01	na	—	na	na	0.02	na	na	—	—	3.08	na
	195[6][n]	0.1109	0.015	0	na	na	0.0486	0.0119	na	na	na	0.0183	3.56	—
	195[8][n]	0.113	0.013	0	na	na	0.0379	0.0109	na	na	na	0.0163	3.28	0.0019
	197	0.025	1.35	na	—	na	na	na	na	—	—	0.01	—	na
	198[n]	0.0201	0.07	0	na	na	0.0259	0.022	na	na	na	0.0385	12.18	0.0011
	201	0.11	22.12	0.015	0.144	na	0.15	na	0.1	0.015	0.015	1.06	—	na
	204	0.07	12.82	na	0.0188	na	na	na	na	0.01	0.0199	0.22	—	na
	207	0.11	23.47	na	0.08	na	na	na	na	0.02	0.0095	1.08	0.03	na
	816	96.92	na	na	na	3.08	na	na	na	na	na	na	na	na
	891	0.13	na	na	na	na	na	na	na	na	0.009	na	10.43	na
	891[n]	0.134	0.015	0.0008	na	na	0.0193	0.0132	na	na	na	0.0345	11.08	—
	893	0.05	0.17	na	na	na	na	0.04	na	na	na	0.03	12.3	na
	893[n]	0.0659	0.1	0.0004	na	na	0.0736	0.0179	na	na	na	0.064	15.77	0.0028
	895	0.05	0.23	na	na	na	na	0.02	na	na	0.02	na	10.55	na
	895[n]	0.0615	0.27	0.0001	na	na	0.0759	0.0152	na	na	na	0.0522	13.12	0.0164
	897	0.16	—	na	—	na	na	0.01	na	na	—	na	7.27	na
	897[n]	0.1549	—	0.0001	na	na	—	0.01	na	na	na	0.3311	6.45	—
	910	0.13	—	na	—	na	na	—	na	na	—	na	11.02	na
	910[n]	0.1439	0.014	0.0008	na	na	—	0.0132	na	na	na	0.0385	11.35	—
	1220	0.01	—	na	—	na	na	0.01	na	na	—	na	3.24	na
	1220[n]	0.0156	0.012	0.0002	na	na	0.1059	0.0062	na	na	na	0.0107	3.23	0.0108
	1253	0.15	0.02	na	na	na	na	0.03	na	na	0.01	na	4.81	na
	1440[n]	0.104	0.158	0	na	na	0.1991	0.0087	na	na	na	0.0198	4.14	0.0027
	1473	0.02	0.26	na	—	na	na	0.02	na	na	—	—	4.00	na

continued

Table A2.1 (continued)

Artifact Type**	ID No.	Composition (weight percent)												
		Ag	As	Au	Bi	Cu	Fe	In	Mg	Ni	Pb	Sb	Sn	Zn
	1473[n]	0.0628	8.69	0.0038	na	na	—	0	na	na	na	0.1611	—	—
	1474[n]	0.0628	8.69	0.0038	na	na	—	0	na	na	na	0.1611	—	—
	1475	0.1	12.9	na	na	na	na	na	na	na	na	0.32	na	na
	1479[n]	0.018	0.116	0	na	na	0.0182	0.0029	na	na	na	0.0066	2.34	0.0004
	1484	0.1	0.1	na	na	na	na	0.01	na	na	0.03	na	3.49	na
	1484[n]	0.103	0.125	0.0022	na	na	0.023	0.0046	na	na	na	0.0209	3.64	—
	1485	0.05	13.8	na	0.0081	na	na	na	na	0.014	0.01	0.21	—	na
	1487[n]	0.0183	0.155	0	na	na	0.0282	0.0034	na	na	na	0.0079	1.62	0.0008
	1493[n]	0.1419	0.007	0	na	na	0.0318	0.0008	na	na	na	0.0267	2.12	0.0016
	1499	0.02	0.19	na	na	na	na	na	na	na	0.008	na	7.12	na
	1499[n]	0.0242	0.217	0	na	na	0.0394	0.0191	na	na	na	0.0265	7.73	0.0025
	1526[n]	0.0234	0.867	0	na	na	0.2748	0.0004	na	na	na	0.0041	—	0.0291
	1532[n]	0.0403	1.191	0.0001	na	na	0.0217	0.0001	na	na	na	0.0044	—	0.0032
	1595	0.05	1.23	na	—	na	na	na	na	na	—	—	0.02	na
	1631[n]	0.0863	0.012	0	na	na	0.0366	0.0094	na	na	na	0.0186	5.38	—
	1795	0.03	1.94	na	—	na	na	na	na	na	—	—	—	na
	1798	0.06	1.11	na	0.0015	na	na	na	na	0.0015	0.0015	—	—	na
	1860[n]	0.0188	0.131	0	na	na	0.0295	0.0097	na	na	na	0.0228	8.54	0.0016
	2183	0.1	0.11	na	—	na	na	0.06	na	na	—	—	6.38	na
	2387[n]	0.0601	0.023	0.0003	na	na	0.0217	0.0218	na	na	na	0.0542	18.1	0.001
	2388[n]	0.0598	0.126	0.0003	na	na	0.0486	0.0428	na	na	na	0.0638	19.98	0.0019
	2492[n]	0.0399	0.179	0	na	na	0.0257	0.0097	na	na	na	0.0369	8.37	0.0032
	2495	0.03	na	na	na	na	na	na	na	na	na	na	4.78	na
	2495[n]	0.0364	0.038	0	na	na	0.0261	0.0069	na	na	na	0.0149	4.58	—
	2539	0.31	0.1	na	—	na	na	—	na	na	0.01	—	5.67	na
	2539[n]	0.1982	0.033	0	na	na	0.1069	0.007	na	na	na	0.0314	7.01	0.0058
	2571	0.03	0.21	na	—	na	na	—	na	na	—	—	2.81	na
	2571[n]	0.0406	0.14	0	na	na	0.0287	0.0045	na	na	na	0.0237	3.33	0.0016
	2589	0.04	0.49	na	—	na	na	na	na	na	—	—	0.17	na
	2650[n]	0.5058	0.358	0.0001	na	na	0.0465	0.0139	na	na	na	0.0931	8.47	—

continued

Table A2.1 (continued)

Artifact Type**	ID No.	Composition (weight percent)												
		Ag	As	Au	Bi	Cu	Fe	In	Mg	Ni	Pb	Sb	Sn	Zn
	2724	0.06	1.00	na	—	na	na	na	na	na	—	—	0.04	na
	Fa12	0.14	—	na	—	na	na	na	na	na	—	—	2.23	na
	Fx6[n]	0.0975	0.025	0.0011	na	na	—	0.0202	na	na	na	0.0524	13.06	0.0026
	H1[n]	0.0859	1.849	0.0005	na	na	0.125	0.0001	na	na	na	0.0099	—	0.0055
Copper Button	2634	0.04	0.02	na	—	na	na	na	na	na	—	0.059	—	na
Copper Hoe	365	0.004	—	—	—	na	na	na	na	—	—	0.05	—	na
Alloy Hoe	32b	0.02	2.3	na	0.0007	na	na	na	na	0.0053	0.0013	0.0079	na	na
	2395	0.03	1.63	na	0.0032	na	na	na	na	0.0065	0.0032	0.0071	—	na
Copper Needle	80	0.04	na	na	0.0022	na	na	na	na	0.002	0.0022	0.002	—	na
	83	0.0086	0.16	na		na	na	na	na	—	—	0.009	0.035	na
	806	0.18	—	—	—	na	—	na	—	—	—	0.02	—	na
	2459	0.11	0.05	—	—	na	0.018	na	—	—	—	0.03	—	na
	2576	0.14	—	—	—	na	0.06	na	0.001	0.0024	0.0024	—	—	na
	2575	0.07	—	na	0.035	na	na	na	na	0.058	—	—	—	na
	2577	0.027	—	na	0.027	na	na	na	na	0.027	—	—	—	na
	2677	0.07	—	—	—	na	0.067	na	—	0.07	0.0074	—	—	na
	2689	0.1	0.04	—	—	na	—	na	—	—	—	0.009	—	na
	Fx14	0.06	na	na	na	na	0.03	na	0.0015	na	0.0119	na	na	na
	Fx15	0.039	na	0.28	na	na	na	na	—	—	0.0098	0.01	—	na
Alloy Needle	74	0.04	1.61	na	0.0026	na	na	na	na	na	0.0026	0.0065	—	na
	98	0.03	2.3	na	0.0016	na	na	na	na	0.0016	0.0016	0.0031	0.03	na
	804a1	0.03	2.17	na	—	na	na	na	na	na	—	0.0097	—	na

continued

Table A2.1 (continued)

Artifact Type**	ID No.	Composition (weight percent)												
		Ag	As	Au	Bi	Cu	Fe	In	Mg	Ni	Pb	Sb	Sn	Zn
	804a2	0.03	2.12	na	—	na	na	na	na	na	—	0.019	—	na
	804a3	0.03	2.13	na	na	na	na	na	na	na	na	na	na	na
	804a4	0.03	2.13	na	0.0016	na	na	na	na	na	0.0032	—	na	na
	804a5	0.03	2.13	na	—	na	na	na	na	na	—	0.098	0.07	na
	2352	0.04	0.26	na	na	na	na	na	na	na	—	na	11.81	na
	2352[n]	0.0391	0.182	0	na	na	0.0147	0.0162	na	na	na	0.0348	13.12	0.0011
Copper														
Open ring	36a	0.75	na	na	na	na	0.01	na	0.001	na	0.09	na	na	na
	997	na	na	na	na	na	0.01	na	na	na	na	na	na	na
	1664	0.18	—	—	—	na	0.015	na	—	—	0.0677	—	0.015	na
Alloy														
Open ring	36d[n]	0.0335	0.039	0.001	na	na	—	0.0232	na	na	na	0.0566	9.57	—
	39b[n]	0.1429	0.107	0.0001	na	na	—	0.0195	na	na	na	0.0435	12.44	—
	40a	0.01	0.04	na	na	na	na	0.02	na	na	0.03	na	8.89	na
	40a[n]	0.0136	0.157	—	na	na	—	0.0145	na	na	na	0.0308	10.16	—
	41a[n]	0.0084	0.034	0	na	na	—	0.0038	na	na	na	0.0352	10.41	0.0047
	45a[n]	0.0186	0.119	0	na	na	0.0179	0.0187	na	na	na	0.0308	10.49	0.0004
	46b[n]	0.0202	0.186	0	na	na	0.0165	0.0153	na	na	na	0.0377	10.08	—
	532[n]	0.0778	0.03	0.0001	na	na	0.011	0.005	na	na	na	0.0467	9.91	0.0011
	535[n]	0.1079	0.079	0	na	na	0.0081	0.0179	na	na	na	0.032	9.91	—
	541[n]	0.014	0.076	0	na	na	0.07	0.0102	na	na	na	0.0265	8.89	0.0017
	578[n]	0.1119	0.055	—	na	na	0.0065	0.0241	na	na	na	0.0468	10.08	0.0006
	598[n]	0.0128	0.119	—	na	na	0.0048	0.0177	na	na	na	0.031	11.08	0.0006
	604[n]	0.0605	0.03	0	na	na	0.0239	0.0076	na	na	na	0.0376	8.89	—
	620	0.01	0.19	na	—	na	na	0.06	na	na	—	—	8.5	na
	631[n]	0.0356	0.076	0.0001	na	na	0.0252	0.006	na	na	na	0.0205	6.29	0.0014
	635[n]	0.0154	0.123	—	na	na	0.0352	0.0276	na	na	na	0.0361	12.44	0.001
	643	0.26	0.19	na	—	na	na	0.03	na	na	0.23	na	7.46	na
	643[n]	0.2301	0.198	0.0005	na	na	—	0.0289	na	na	na	0.1321	6.11	—

continued

Table A2.1 (continued)

Artifact Type**	ID No.	Composition (weight percent)												
		Ag	As	Au	Bi	Cu	Fe	In	Mg	Ni	Pb	Sb	Sn	Zn
	682a	0.19	0.07	na	—	na	na	0.03	na	na	—	na	13.38	na
	682a[n]	0.1871	0.076	0.0003	na	na	—	0.0293	na	na	na	0.0533	14.75	—
	682b	0.04	0.49	na	—	na	0.02	na	na	na	—	na	9.26	na
	682b[n]	0.0385	0.318	0	na	na	—	0.0152	na	na	na	0.0425	10.93	—
	873	0.17	0.05	na	na	na	na	na	na	na	0.009	na	15.16	na
	873[n]	0.1871	0.104	0.0001	na	na	—	0.02	na	na	na	0.0414	17.77	—
	874	93.43	na	0.29	na	6.28	na	na	na	na	na	na	na	na
	1653[n]	0.1059	0.032	0	na	na	0.0229	0.0225	na	na	na	0.0488	11.85	0.0022
	1657[n]	0.1469	0.022	0	na	na	0.0169	0.0063	na	na	na	0.0163	5.84	0.0006
	1665	0.09	na	na	na	na	na	na	na	na	0.01	na	10.27	na
	1665[n]	0.1159	0.046	0	na	na	0.0327	0.0164	na	na	na	0.0388	12.44	0.0024
	1666[n]	0.0851	0.032	0.0006	na	na	—	0.0169	na	na	na	0.0431	11.08	0.0019
	1667[n]	0.0417	0.03	0.0001	na	na	—	0.0161	na	na	na	0.0469	8.07	0.0014
	1668[n]	0.0086	0.019	0.0033	na	na	0.0091	0.004	na	na	na	0.0333	7.1	0.0009
	1673[n]	0.064	0.024	0.0024	na	na	—	0.0048	na	na	na	0.0282	6.21	—
	1718	0.05	0.13	na	na	na	na	na	na	na	0.02	na	10.94	na
	1718[n]	0.0598	0.111	0	na	na	—	0.0325	na	na	na	0.0521	12.78	0.0055
	1737[n]	0.0538	0.038	0.0001	na	na	—	0.0204	na	na	na	0.0521	16.86	0.0012
	1744[n]	0.0313	0.046	0.0002	na	na	—	0.0087	na	na	na	0.0296	10.58	0.0011
	1746[n]	0.0809	0.074	0	na	na	0.0325	0.0156	na	na	na	0.0306	11.35	—
	1747	0.04	na	na	na	na	na	0.02	na	na	0.03	na	12.85	na
	1748[n]	0.031	0.048	0.0002	na	na	0.0173	0.0121	na	na	na	0.0395	13.64	—
	2313	0.05	0.02	na	0.008	na	na	—	na	na	0.02	na	10.43	na
	2399[n]	1.38	6.87	—	na	na	—	0	na	na	na	—	—	—
	2419[n]	0.0227	0.225	0	na	na	0.0703	0.0179	na	na	na	0.0345	9.75	0.0007
	2421[n]	1.13	4.76	—	na	na	—	0	na	na	na	—	—	—
	Fa8	0.06	0.2	na	—	na	na	—	na	na	—	na	7.95	na
Alloy														
Ornament	1865	0.16	0.32	na	—	na	na	na	na	na	—	0.087	1.27	na
	1865[n]	0.1429	0.282	0.0004	na	na	—	0.0013	na	na	na	0.0068	1.09	0.0024

continued

Table A2.1 (continued)

Artifact Type**	ID No.	Composition (weight percent)												
		Ag	As	Au	Bi	Cu	Fe	In	Mg	Ni	Pb	Sb	Sn	Zn
Alloy														
Pin bell	2454b	0.04	0.77	na	—	na	na	—	na	na	—	na	0.08	na
Silver														
Sheet	20	99.12	na	0.13	na	0.75	na	na	na	na	na	na	0.01	na
	231	99.46	na	0.011	na	0.53	na	na	na	na	na	na	na	na
	235	98.37	na	na	na	1.63	na	na	na	na	na	na	na	na
	2627	99.99	na	na	na	0.0087	na	na	na	na	na	na	na	na
	2628	99.98	na	na	na	0.019	na	na	na	na	na	na	na	na
Alloy														
Sheet	19	96.33	na	na	na	3.67	na	na	na	na	na	na	na	na
	21	85.22	na	0.072	na	14.7	na	na	na	na	na	na	na	na
	23b	26.1	na	0.024	na	73.88	na	na	na	na	na	na	na	na
	229	14.2	—	0.015	—	na	—	na	—	—	0.0296	—	—	na
	234	33.9	na	na	na	66.1	na	na	na	na	na	na	na	na
	885	70.5	na	na	na	29.5	na	na	na	na	na	na	na	na
	2586	95.22	na	na	na	4.78	na	na	na	na	na	na	na	na
	Fx	93.33	na	0.034	na	6.67	na	na	na	na	na	na	na	na
	Fx1	93.3	na	0.28	na	6.43	na	na	na	na	na	na	na	na
	Fx2a	94.4	na	0.011	na	5.5	na	na	na	na	na	na	0.08	na
	Fx4	88.84	na	0.059	na	11.1	na	na	na	na	na	na	na	na
	Fx5	0.16	—	na	—	na	na	na	na	na	—	0.102	7.56	na
	Fx5^n	0.1919	0.018	0.0005	na	na	—	0.0106	na	na	na	0.0308	8.37	0.0034
Copper														
Tweezer	5	0.13	na	na	0.0013	na	na	na	na	na	0.0013	0.0106	0.008	na
	224	0.63	—	—	—	na	0.029	na	0.0097	—	—	—	—	na
	225	0.13	na	na	0.0007	na	na	na	na	0.0015	0.0029	0.0161	—	na
	2345	0.06	na	na	na	na	0.02	na	na	na	na	na	na	na

continued

Table A2.1 (continued)

Artifact Type**	ID No.	Composition (weight percent)												
		Ag	As	Au	Bi	Cu	Fe	In	Mg	Ni	Pb	Sb	Sn	Zn
	2346	0.15	na	na	0.0013	na	0.01	na	0.001	na	0.09	na	na	na
	2515	0.17	0.35	na	0.0015	na	na	na	na	na	0.0015	0.012	0.01	na
	2657b	0.03	—	na	—	na	na	na	na	na	—	—	—	na
	2678	0.11	na	na	na	na	0.01	na	na	na	na	na	na	na
	2679	0.098	—	—	—	na	0.102	na	—	—	—	—	—	na
	2686	0.14	0.13	na	0.0022	na	0.0423	na	0.011	—	0.0066	0.02	—	na
Alloy														
Tweezer	2	0.08	—	na	—	na	na	na	na	na	—	—	2.54	na
	3	0.03	2.7	na	—	na	na	na	na	na	—	—	—	na
	4	0.03	3.02	na	0.0024	na	na	na	na	0.017	—	0.04	—	na
	7	0.03	0.06	na	na	na	0.01	na	na	na	0.02	na	10.68	na
	8	0.11	2.25	na	na	na	0.04	na	na	na	0.02	0.04	10.06	na
	8^n	0.1191	0.089	0.0003	na	na	0.0185	0.0283	na	na	na	0.0483	10.41	—
	9	0.1	0.39	na	na	na	0.13	na	na	na	na	na	9.93	na
	9^n	0.1069	0.628	0.0004	na	na	0.0225	0.0201	na	na	na	0.0881	11.26	0.0015
	11	0.14	1.19	na	na	na	na	na	na	na	0.01	0.16	9.09	na
	11^n	0.1361	1.4	0.0046	na	na	0.1159	0.0125	na	na	na	0.175	8.81	—
	12	0.09	na	na	na	na	na	na	na	na	0.009	na	4.34	na
	12^n	0.0959	0.017	0.0006	na	na	0.0489	0.0054	na	na	na	0.0291	5.72	—
	13	0.17	1.14	na	na	na	na	na	na	na	na	0.16	8.96	na
	13^n	0.1429	1.449	0.0003	na	na	0.0483	0.0127	na	na	na	0.1641	8.43	—
	32a	0.05	0.78	na	na	na	na	na	na	na	0.03	0.09	5.87	na
	$32a^n$	0.0561	0.766	0.0004	na	na	—	0.0046	na	na	na	0.125	5.02	—
	339/76	0.14	0.79	na	na	na	na	na	na	na	0.01	na	9.55	na
	$339/76^n$	0.1449	1.03	0.0004	na	na	—	0.0094	na	na	na	0.0451	9.57	—
	939/76	24.3	na	0.009	na	75.69	na	na	na	na	na	na	na	na
	2017	0.09	na	na	na	na	na	na	na	na	0.01	na	9.6	na
	2017^n	0.0879	0.033	0.0002	na	na	—	0.0123	na	na	na	0.0293	10.76	0.0043
	2343	0.1	1.21	na	0.02	na	na	—	na	na	0.23	0.17	11.76	na

continued

Table A2.1 (continued)

Artifact Type**	ID No.	Composition (weight percent)												
		Ag	As	Au	Bi	Cu	Fe	In	Mg	Ni	Pb	Sb	Sn	Zn
	2343[n]	0.0953	1.211	0.0017	na	na	—	0.0146	na	na	na	0.1581	10.66	—
	2344	0.12	2.03	na	0.01	na	na	na	na	na	0.06	na	8.84	na
	2344[n]	0.1219	2.208	0.0012	na	na	0.0499	0.0127	na	na	na	0.1932	9.81	—
	2513	0.07	0.28	na	0.02	na	na	na	na	na	na	0.01	10.34	na
	2513[n]	0.0684	0.155	0.0006	na	na	—	0.0264	na	na	na	0.0718	14.15	0.0009
	2516	0.12	0.19	na	na	na	na	na	na	na	0.02	0.04	8.18	na
	2516[n]	0.1169	0.135	0.0005	na	na	0.0798	0.0092	na	na	na	0.0729	9.14	0.0043
	2517	0.08	0.12	na	na	na	na	na	na	na	0.5	na	9.73	na
	2517[n]	0.0818	0.479	0.0005	na	na	0.0152	0.0105	na	na	na	0.0466	10.08	—
	2518	55.9	na	na	na	44.1	na	na	na	na	na	na	na	na
	2528a	0.03	0.15	na	na	na	na	na	na	na	na	na	11.02	na
	2528a[n]	0.0282	0.161	0.0003	na	na	0.1079	0.0072	na	na	na	0.0397	10.58	0.0037
	2528b	0.03	0.14	na	na	na	na	na	na	na	0.01	na	10.7	na
	2528b[n]	0.0313	0.179	0.0003	na	na	—	0.0088	na	na	na	0.0431	12.02	0.0017
	2556[n]	0.1089	1.871	—	na	na	—	0.0098	na	na	na	0.3304	5.06	—
	2617	0.06	4.43	na	—	na	na	na	na	na	—	0.028	0.09	na
	2647	0.13	0.18	na	—	na	na	—	na	na	0.01	—	10.11	na
	2647[n]	0.1671	0.085	0.0001	na	na	—	0.0083	na	na	na	0.0455	13.12	—
	2656	0.1	0.93	na	—	na	na	0.11	na	na	0.01	0.24	5.52	na
	2680	0.1	0.76	na	na	na	na	na	na	na	0.06	0.1	6.36	na
	2680[n]	0.1109	1.079	0.001	na	na	0.0179	0.0051	na	na	na	0.14	7.99	—
	2682	0.11	0.75	na	na	na	na	na	na	na	0.01	0.13	6.46	na
	2682[n]	0.0828	0.875	0.0011	na	na	—	0.0204	na	na	na	0.1409	5.74	—
	2683	0.13	2.43	na	na	na	na	na	na	na	0.01	0.1	6.02	na
	2683[n]	0.1489	2.761	0.0028	na	na	0.0256	0.01	na	na	na	0.1449	7.36	—
	2684	0.06	0.62	na	—	na	na	0.01	na	na	—	na	7.93	na
	2684[n]	0.0518	0.56	0.0012	na	na	—	0.0146	na	na	na	0.1622	7.49	0.0018
	2687	0.07	0.81	na	—	na	na	na	na	na	0.5	0.12	5.69	na
	2687[n]	0.0665	0.83	0.0012	na	na	—	0.0064	na	na	na	0.1309	5.4	—
	2688[n]	0.0847	0.875	0.001	na	na	—	0.0162	na	na	na	0.166	10.58	—

Note: Analyses carried out by atomic absorption spectrometry and neutron activation (^{n} indicates analysis by neutron activation). A dash indicates element not detected in quantitative analysis (usually present in qualitative); "na" indicates element not analyzed (not detected in qualitative analyses for most cases; see Hosler 1986).

* See Hosler (1986) for qualitative analyses of these objects.

** Hosler 1988a indicates subtype designations for most of these objects.

*** See Hosler, Lechtman, and Holm (1990) for analyses of other type 1a axe-monies.

Notes

1. The Perspective and the Region

1. Styles of taking data, interpretations of laboratory information, and the questions that generate research inevitably incorporate and reproduce dominant paradigms, intellectual fads, and biases of class and gender. However, the composition of a 5% tin bronze axe is normally invariant, and the microstructure of the same axe unambiguously reveals how it was made, provided it has not been deformed after deposition and that corrosion is minimal. In other words, laboratory-analytical data can be replicated and will yield the same results long after paradigms have shifted.

2. The equation of technology with tools is discussed in Hosler (1986); also see Childe (1944, 1983) and Adams (1975). Pfaffenberger (1988) provides a useful summary and criticism of this point of view. Ingold (1990) examines the notion in western culture that technical activity means tool-using, arguing that western societies confuse the "technical" with the mechanical. Unfortunately Cotterell and Kamminga's (1990) recent commendable examination of the mechanics of preindustrial technology contributes little to alter this equation, since almost all of their examples represent tool and machine inventories.

3. Staudenmaier (1985) presents a lucid review of the emergence of the history of technology as a discipline, and traces the contextualist school's interest in examining technology in its social milieu. M. R. Smith (1977) provides a stellar example of this perspective.

4. Several notable exceptions exist, for example Childs (1986), De Atley (1986), Hosler (1986), Killick (1990), Lechtman and Steinberg (1970), and Steinberg (1973).

5. A large body of literature exists on these topics; the brief selection listed here includes works that focus primarily on production technologies. Ceramics: London (1991), Steponaitis (1984), Stimmell and Stromberg (1986), van As (1989), Vandiver and Koehler (1986). Cloth: Bolland (1991), Conklin (1979), Frame (1986), Paul and Niles (1984), Rowe (1980). Stone: McAnany (1989), Shafer and Hester (1991), Torrence (1984), Yerkes (1989), Young and Bonnichsen (1984).

6. See, for example, Braun (1983); Hosler (1986); Schiffer (1992); Schiffer and Skibo (1987); Skibo and Schiffer (1987).

7. The summer 1990 issue of the *Archaeological Review from Cambridge* (vol. 9, no. 1) deals with the topic "Technology in the Humanities," and the contributions give a sense of the ways in which contemporary scholars who are not archaeologists are approaching it. Many contributors, including Tim Ingold and Pierre Lemonnier, are concerned with the production of material culture and the ways in which technology and technical relations are embedded in social relations. Few scholars are examining the physical aspects of materials and techniques in the production of material culture. Herbert (1984) has carried out a provocative study of the role of copper in traditional African societies from the point of view of an economic historian, but without the added perspective provided by technical studies. Franklin, Berthrong, and Chan (1985) have attempted to explore the relations between Shang metallurgy and cosmology. Niessen (1991), in considering alternative approaches to studying weaving technology among contemporary craftspeople, raises the notion of "insistence," that is, that technologies are cultural expressions with homologies in other areas of culture.

8. A copper-gold metallurgy based on lost-wax casting did emerge in Oaxaca around or after A.D. 1200, and was apparently closely related to the technologies of lower Central America and Colombia. The technology of Oaxacan metallurgy has not been studied. In addition, similar metallurgical technologies, also related to those in lower Central America and Colombia, developed in the southeastern Maya area around this time.

9. Otto Schöndube B., personal communication 1991.

10. All translations from Spanish are by the author, unless otherwise noted.

11. Arturo Oliveros, personal communication 1991.

12. Included also are data from my examinations of collections in the Museo Nacional de Antropología e Historia in Mexico City, The American Museum of Natural History in New York City, and others.

13. Data from the qualitative analyses are on file at the Center for Materials Research in Archaeology and Ethnology at MIT; quantitative data are presented in table 4.9.

2. Resources, Metals, and Alloys

1. Artifacts recovered that definitively reflect processing include four small ingots from the Postclassic occupation at Lamanai in Belize (see chapter 7), several pieces of smelted material (Grinberg, Ru- binovich, and Gasca 1986; Hosler and Stresser-Péan 1992) and an ingot (Hosler and Stresser-Péan 1992) from the Huastec site of Vista Hermosa, and small pieces of slag from Amapa in Nayarit (Meighan 1976) and from several other sites.

2. I want to emphasize that, even if such data on ores and slags were available, it would not allow matching of artifact to source in the same fashion possible with materials such as obsidian. For decades scholars working in Old World metallurgies have attempted to do so with little success.

3. Gold is the only metal that is not considered here but that smiths used in the prehispanic era; gold artifacts were unavailable for study.

4. Amador Osoria, personal communication 1985.

5. Ulrich Petersen, personal communication 1985.

6. I excluded the quantitative analyses of a few metal objects in the RMG assemblage said to be from Oaxaca.

7. Ulrich Petersen, personal communication 1986.

8. There is no commonly accepted terminology that describes the binary alloys of copper and arsenic. Metallurgists refer to all such alloys as arsenical copper, regardless of the amount of arsenic alloyed with copper. Lechtman (1981) introduced the term "arsenic bronze" to refer to alloys of copper and arsenic whose mechanical properties resemble those of tin bronze. Throughout this book I adhere to a terminology that relates the arsenic concentration of a copper-arsenic alloy to the mechanical properties of the alloy, in the most general sense: arsenical copper (< 0.1% As); low-arsenic copper-arsenic alloy (0.1–0.5% As); arsenic bronze (0.5–10% As). Alloys containing more than 10 percent arsenic are arsenic bronzes, but they rapidly become too brittle to work cold. Such alloys were used in West Mexico for casting objects such as bells. These alloys, and the objects cast from them, are a rich silver color.

Arsenical coppers are impure coppers whose electrical properties are markedly affected by the presence of arsenic but whose mechanical properties are similar to those of copper. Mechanical properties of copper-arsenic alloys, such as hardness and malleability, begin to change appreciably with arsenic concentrations of about 0.5 weight percent. At these relatively low arsenic levels, the overall strength of the alloy increases, with gains in hardness approaching 20% when the alloy is cold-worked (Heather Lechtman, personal communication 1991). At arsenic concentrations of about 0.5 weight percent and higher, copper-arsenic alloys can be considered bronzes. Their improved mechanical properties upon working are undergoing systematic study (Heather Lechtman, personal communication 1991; Northover 1989; Budd and Ottaway 1991).

9. These possible alternatives are being explored through experimental investigations into prehistoric methods of producing copper-arsenic alloys by Heather Lechtman at MIT.

10. Heather Lechtman, personal communication 1990.

11. Ulrich Petersen, personal communication 1986.

12. It is conceivable, because tetrahedrites associate with chalcopyrite in West Mexico and because tennantite, the arsenic-rich end of this series, is occasionally reported, that these mixed, associated ore minerals were smelted directly, explaining the presence of the tin but not the nickel in the metal.

13. Warren (1989) is a revised republication of Warren (1968).

14. Archives: Industrial Minera México, San Luis Potosí, Mexico.

15. Nancy Troike, manuscript on file, Instituto de Investigaciones Antropológicas, Universidad Nacional Autónoma de México.

3. Period 1 of West Mexican Metalworking: A.D. 600 to A.D. 1200/1300

1. Complete results of hardness tests, qualitative chemical analyses, and additional metallographic studies appear in Hosler (1986).

2. The terms working, cold working, and hammering refer to the plastic deformation of metal while cold, that is, at ambient temperatures. Hot work refers to plastic deformation while the metal is hot. Annealing involves heating the metal after cold work has reduced its plasticity; the metal recrystallizes when it is heated and regains its plastic properties.

3. In chapter 5, I refer to sheet metal objects made from alloys of copper-silver-gold. These are not common in West Mexico but do sometimes appear in private and museum collections. It is difficult to estimate the relative proportion of such alloy objects in view of the fact that so many were melted down by the Spaniards.

4. Clement Meighan, personal communication 1991.

5. Copper-silver alloys may appear slightly earlier than this at Infiernillo sites, but the dates are uncertain.

6. Robert Rose, personal communication 1987.

7. The design of some instruments such as the Aztec *huehuetl* or two-toned drum was such that they invariably produced the same two notes (Stevenson 1968).

8. The picture from the Infiernillo burials is intriguing, since fewer bells were recovered proportionate to other artifact classes. This may indicate that bells were reserved only for extremely high-status individuals interred at major centers, that bells were uncommon in the Infiernillo region, or a sampling problem. Large numbers of bells have been reported from sites near the Infiernillo area, however.

9. David Parks, personal communication 1986.

10. Regis Pelloux, personal communication 1985.

11. That value was experimentally determined by Matthew Lewis and Melissa Krawizcki through an ingenious experiment with one of Mr. Lewis's beard hairs.

12. Since we do not know exactly how the tweezers were held, force was applied at two different positions on the model and stresses resulting from the loading at each location were simulated. A contact force of 100 to 200 grams generated at the tips of the tweezer blades was found sufficient to close these tweezers and to pull a hair.

13. Tweezer no. 2345 was functional even though the metal exceeded its yield strength. The hinge metal experienced cold work as it was bent during shaping, producing a phenomenon known as residual stress. As a result the real stresses the tweezer hinge can tolerate are substantially higher than the theoretical yield strength of the metal.

14. The axe that was analyzed comes from Peñitas (Nayarit) and dates to A.D. 1000–1200 (Clement Meighan, personal communication 1990).

15. The Vickers Hardness Number (VHN) is a widely used measure of the hardness of metal. The VHN is determined by indenting the metal's surface with a diamond indenter; hardness is a function of the load applied and the size of the indenter. When annealed copper is cold-rolled (a method not unlike cold-hammering) and thickness is reduced to about 75%, hardness can rise to about 131 VHN. These numbers acquire some meaning when compared to data for other metals: cast copper measures about 50 VHN, cold-worked copper-tin bronze tools (containing 5–8% Sn) measure from 200 to 300 VHN, and high-carbon steel tools range from 300 to 400 VHN, although steel can be made far harder.

16. At Amapa 17 needles were reported excavated, but study of that collection using microscopic techniques shows that many objects reported as "wire" are needles, and that at least 16 must be added to Meighan's original count.

17. There may be two examples of the loop eye design from Amapa; since the eye portion of the needle was broken I could not be absolutely certain.

4. Origins of Period 1 West Mexican Metallurgy

1. Hosler, field notes 1991.

2. Olaf Holm, personal communication 1981.

3. Karen Stothert, personal communication 1987.

4. Karen Stothert, personal communication 1988.

5. John Murra, personal communication 1992.

6. Clemencia Plazas, personal communication 1991.

7. Data provided by Clemencia Plazas, Director of the Museo del Oro, Bogotá.

8. Juana Sáenz, personal communication 1985.

9. Hosler, field notes 1984.

10. Hosler, field notes 1983; Clemencia Plazas, personal communication 1993.

11. A few perforated-eye needles made from gold and *tumbaga* do appear in Colombia and Costa Rica.

12. Patricia Plunkett reports three bells from Oaxaca dating to between A.D. 350 and A.D. 850 in her doctoral dissertation. Dr. Plunkett has later placed these bells after A.D. 1100 (Patricia Plunkett, personal communication 1991).

13. Dates for the Bay Islands material derive from Paul Healy's work in Honduras, which shows that the Classic Period extends to approximately A.D. 1000 in that area. Based on her own research in the region, Rosemary Joyce suggests that A.D. 900 to A.D. 1100 constitute good bracketing dates for these objects (Rosemary Joyce, personal communication 1991). Copper objects are also reported from the following sites dating to the Early Postclassic Period: Santa Bárbara (Honduras), Los Naranjos (Honduras), Comaygua (Honduras), and Las Flores Bolsa (Honduras). Bracketing dates of A.D. 900 to A.D. 1200 are given for the material from Los Naranjos (Joyce 1986).

14. Marie Areti Hers, personal communication 1991.

15. Heather Lechtman, personal communication 1991.

16. Karen Bruhns, personal communication 1991.

17. Heather Lechtman, personal communication 1990.

18. Jorge Marcos, personal communication 1988.

19. Olaf Holm, personal communication 1981.

20. The Museo Antropológico also possesses a large volume of copper-silver sheet in fragmentary condition.

5. The Florescence of West Mexican Metallurgy: A.D. 1200/1300 to the Spanish Invasion

1. Hosler, field notes 1991.

2. Hosler, field notes 1991.

3. These ingots appear to be made of tin bronze and are housed in the collections of the Museo Nacional de Antropología e Historia in Mexico City.

4. Hosler, field notes 1991.

5. Hosler, field notes 1991.

6. Helen Pollard, personal communication 1992.

7. Helen Pollard, personal communication 1992.

8. The study corpus and the analytic methods used in this investigation are described in appendix 1.

9. I was able visually to examine material ascribed to Calixlahuaca housed in the American Museum of Natural History, in New York City, and in the Museo Nacional de Antropología e Historia, in Mexico City.

10. See Hosler (1986) for a fuller discussion of this point.

11. Heather Lechtman, personal communication 1988.

12. Some of the RMG copper objects (especially the rings and perforated-eye needles) may have been made during Period 1 (see table 3.1).

13. Heather Lechtman, personal communication 1989.

14. Helen Pollard, personal communication 1991.

15. I determined that the axes illustrated were metal rather than stone by comparing the blade shapes of archaeologically known stone axes with the metal RMG axes.

16. Heather Lechtman, personal communication 1985.

17. Hosler, field notes 1989.

18. Hosler, field notes 1982.

6. Period 2: Origins and Transformations

1. In Mexico, the earliest dates for copper-arsenic bronze could be at Tomatlán (Jalisco) at about A.D. 800. However, researchers who reported those compositional data advise that analytical results for arsenic should be treated cautiously (Joseph Mountjoy, personal communication 1985).

2. Heather Lechtman, personal communication 1993.

3. Hosler, field notes 1983.

4. William Root, notes on file at the Center for Materials Research in Archaeology and Ethnology, MIT.

5. Heather Lechtman, personal communication 1989.

6. Two are in the Regional Museum of Guadalajara (RMG) and two are in the museum in Taxco.

7. Also, Heather Lechtman, personal communication 1988.

7. The Dissemination of West Mexican Metallurgy

1. The possibility that the alloy derived from melting down objects made from tin bronze is treated later in this chapter.

2. They are found in the collections of the Museo Nacional de Antropología e Historia in Mexico City and were recovered at San Miguel Tecuizapa.

3. One object classified as an ornament is made from recycled metal; however, this object was a piece of wire subsequently rather inelegantly reshaped into a wire circlet.

4. David Pendergast, personal communication 1984.

5. Analytic data are presented in Hosler and Stresser-Péan (1992).

6. The raw materials were available at the mines near the site; a large deposit of cassiterite is shown at San Antonio, in San Luis Potosí, to the south of Vista Hermosa, and tin metal may have been won from this ore. According to the UNAM maps discussed in chapter 2 and other sources (Panczner 1987), copper deposits, as well as deposits of arsenopyrite, also are present in the region.

7. Tin in the copper-tin and copper-arsenic-tin alloy bells ranges from 1.08 to 4.9 weight percent. In the copper-arsenic alloy bells, arsenic concentration ranges from 0.52 to 2.2 weight percent.

8. One type 10b tin bronze bell was also recovered and was probably imported from West Mexico.

9. Guy Stresser-Péan, unpublished data.

10. Chemical analyses show that the Tamtok bell is a copper-arsenic-lead alloy (Hosler and Stresser-Péan 1992).

11. Also, William Root, notes on file at the Center for Materials Research in Archaeology and Ethnology, MIT.

12. Hosler, field notes 1982.

13. Also, William Root, notes on file at the Center for Materials Research in Archaeology and Ethnology, MIT.

14. Also, J. Charles Kelley, personal communication 1992.

15. Also, Michael E. Smith, personal communication 1992.

16. Significant numbers of metal objects have been reported from the Valley of Toluca. Although the technology has not been studied, Quezada (1972) remarks that it resembles the metallurgy of the Tarascan region. The limited material I have examined from the Valley of Toluca in the American Museum of Natural History and the Museo Nacional de Antropología e Historia bears out her observations. Root (in Lothrop 1952) analyzed a few bells from Calixlahuaca that were made from alloys of copper and tin, but others he studied were made from copper-lead alloys. Another copper-lead alloy bell was studied by Flores de Aguirrezabal and Quijada (1980).

17. Helen Pollard, personal communication 1991.

8. The Sounds and Colors of Sacred Power

1. See chapter 2 for a description of the *Relaciones Geográficas*. Ajuchitlán, located in northern Guerrero, contains many copper deposits according to the *Relación,* and some were probably exploited at the time of the Spanish invasion. Geological maps also show copper deposits in that area.

2. Also, Kenneth Hale, personal communication 1992.

3. Kenneth Hale, personal communication 1992.

4. I would like to thank R. Tom Zuidema for bringing this to my attention.

5. Louise Burkhart, personal communication 1993.

6. This statement is based on their color. We have no chemical analytical data for such tweezers.

7. Hosler, field notes 1983.

8. Enrique Mayer, personal communication 1986.

9. I am indebted to R. Tom Zuidema for this suggestion.

10. *Psalmodia Christiana* is a hymnbook composed by Fray Bernardino de Sahagún in Nahuatl as a means of Christianizing the native peoples of Mexico. He began composing it sometime between 1558 and 1561; it was first published in 1583. *Cantares Mexicanos* is one of the chief sources of Aztec poetry, and was composed in Nahuatl between 1550 and 1580; the poetry deals with the conquest and its aftermath, and provides evidence for a kind of revitalization movement. The songs call ghost-warriors down from paradise to aid the warrior-singers in overwhelming the enemy indigenous groups, such as the Tlaxcalans and the Chalcans, who collaborated with the Spaniards in the conquest.

11. This poem was translated from a Nahuatl poem in *Psalmodia Christiana* by Louise Burkhart.

12. Spicer (1964) observes that the religious complex in which flowers and blood are equivalent is typical of Uto-Aztecan groups in northwest Mexico (Cora, Huichol, Tarahumara, Yaqui, Mayo, and Pima). The Aztec diverged from these groups' predecessors. Burkhart (1992), citing an unpublished article by Jane Hill (1987), says that Hill has traced this flowery world complex throughout the Uto-Aztecan language family. I was unable to consult Hill's article, but we may know too little about Tarascan to exclude Tarascan a priori. Burkhart also has commented that there is some evidence for such a complex among Maya speakers (personal communication 1993); R. Tom Zuidema has noted a similar interest in shimmering qualities in Incaic thought and reproduced in certain textiles (personal communication 1993).

13. The other root in this word, *ayaui*, means to drizzle, or for there to be fog (Campbell 1985: 22).

14. Isabel Kelly, personal communication 1981.

15. Broda (1970) points out that sacrificial flaying has a wide distribution in the Americas, and does not always connote fertility.

16. These same concepts—metal, bells, and sound—are related in Mixtec as well, as the following terms indicate:

kaa/saa: Metal, fundir metal, sonar alto, claro, metalico. [metal, to cast or smelt metal, to sound loud, metallic.]
kaa: Sonajas de los indios. [indian rattles or bells.]
kadzi: Sonar sonajas de metal. [to ring metal bells.]

17. Kenneth Hale, personal communication 1993.

18. Another example comes from the Quiché, in highland Guatemala, where Termer (1928) describes a dance called *el Baile de la Culebra,* in which women are symbolically robbed, possessed, and then symbolically ingest snakes. This is the only explicit reference to sexuality that I found.

19. For a recent example of the incorporation of a foreign technology for some of these ends, see Clark and Blake (1994).

References

Acuña, René (editor). 1987. *Relaciones geográficas del siglo XVI: Michoacán*. Universidad Nacional Autónoma de México, Mexico.

Adams, Richard Newbold. 1975. *Energy and Structure: A Theory of Social Power*. University of Texas Press, Austin.

Alva, Walter, and Christopher B. Donnan. 1993. *Royal Tombs of Sipán*. Fowler Museum of Cultural History, Los Angeles.

Ambrosetti, Juan B. 1904. El bronce en la región Calchaquí. *Anales del Museo Nacional de Buenos Aires* (Serie 3) 4:163–314.

Anonymous. 1991. *Diccionario grande de la lengua de Michoacán*. 2 vols. Introduction and notes by B. Warren. FIMAX, Morelia.

Arsandaux, H., and Paul Rivet. 1921. Contribution à l'étude de la métallurgie mexicaine. *Journal de la Société des Américanistes de Paris* 13:261–280.

——— 1923. Nouvelle note sur la métallurgie mexicaine. *L'Antropologie* 33:63–85.

Bahr, Donald, Juan Gregorio, David López, and Albert Alvarez. 1974. *Pima Shamanism and Staying Sickness*. University of Arizona Press, Tucson.

Barrera, Tomás. 1931. *Zonas mineras de los estados de Jalisco y Nayarit*. Boletín Instituto Geológico de México no. 51. Universidad Nacional Autónoma de México, Mexico.

Barrett, Elinore M. 1981. The King's Copper Mine: Inguarán in New Spain. *The Americas* 38(1):1–29.

Battelle Memorial Institute. n.d. Metallurgical Evaluation of Pre-Columbian Copper Artifacts. Unpublished laboratory report, on file at the Department of Anthropology, Smithsonian Institution.

Baudez, Claude-François. 1992. The Maya Snake Dance: Ritual and Cosmology. *Res* 21:37–52.

Bell, Betty. 1971. Archaeology of Nayarit, Jalisco, and Colima. In *Handbook of Middle American Indians*, vol. 11, edited by G. F. Ekholm and I. Bernal, pp. 694–753. University of Texas Press, Austin.

Benzoni, Girolamo. 1857. *History of the New World*. Hakluyt Society, no. 21.

Berdan, Frances F., Richard E. Blanton, Elizabeth Hill Boone, Mary G. Hodge, Michael E. Smith, and Emily Umberger. In press. *Aztec Imperial Strategies*. Dumbarton Oaks, Washington, D.C.

Bergsøe, Paul. 1937. The Metallurgy and Technology of Gold and Platinum among the Pre-Columbian Indians. *Ingeniørvidenskabelige Skrifter* A44:1–45.

Berlin, Heinrich. 1956. *Late Pottery Horizons of Tabasco, Mexico*. Contributions to American Anthropology and History no. 59. Carnegie Institute of Washington, Washington, D.C.

Berrocal L., G., and F. Querol S. 1991. Geological Description of the Cuale District Ore Deposits, Jalisco, Mexico. In *Economic Geology, Mexico*, edited by G. P. Salas, pp. 355–363. The Geology of North America. Geological Society of America, Boulder.

Bierhorst, John. 1985. *Cantares Mexicanos: Songs of the Aztecs*. Stanford University Press, Stanford.

Bird, Junius B. 1967–1968. Treasures from the Land of Gold. *Arts in Virginia*. 8 (1–2):21–23.

——— 1979. Legacy of the Stingless Bee. *Natural History* 88(9):49–51.

Bolland, Rita. 1991. *Tellem Textiles: Archaeological Finds from Burial Caves in Mali's Bandiagara Cliff*. Royal Tropical Institute, Amsterdam.

Boman, Éric. 1908. *Antiquités de la région andine de la République argentine et du désert d'Atacama,* vol. 2. Imprimerie Nationale, Paris.

Brand, Donald D. 1951. *Quiroga: A Mexican Municipio.* United States Printing Office, Washington, D.C.

Braun, David P. 1983. Pots as Tools. In *Archaeological Hammers and Theories,* edited by J. A. Moore and A. S. Keene, pp. 107–134. Academic Press, New York.

Bray, Warwick. 1977. Maya Metalwork and Its External Connections. In *Social Process in Maya Prehistory,* edited by N. Hammond, pp. 365–403. Academic Press, New York.

1978. *The Gold of El Dorado.* Catalogue of an exhibition held at The Royal Academy, Piccadilly, London, 21 November 1978 to 18 March 1979. Times Newspapers, London.

1981. Gold Work. In *Between Continents, Between Seas: Precolumbian Art of Costa Rica,* pp. 153–166. H. N. Abrams, New York.

1985. Ancient American Metallurgy: Five Hundred Years of Study. In *The Art of Precolumbian Gold: The Jan Mitchell Collection,* pp. 76–84. Metropolitan Museum of Art, New York.

Broda, Johanna. 1970. Tlacxipehualiztli: A Reconstruction of an Aztec Calendar Festival from 16th Century Sources. *Revista Española de Antropología Americana* 5:197–274.

1971. Las fiestas aztecas de los dioses de la lluvia. *Revista Española de Antropología Americana* 5:245–327.

Bruhns, Karen O. 1989. Intercambio entre la costa y la sierra en el formativo tardío: nuevas evidencias del Azuay. In *Relaciones interculturales en el área ecuatorial del Pacífico durante la época precolombina,* edited by J. F. Bouchard and M. Guinea, pp. 57–74. British Archaeological Reports, Oxford.

Bruhns, Karen O., James H. Burton, and George R. Miller. 1990. Excavations at Pirincay in the Paute Valley of Southern Ecuador, 1985–1988. *Antiquity* 64:221–233.

Brumfiel, Elizabeth M. 1987. Elite and Utilitarian Crafts in the Aztec State. In *Specialization, Exchange, and Complex Societies,* edited by E. M. Brumfiel and T. K. Earle, pp. 102–118. Cambridge University Press, Cambridge.

Brundage, Burr C. 1982. *The Phoenix of the Western World: Quetzalcoatl and the Sky Religion.* University of Oklahoma Press, Norman.

1985. *The Jade Steps: A Ritual Life of the Aztecs.* University of Utah Press, Salt Lake City.

1988. *The Fifth Sun: Aztec Gods, Aztec World.* University of Texas Press, Austin.

Brush, Charles F. 1962. Pre-Columbian Alloy Objects from Guerrero, Mexico. *Science* 138:1336–1337.

Budd, P., and B. S. Ottaway. 1991. The Properties of Arsenical Copper Alloys: Implications for the Development of Eneolithic Metallurgy. In *Archaeological Sciences 1989,* edited by P. Budd, B. Chapman, C. Jackson, R. Janaway, and B. Ottaway, pp. 132–142. Oxbow Books, Oxford.

Burger, Richard L. 1984. Archaeological Areas and Prehistoric Frontiers: The Case of Formative Peru and Ecuador. In *Social and Economic Organization in the Prehispanic Andes,* edited by D. F. Browman, R. L. Burger, and M. A. Rivera, pp. 33–71. British Archaeological Reports, Oxford.

Burkhart, Louise M. 1992. Flowery Heaven: The Aesthetic of Paradise in Nahuatl Devotional Literature. *Res* 21:89–109.

Bushnell, Geoffrey H. S. 1951. *The Archaeology of the Santa Elena Peninsula in South-West Ecuador.* Cambridge University Press, Cambridge.

Butler, Mary. 1959. Spanish Contact at Chipal. *Mitteilungen aus dem Museum für Völkerkunde und Vorgeschichte in Hamburg* 25:28–35.

Buys, Jozef, and Victoria Domínguez. 1988. Un cementerio de hace 2000 años: Jardín del Este. In *Quito antes de Benalcázar*, edited by I. Cruz C., pp. 31–50, 67–73. Centro Cultural Artes, Quito.

Buys, Jozef, and Michael Muse. 1987. Arqueología de asentamientos asociados a los campos elevados de Peñón del Río, Guayas, Ecuador. In *Pre-Hispanic Agricultural Fields in the Andean Region,* edited by W. M. Denevan, K. Mathewson, and G. Knapp, pp. 225–248. British Archaeological Reports, Oxford.

Cabrera C., Rubén. 1976. *Arqueología en el bajo Balsas, Guerrero y Michoacán. Presa La Villita*. Tesis de Maestría, Escuela Nacional de Antropología e Historia, Mexico.

——— 1986. El desarrollo cultural prehispánico en la región del bajo río Balsas. In *Arqueología y Etnohistoria del Estado de Guerrero*, pp. 117–151. Instituto Nacional de Antropología e Historia, Mexico.

——— 1988. Nuevos resultados de Tzintzuntzan, Michoacán, en su décima temporada de excavaciones. In *Primera reunión sobre las sociedades prehispánicas en el centro occidente de México*, pp. 193–218. Instituto Nacional de Antropología e Historia, Mexico.

Caley, Earle R., and Dudley T. Easby, Jr. 1964. New Evidence of Tin Smelting and the Use of Metallic Tin in Preconquest Mexico. In *35th International Congress of Americanists,* vol. 1, pp. 507–517. Mexico.

Campbell, R. Joe. 1985. *A Morphological Dictionary of Classical Nahuatl: A Morpheme Index to the Vocabulario en Lengua Mexicana y Castellana of Fray Alonso de Molina*. Hispanic Seminary of Medieval Studies, Madison.

Carriveau, Gary W. 1978. Application of Thermoluminescence Dating Techniques to Prehistoric Metallurgy. In *Application of Science to the Examination of Works of Art,* edited by W. J. Young, pp. 59–67. Museum of Fine Arts, Boston.

Caso, Alfonso. 1965. Lapidary Work, Goldwork, and Copperwork from Oaxaca. In *Handbook of Middle American Indians,* vol. 3, edited by G. R. Willey, pp. 896–930. University of Texas Press, Austin.

Chadwick, Robert. 1971. Archaeological Synthesis of Michoacan and Adjacent Regions. In *Handbook of Middle American Indians,* vol. 11, edited by G. F. Ekholm and I. Bernal, pp. 657–693. University of Texas Press, Austin.

Charles, James A. 1980. The Coming of Copper and Copper-Base Alloys and Iron: A Metallurgical Sequence. In *The Coming of the Age of Iron,* edited by T. A. Wertime and J. D. Muhly, pp. 151–181. Yale University Press, New Haven.

Childe, V. Gordon. 1944. Archaeological Ages as Technological Stages. *Journal of the Royal Anthropological Institute of Great Britain and Ireland* 74:7–24.

——— 1983. *Man Makes Himself.* Meridian, New York.

Childs, Susan Terry. 1986. *Style in Technology: A View of African Early Iron Age Iron Smelting through Refractory Ceramics.* Ph.D. Dissertation, Boston University, Boston.

Civrieux, Marc de. 1974. *Religión y magia Kari'ña*. Instituto de Investigaciones Históricas, Caracas.

Clark, James C. (editor and translator). 1938. *Codex Mendoza: The Mexican Manuscript Known as the Collection of Mendoza Preserved in the Bodleian Library, Oxford*. 3 vols. Waterlow and Sons, London.

Clark, John E., and Michael Blake. 1994. The Power of Prestige: Competitive Generosity and the Emergence of Rank Societies in Lowland Mesoamerica. In *Factional Competition and*

Political Development in the New World, edited by E. M. Brumfiel and J. W. Fox, pp. 17–30. Cambridge University Press, Cambridge.

Coe, Michael D. 1965. *The Jaguar's Children: Pre-Classic Central Mexico.* Museum of Primitive Art, New York.

Coggins, Clemency Chase. 1984. *The Cenote of Sacrifice: Catalogue.* In *Cenote of Sacrifice: Maya Treasures from the Sacred Well at Chichén Itzá,* edited by C. C. Coggins and O. C. Shane, pp. 23–166. University of Texas Press, Austin.

Collier, Donald, and John V. Murra. 1943. *Survey and Excavations in Southern Ecuador.* Field Museum of Natural History, Chicago.

Conklin, William J. 1979. Moche Textile Structures. In *The Junius B. Bird Pre-Columbian Textile Conference,* pp. 165–184. Textile Museum and Dumbarton Oaks, Washington, D.C.

Cordy-Collins, Alana. 1990. Fonga Sigde, Shell Purveyor to the Chimu Kings. In *The Northern Dynasties: Kingship and Statecraft in Chimor,* edited by A. Cordy-Collins and M. E. Moseley, pp. 393–417. Dumbarton Oaks, Washington, D.C.

Costin, Cathy, Timothy Earle, Bruce Owen, and Glenn Russell. 1989. Impact of Inca Conquest on Local Technology in the Upper Mantaro Valley, Peru. In *What's New? A Closer Look at the Process of Innovation,* edited by S. E. van der Leeuw and R. Torrence, pp. 107–139. Unwin Hyman, London.

Cotterell, Brian, and Johan Kamminga. 1990. *Mechanics of Pre-Industrial Technology: An Introduction to the Mechanics of Ancient and Traditional Material Culture.* Cambridge University Press, Cambridge.

Covarrubias V., Manuel. 1961. Notas para el estudio de la arqueología de la costa de Jalisco. *Eco* 7:4–7.

Covey, Cyclone (translator). 1961. *Cabeza de Vaca's Adventures in the Unknown Interior of America.* University of New Mexico Press, Albuquerque.

Craine, Eugene R., and Reginald C. Reindorp. 1970. *The Chronicles of Michoacan.* University of Oklahoma Press, Norman.

Crossin, Richard S. 1967. The Breeding Biology of the Tufted Jay. *Proceedings of the Western Foundation of Vertebrate Zoology* 1(5):264–299.

Crumrine, N. Ross. 1977. *The Mayo Indians of Sonora: A People Who Refuse to Die.* University of Arizona Press, Tucson.

Dahlgren de Jordan, Barbro. 1979. *La Mixteca: su cultura e historia prehispánicas.* 3rd ed. Dirección General de Educación y Bienestar Social, Oaxaca.

De Atley, Suzanne P. 1986. Mix and Match: Traditions of Glaze Paint Preparation at Four Mile Ruin, Arizona. In *Technology and Style,* edited by W. D. Kingery, pp. 297–329. Ceramics and Civilization, vol. 4. American Ceramic Society, Columbus.

Delgado, Agustin. 1965. *Archaeological Reconnaissance in the Region of Tehuantepec, Oaxaca, Mexico.* New World Archaeological Foundation, Provo.

Díaz del Castillo, Bernal. 1939. *Historia verdadera de la conquista de la Nueva España.* 3 vols. Editorial Pedro Robredo, Mexico.

Doyon, León G. 1988. Tumbas da la nobleza en La Florida. In *Quito Antes de Benalcázar,* edited by I. Cruz C., pp. 51–66, 86–100. Centro Cultural Artes, Quito.

Duque G., Luis. 1964. *Exploraciones arqueológicas en San Agustín.* Instituto Colombiano de Antropología, Bogotá.

Durán, Fray Diego. 1967. *Historia de las indias de Nueva España e islas de la tierra firme.* 2 vols. Editorial Porrúa, Mexico.

Dutton, Bertha P., and Hulda R. Hobbs. 1943. *Excavations at Tajumulco, Guatemala*. University of New Mexico Press, Santa Fe.

Earle, Timothy. 1989. The Evolution of Chiefdoms. *Current Anthropology* 30:84–88.

Easby, Dudley T., Earle R. Caley, and Khosrow Moazed. 1967. Axe-Money: Facts and Speculation. *Revista Méxicana de Estudios Antropológicos* 21:107–148.

Edwards, Clinton R. 1969. Possibilities of Pre-Columbian Maritime Contacts Among New World Civilizations. *Mesoamerican Studies* 4:3–10.

Ekholm, Gordon F. 1942. *Excavations at Guasave, Sinaloa, Mexico*. American Museum of Natural History, New York.

Escalera U., Andrés, and María Angeles Barriuso P. 1978. Estudio científico de los objetos de metal de Ingapirca (Ecuador). *Revista Española de Antropología Americana* 8:19–47.

Estrada, Emilio. 1955. Balsa and Dugout Navigation in Ecuador. *The American Neptune* 15:142–149.

Estrada, Jenny (editor). 1988. *La balsa en la historia de la navegación ecuatoriana: compilación de crónicas, estudios, gráficas y testimonios*. Instituto de Historia Marítima, Guayaquil.

Farabee, William C. 1922. *Indian Tribes of Eastern Peru*. Peabody Museum, Cambridge.

Fester, Gustavo A. 1962. Copper and Copper Alloys in Ancient Argentina. In *Chymia: Annual Studies in the History of Chemistry*, vol. 8, pp. 21–31. University of Pennsylvania Press, Philadelphia.

Flores, Teodoro. 1946. *Geología minera de la región NE. del estado de Michoacán*. Boletín Instituto Geológico de México no. 52. Universidad Nacional Autónoma de México, Mexico.

Flores de Aguirrezabal, María Dolores, and César A. Quijada L. 1980. Distribución de objetos de metal en el occidente de México. In *Rutas de intercambio en Mesoamérica y el norte de México*, vol. 2, pp. 83–92. Sociedad Mexicana de Antropología, XVI Reunión de Mesa Redonda, Saltillo.

Foshag, William F., and C. Fries, Jr. 1942. Tin Deposits of the Republic of Mexico. In *U.S. Geological Survey Bulletin*, vol. 935-C, pp. 99–176. United States Geological Survey, Washington, D.C.

Frame, Mary. 1986. Nasca Sprang Tassels: Structure, Technique, and Order. *Textile Museum Journal* 25:67–82.

Franklin, Ursula, John Berthrong, and Alan Chan. 1985. Metallurgy, Cosmology, Knowledge: The Chinese Experience. *Journal of Chinese Philosophy* 12:333–370.

Friedrich, Paul. 1984. Tarascan: From Meaning to Sound. In *Supplement to the Handbook of Middle American Indians*, vol. 2, edited by M. S. Edmonson, pp. 56–82. University of Texas Press, Austin.

Furst, Peter T. 1965a. West Mexican Tomb Sculpture as Evidence for Shamanism in Prehispanic Mesoamerica. *Antropológica* 15:29–80.

———. 1965b. West Mexico, the Caribbean and Northern South America: Some Problems in New World Interrelationships. *Antropológica* 14:1–37.

———. 1974. Shamanistic Survivals in Mesoamerican Religion. In *41st International Congress of Americanists*, vol. 3, pp. 149–157. Mexico.

Gann, Thomas W. F. 1918. *The Maya Indians of Southern Yucatan and Northern British Honduras*. United States Government Printing Office, Washington, D.C.

Gann, Thomas W. F., and Mary Gann. 1939. *Archaeological Investigations in the Corozal District of British Honduras*.

United States Government Printing Office, Washington, D.C.

Goldstein, Paul. 1989a. *Omo, a Tiwanaku Provincial Center in Moquegua, Peru*. Ph.D. Dissertation, University of Chicago, Chicago.

——— 1989b. The Tiwanaku Occupation of Moquegua. In *Ecology, Settlement, and History in the Osmore Drainage, Peru*, edited by D. S. Rice, C. Stanish, and P. R. Scarr, pp. 219–256. British Archaeological Reports, Oxford.

González, Alberto R. 1979. Pre-Columbian Metallurgy of Northwest Argentina: Historical Development and Cultural Process. In *Pre-Columbian Metallurgy of South America*, edited by E. P. Benson, pp. 133–202. Dumbarton Oaks, Washington, D.C.

González R., Jenaro. 1956. *Riqueza minera y yacimientos minerales de México*. 3rd ed. 20th International Geological Congress, Mexico.

Goossens, Pierre J. 1972a. Metallogeny in the Ecuadorian Andes. *Economic Geology* 67:458–468.

——— 1972b. *Los yacimientos e indicios de los minerales metálicos y no metálicos de la república del Ecuador*. Universidad de Guayaquil, Guayaquil.

Gordon, Robert B. 1985. Laboratory Evidence of the Use of Metal Tools at Machu Picchu (Peru) and Environs. *Journal of Archaeological Science* 12:311–327.

Gorenstein, Shirley. 1985. *Acámbaro: Frontier Settlement on the Tarascan-Aztec Border*. Vanderbilt University Press, Nashville.

Gossen, Gary H. 1974. *Chamulas in the World of the Sun: Time and Space in a Maya Oral Tradition*. Harvard University Press, Cambridge.

Grinberg, Dora M. 1989. Tecnologías metalúrgicas tarascas. *Ciencia y Desarrollo* 15(89):37–52.

Grinberg, Dora M., and Francisca Franco V. 1987. Estudio de cuatro cascabeles de falso alambre provenientes de las excavaciones del tren subterráneo de la Ciudad de México. *Antropología y Técnica* 2:143–151.

Grinberg, Dora M., R. E. Rubinovich, and A. A. Gasca. 1986. Intentional Production of Bronze in Mesoamerica. In *Precolumbian American Metallurgy*, pp. 57–65. 45th International Congress of Americanists, Bogotá.

Haemig, Paul D. 1979. Secret of the Painted Jay. *Biotropica* 11(2):81–87.

Healy, Paul F. 1988. Music of the Maya. *Archaeology* 41(1):24–31.

Hearne, Pamela. 1992. The Story of the River Gold. In *River of Gold: Precolumbian Treasures from Sitio Conte*, edited by P. Hearne and R. J. Sharer, pp. 1–21. The University Museum, University of Pennsylvania, Philadelphia.

Helms, Mary W. 1979. *Ancient Panama: Chiefs in Search of Power*. University of Texas Press, Austin.

Herbert, Eugenia W. 1984. *Red Gold of Africa: Copper in Precolonial History and Culture*. University of Wisconsin Press, Madison.

Hers, Marie-Areti. 1989. *Los toltecas en tierras chichimecas*. Instituto de Investigaciones Estéticas, Universidad Nacional Autónoma de México, Mexico.

——— 1990. Los objetos de cobre en la cultura Chalchihuites. In *Homenaje a Federico Sescosse: un hombre, un destino y un lugar*, pp. 45–60. Gobierno del Estado de Zacatecas.

Heyden, Doris. 1986. Metaphors, *Nahualtocaitl*, and Other "Disguised" Terms among the Aztecs. In *Symbol and Meaning beyond the Closed Community: Essays in Mesoamerican*

Ideas, edited by G. H. Gossen. Institute for Mesoamerican Studies, Albany.

Hill, Jane H. 1987. The Flowery World of Old Uto-Aztecan. Paper presented at the 86th Annual Meeting of the American Anthropological Association, Chicago.

Holm, Olaf. 1963. Copper Needles from Manabí, Ecuador. *Ethnos* 28(2–4):177–187.

——— 1978. Hachas monedas del Ecuador. In *El hombre y la cultura andina,* vol. 1, pp. 347–361. Editora Lasontay, Lima.

Hosler, Dorothy. 1986. *The Origins, Technology, and Social Construction of Ancient West Mexican Metallurgy.* Ph.D. Dissertation, University of California, Santa Barbara.

——— 1988a. Ancient West Mexican Metallurgy: A Technological Chronology. *Journal of Field Archaeology* 15:191–217.

——— 1988b. Ancient West Mexican Metallurgy: South and Central American Origins and West Mexican Transformations. *American Anthropologist* 90:832–855.

——— 1988c. The Metallurgy of Ancient West Mexico. In *The Beginning of the Use of Metals and Alloys,* edited by R. Maddin, pp. 328–343. MIT Press, Cambridge.

Hosler, Dorothy, Heather Lechtman, and Olaf Holm. 1990. *Axe-Monies and Their Relatives.* Dumbarton Oaks, Washington, D.C.

Hosler, Dorothy, and Guy Stresser-Péan. 1992. The Huastec Region: A Second Locus for the Production of Bronze Alloys in Ancient Mesoamerica. *Science* 257:1215–1220.

Hunt, Eva. 1977. *The Transformation of the Hummingbird: Cultural Roots of a Zinacantecan Mythical Poem.* Cornell University Press, Ithaca.

Ichon, Alain. 1990. *La religión de los totonacas de la sierra.* Instituto Nacional Indigenista, Mexico.

Ingersoll, Daniel W., Jr., and Gordon Bronitsky (editors). 1987. *Mirror and Metaphor: Material and Social Constructions of Reality.* University Press of America, Lanham.

Ingold, Tim. 1990. Society, Nature and the Concept of Technology. *Archaeological Review from Cambridge* 9:5–17.

Izikowitz, Karl G. 1935. *Musical and Other Sound Instruments of the South American Indians: A Comparative Ethnographical Study.* Elanders Boktryckeri Aktiebolag, Göteborg.

Jansen, Maarten. 1990. The Search for History in the Mixtec Codices. *Ancient Mesoamerica* 1:99–112.

Jijón y Caamaño, Jacinto. 1940–1945. *El Ecuador interandino y occidental antes de la conquista castellana.* 5 vols. Editorial Ecuatoriana, Quito.

Jiménez M., Wigberto. 1948. Historia antigua de la zona tarasca. In *El occidente de México,* pp. 146–157. Sociedad Mexicana de Antropología, Cuarta Reunión de Mesa Redonda, Mexico.

Johnson, Allen W., and Timothy Earle. 1987. *The Evolution of Human Societies: From Foraging Group to Agrarian State.* Stanford University Press, Stanford.

Joyce, Rosemary A. 1986. Terminal Classic Interaction on the Southeastern Maya Periphery. *American Antiquity* 51(2):313–329.

Karttunen, Frances. 1992. *An Analytical Dictionary of Nahuatl.* University of Oklahoma Press, Norman.

Kelley, J. Charles. 1986. The Mobile Merchants of Molino. In *Ripples in the Chichimec Sea: New Considerations of Southwestern-Mesoamerican Interactions,* edited by F. J. Mathien and R. H. McGuire, pp. 81–104. Southern Illinois University Press, Carbondale.

Kelly, Isabel T. 1945. *Excavations at Culiacán, Sinaloa*. University of California Press, Berkeley.

1947. *Excavations at Apatzingán, Michoacán*. Viking Fund, New York.

1949. *The Archaeology of the Autlán-Tuxcacuesco Area of Jalisco*, vol. 2. University of California Press, Berkeley.

1980. *Ceramic Sequence in Colima: Capacha, an Early Phase*. University of Arizona Press, Tucson.

1985. Some Gold and Silver Artifacts from Colima. In *The Archaeology of West and Northwest Mesoamerica*, edited by M. S. Foster and P. G. Weigand, pp. 153–179. Westview Press, Boulder.

Killick, David John. 1990. *Technology in Its Social Setting: Bloomery Iron Smelting at Kasungu, Malawi, 1860–1940*. Ph.D. Dissertation. Yale University, New Haven.

Kingsborough, Edward K., Viscount. 1964–1967. *Antigüedades de México, basadas en la recopilación de Lord Kingsborough*. 4 vols. Secretaría de Hacienda y Crédito Público, Mexico.

Kroeber, Alfred L., and William D. Strong. 1924. *The Uhle Collections from Chincha*. University of California Press, Berkeley.

Landa, Fray Diego de. 1978. *Relación de las cosas de Yucatán*. 11th edition. Editorial Porrúa, Mexico.

Lathrap, Donald W. 1982. Complex Iconographic Features Shared by Olmec and Chavin and Some Speculations on Their Possible Significance. In *Primer simposio de correlaciones antropológicas andino-mesoamericano*, pp. 301–327. Escuela Superior Politécnica del Litoral, Guayaquil.

Leach, Edmund R. 1958. Magical Hair. *Journal of the Royal Anthropological Institute of Great Britain and Ireland* 88:147–164.

Lechtman, Heather. 1970. The Gilding of Metals in Pre-Columbian Peru. In *Application of Science in Examination of Works of Art*, edited by W. J. Young, pp. 38–52. Museum of Fine Arts, Boston.

1976. A Metallurgical Site Survey in the Peruvian Andes. *Journal of Field Archaeology* 3:1–42.

1977. Style in Technology: Some Early Thoughts. In *Material Culture: Styles, Organization, and Dynamics of Technology*, edited by H. Lechtman and R. S. Merrill, pp. 3–20. West Publishing, Saint Paul.

1979. Issues in Andean Metallurgy. In *Pre-Columbian Metallurgy of South America*, edited by E. P. Benson, pp. 1–40. Dumbarton Oaks, Washington, D.C.

1980. The Central Andes: Metallurgy without Iron. In *The Coming of the Age of Iron*, edited by T. A. Wertime and J. D. Muhly, pp. 267–334. Yale University Press, New Haven.

1981. Copper-Arsenic Bronzes from the North Coast of Peru. *Annals of the New York Academy of Sciences* 376:77–121.

1984. Andean Value Systems and the Development of Prehistoric Metallurgy. *Technology and Culture* 25:1–36.

1985. The Manufacture of Copper-Arsenic Alloys in Prehistory. *Historical Metallurgy* 19(1):141–142.

1988. Traditions and Styles in Central Andean Metalworking. In *The Beginning of the Use of Metals and Alloys*, edited by R. Maddin, pp. 344–378. MIT Press, Cambridge.

1991. The Production of Copper-Arsenic Alloys in the Central Andean Culture Area: Highland Ores and Coastal Smelters? *Journal of Field Archaeology* 18:43–76.

1993. Technologies of Power: The Andean Case. In *Configurations of Power*, edited by J. S. Henderson and P. J. Netherly, pp. 244–280. Cornell University Press, Ithaca.

Lechtman, Heather, and Robert S. Merrill (editors). 1977. *Material Culture: Styles, Organization, and Dynamics of Technology.* West Publishing, Saint Paul.

Lechtman, Heather, and Arthur Steinberg. 1970. Bronze Joining: A Study in Ancient Technology. In *Art and Technology: A Symposium on Classical Bronzes,* edited by S. Doeringer, D. Mitten, and A. Steinberg, pp. 5–35. Cambridge: MIT Press.

Lemonnier, Pierre. 1976. La Description des chaînes opératoires: contribution à l'analyse des systèmes techniques. *Techniques et Culture* 1:100–151.

——— 1986. The Study of Material Culture Today: Toward an Anthropology of Technical Systems. *Journal of Anthropological Archaeology* 5:147–186.

Leone, Mark. 1973. Archaeology as the Science of Technology: Mormon Town Plans and Fences. In *Research and Theory in Current Archaeology,* edited by C. L. Redman, pp. 125–150. John Wiley, New York.

——— 1977. The Role of Primitive Technology in Nineteenth Century American Utopias. In *Material Culture: Styles, Organization, and Dynamics of Technology,* edited by H. Lechtman and R. S. Merrill, pp. 87–107. West Publishing, Saint Paul.

Lister, Robert H. 1949. *Excavations at Cojumatlán, Michoacán, Mexico.* University of New Mexico Press, Albuquerque.

Litvak K., Jaime. 1968. Excavaciones de rescate en la presa de La Villita. *Boletín del Instituto Nacional de Antropología e Historia* 31:28–30.

London, Gloria A. 1991. Standardization and Variation in the Work of Craft Specialists. In *Ceramic Ethnoarchaeology,* edited by W. A. Longacre, pp. 182–204. University of Arizona Press, Tucson.

Long, Stanley, and Marcia Wire. 1966. Excavations at Barra de Navidad, Jalisco. *Antropológica* 18:3–81.

Lorinczi, G. I., and J. C. Miranda V. 1978. Geology of the Massive Sulfide Deposits of Campo Morado, Guerrero, Mexico. *Economic Geology* 73:180–191.

Lothrop, Samuel K. 1952. *Metals from the Cenote of Sacrifice, Chichen Itza, Yucatan.* With sections by W. C. Root and T. Proskouriakoff. Peabody Museum, Cambridge.

Lowe, Gareth W. 1959. *The Chiapas Project, 1955–1958. Report of the Field Director.* New World Archaeological Foundation, Orinda.

Lumholtz, Carl. 1973. *Unknown Mexico: A Record of Five Years' Exploration among the Tribes of the Western Sierra Madre; in the Tierra Caliente of Tepic and Jalisco; and among the Tarascos of Michoacan.* 2 vols. Reprinted, A.M.S. Press, New York. Originally published 1902, Charles Scribner's Sons, New York.

Lyons, James I. 1988. Geology and Ore Deposits of the Bolaños Silver District, Jalisco, Mexico. *Economic Geology* 83:1560–1582.

Macías, Martha C. 1990. La metalurgia en el México prehispánico: Oaxaca, tierra de orfebres. In *Mesoamerica y norte de México, siglo IX–XII,* vol. 1, edited by F. Soldi M., pp. 179–193. Instituto Nacional de Antropología e Historia, Mexico.

Macías G., Angelina. 1989. La cuenca de Cuitzeo. In *Historia general de Michoacán,* vol. 1, pp. 171–190. Gobierno de Michoacán, Morelia.

——— 1990. *Huandacareo: lugar de juicios, tribunal.* Instituto Nacional de Antropología e Historia, Mexico.

Maddin, Robert, Tamara Stech Wheeler, and James D. Muhly. 1980. Distinguishing Artifacts Made of Native Copper. *Journal of Archaeological Science* 7:211–225.

Maldonado C., Rubén. 1980. *Ofrendas asociadas a entierros del Infiernillo en el Balsas: estudio y experimentación con tres métodos de taxonomía numérica.* Instituto Nacional de Antropología e Historia, Mexico.

Marcos, Jorge G. 1978. Cruising to Acapulco and Back with the Thorny Oyster Set: A Model for a Lineal Exchange System. *Journal of the Steward Anthropological Society* 9:99–132.

——— 1981. Informe sobre el área ceremonial del complejo Manteño-Huancavilca de la Loma de los Cangrejitos, valle de Chanduy, Ecuador. *El Arquitecto* 1(5):54–63.

——— 1986. Breve prehistória del Ecuador. In *Arqueología de la costa Ecuatoriana: nuevos enfoques,* edited by J. Marcos, pp. 25–50. Biblioteca Ecuatoriana de Arqueología, vol. 1. Corporación Editora Nacional, Quito.

Marcos, Jorge G., and Presley Norton. 1981. Interpretación sobre la arqueología de la Isla de La Plata. *Miscelània Antropológica Ecuatoriana* 1:136–154.

Martí, Samuel, and Gertrude P. Kurath. 1964. *Dances of Anáhuac: The Choreography and Music of Precortesian Dances.* Aldine Publishing, Chicago.

Masucci, Maria. 1992. *Ceramic Change in the Guangala Phase, Southwest Ecuador: A Typology and Chronology.* Ph.D. Dissertation, Southern Methodist University, Dallas.

Mathewson, C. H. 1915. A Metallographic Description of Some Ancient Peruvian Bronzes from Machu Picchu. *American Journal of Science* (4th series) 40:525–616.

Matos M., Eduardo. 1987. The Templo Mayor of Tenochtitlan: History and Interpretation. In *The Great Temple of Tenochtitlan: Center and Periphery in the Aztec World,* edited by J. Broda, D. Carrasco, and E. Matos M., pp. 15–60. University of California Press, Berkeley.

Mayer, Eugen F. 1992. *Armas y herramientas de metal prehispánicas en Ecuador.* Philipp von Zabern, Mainz.

McAnany, Patricia A. 1989. Stone-Tool Production and Exchange in the Eastern Maya Lowlands: The Consumer Perspective from Pulltrouser Swamp, Belize. *American Antiquity* 54(2):332–346.

Mead, Charles W. 1915. *Prehistoric Bronze in South America.* American Museum of Natural History, New York.

Meighan, Clement W. 1960. Prehistoric Copper Objects from Western Mexico. *Science* 131:1534.

——— 1969. Cultural Similarities between Western Mexico and Andean Regions. *Mesoamerican Studies* 4:11–25.

——— 1974. Prehistory of West Mexico. *Science* 184:1254–1261.

——— 1976. *The Archaeology of Amapa, Nayarit.* Institute of Archaeology, University of California, Los Angeles.

Meighan, Clement W., and Leonard J. Foote. 1968. *Excavations at Tizapán el Alto, Jalisco.* Latin American Center, University of California, Los Angeles.

Meighan, Clement W., and H. B. Nicholson. 1989. The Ceramic Mortuary Offerings of Prehistoric West Mexico: An Archaeological Perspective. In *Sculpture of Ancient West Mexico: Nayarit, Jalisco, and Colima,* edited by M. Kan, C. Meighan, and H. B. Nicholson, pp. 29–67. University of New Mexico Press, Albuquerque.

Merkel, J. F., I. Shimada, C. P. Swann, and R. Doonan. 1994. Pre-Hispanic Copper Alloy Production at Batán Grande, Peru: Interpretation of the Analytical Data for Ore Samples. In *Archaeometry of Pre-Columbian Sites and Artifacts,* edited by D. A. Scott and P. Meyers, pp. 199–227. The Getty Conservation Institute, Los Angeles.

Métraux, Alfred. 1949. Religion and Shamanism. In *Handbook of South American Indians,* vol. 5, edited by J. H. Steward, pp. 559–599. Smithsonian Institution Bureau of American Ethnology Bulletin no. 143. United States Government Printing Office, Washington, D.C.

Millon, René. 1981. Teotihuacan: City, State, and Civilization. In *Supplement to the Handbook of Middle American Indians,* vol. 1, edited by J. A. Sabloff, pp. 198–243. University of Texas Press, Austin.

Mountjoy, Joseph B. 1969. On the Origin of West Mexican Metallurgy. *Mesoamerican Studies* 4:26–42.

——— 1970. *Prehispanic Culture History and Cultural Contact on the Southern Coast of Nayarit, Mexico.* Ph.D. Dissertation, Southern Illinois University, Carbondale.

——— 1982. *Proyecto Tomatlán de salvamento arqueológico.* Instituto Nacional de Antropología e Historia, Mexico.

——— 1992. Resources for Prehispanic Cultures on the Coast of West Mexico. Paper presented at the Cultural Dynamics of Precolumbian West and Northwest Mesoamerica CISA conference, March 22–24, 1992, Phoenix.

Mountjoy, Joseph B., and Luis Torres M. 1985. The Production and Use of Prehispanic Metal Artifacts in the Central Coastal Area of Jalisco, Mexico. In *The Archaeology of West and Northwest Mesoamerica,* edited by M. S. Foster and P. G. Weigand, pp. 133–152. Westview Press, Boulder.

Mountjoy, Joseph B., and Phillip C. Weigand. 1974. The Prehispanic Settlement Zone at Teuchitlan, Jalisco. In *41st International Congress of Americanists,* vol. 1, pp. 353–363. Mexico.

Muhly, James D. 1988. The Beginnings of Metallurgy in the Old World. In *The Beginning of the Use of Metals and Alloys,* edited by R. Maddin, pp. 2–20. MIT Press, Cambridge.

Murra, John V. 1972. El "control vertical" de un máximo de pisos ecológicos en la economía de las sociedades andinas. In *Documentos para la historia y etnología de Huánuco y la Selva Central,* vol. 2, edited by J. V. Murra, pp. 427–476. Universidad Nacional Hermilio Valdizán, Huánuco.

——— 1975. El tráfico de mullu en la costa del Pacífico. In *Formaciones económicas y políticas del mundo andino,* edited by J. V. Murra, pp. 255–267. Instituto de Estudios Peruanos, Lima.

Neumann, Frank J. 1974. The Rattle-Stick of Xipe Totec: A Shamanic Element in Prehispanic Mesoamerican Religion. In *41st International Congress of Americanists,* vol. 2, pp. 243–251. Mexico.

Nicholson, Henry B. 1960. The Mixteca-Puebla Concept in Mesoamerican Archaeology: A Re-examination. *International Congress of Anthropological and Ethnological Sciences: Men and Cultures; 5th International Congress,* pp. 612–618. Trustees of the University of Pennsylvania, Philadelphia.

Niessen, Sandra A. 1991. Interpreting Weaving Techniques. Paper presented at the annual meeting of the Canadian Anthropological Society (CASCA), panel on "Technology and Interpretation," May 1991, London, Ontario.

Nordenskiöld, Erland F. 1921. *The Copper and Bronze Ages in South America.* Elanders Boktryckeri Aktiebolag, Göteborg.

Northover, J. Peter. 1989. Properties and Use of Arsenic-Copper Alloys. In *Der Anschnitt* (Beiheft 7), edited by A. Hauptmann, E. Pernicka, and G. Wagner, pp. 111–118. Selbstverlag Deutsches Bergbau-Museum, Bochum.

Northover, J. Peter, S. Shalev, and M. Schindler. 1992. Exotic Alloys in Prehistory. Paper presented at the 28th International Symposium on Archaeometry, Los Angeles.

Norton, Presley. 1986. El señorío de Salangone y la liga de mercaderes. *Miscelánia Antropológica Ecuatoriana* 6:131–143.

Oviedo y Valdés, Gonzalo F. de. 1945. *Historia general y natural de las indias, islas y tierra-firma del mar océano.* 13 vols. Editorial Guaranía, Asunción.

Owen, Bruce. 1986. *The Role of Common Metal Objects in the Inka State.* Master's Thesis, University of California, Los Angeles.

Panczner, William D. 1987. *Minerals of Mexico.* Van Nostrand Reinhold Company, New York.

Parga P., J. de J., and J. de J. Rodríguez S. 1991. Geology of the Tizapa Ag, Zn, Pb, Cu, Cd, and Au Massive Polymetallic Sulfides, Zacazonapan, Mexico. In *Economic Geology, Mexico,* edited by G. P. Salas, pp. 373–378. The Geology of North America. Geological Society of America, Boulder.

Paso y Troncoso, Francisco del (editor). 1905–1906. *Papeles de Nueva España.* 2nd series. 9 vols. Sucesores de Rivadeneyra, Madrid.

Pasztory, Esther. 1974. *The Iconography of the Teotihuacan Tlaloc.* Dumbarton Oaks, Washington, D.C.

Paul, Anne, and Susan A. Niles. 1984. Identifying Hands at Work on a Paracas Mantle. *Textile Museum Journal* 23:5–15.

Paulsen, Allison C. 1974. The Thorny Oyster and the Voice of God: *Spondylus* and *Strombus* in Andean Prehistory. *American Antiquity* 39(4):597–607.

1977. Patterns of Maritime Trade between South Coastal Ecuador and Western Mesoamerica, 1500 B.C.–A.D. 600. In *The Sea in the Pre-Columbian World,* edited by E. P. Benson, pp. 141–166. Dumbarton Oaks, Washington, D.C.

Pendergast, David M. 1962a. Metal Artifacts from Amapa, Nayarit, Mexico. *American Antiquity* 27(3):370–379.

1962b. Metal Artifacts in Prehispanic Mesoamerica. *American Antiquity* 27(4):520–545.

1981. Lamanai, Belize: Summary of Excavation Results, 1974–1980. *Journal of Field Archaeology* 8:29–53.

1991. The Southern Maya Lowlands Contact Experience: The View from Lamanai, Belize. In *Columbian Consequences,* edited by D. H. Thomas, pp. 337–354. Smithsonian Institution Press, Washington, D.C.

Pfaffenberger, Bryan. 1988. Fetishised Objects and Humanised Nature: Towards an Anthropology of Technology. *Man* (N.S.) 23:236–252.

Plazas, Clemencia, and Ana María Falchetti. 1985. Cultural Patterns in the Prehispanic Goldwork of Colombia. In *The Art of Precolumbian Gold: The Jan Mitchell Collection,* pp. 46–59. Metropolitan Museum of Art, New York.

Pollard, Helen P. 1987. The Political Economy of Prehispanic Tarascan Metallurgy. *American Antiquity* 52(4):741–752.

1993. *Taríacuri's Legacy: The Prehispanic Tarascan State.* University of Oklahoma Press, Norman.

Poma de Ayala, Felipe Guaman. 1936. *Nueva crónica y buen gobierno.* Facsimile Edition. Institut d'Ethnologie, Paris.

Quezada R., María Noemí. 1972. *Los matlatzincas: época prehispánica y época colonial hasta 1650.* Instituto Nacional de Antropología e Historia, Mexico.

Rapp, George, Jr. 1982. Native Copper and the Beginning of Smelting: Chemical Studies. In *Early Metallurgy in Cyprus, 4000–500 B.C.,* edited by J. D. Muhly, R. Maddin, and V. Karageorghis, pp. 33–38. Pierides Foundation, Nicosia.

Reddy, Junuthula N. 1984. *An Introduction to the Finite Element Method.* McGraw-Hill, New York.

Reichel-Dolmatoff, Gerardo. 1981. Things of Beauty Replete with Meaning: Metals and Crystals in Colombian Indian Cosmology. In *Sweat of the Sun, Tears of the Moon: Gold and Emerald Treasures of Colombia,* pp. 17–33. Natural History Museum of Los Angeles County, Los Angeles.

1988. *Orfebrería y chamanismo: un estudio iconográfico del Museo del Oro.* Compañía Litográfica Nacional, Medellín.

Reiss, Wilhelm, and Alphons Stübel. 1880–1887. *The Necropolis of Ancon in Peru: A Contribution to Our Knowledge of the Culture and Industries of the Empire of the Incas, Being a Result of Excavations Made on the Spot.* A. Asher, Berlin.

Renfrew, Colin. 1969. Trade and Culture Process in European Prehistory. *Current Anthropology* 10:151–169.

1975. Trade as Action at a Distance: Questions of Integration and Communication. In *Ancient Civilization and Trade,* edited by J. A. Sabloff and C. C. Lamberg-Karlovsky, pp. 3–59. University of New Mexico Press, Albuquerque.

Roe, Peter G. 1982. *The Cosmic Zygote: Cosmology in the Amazon Basin.* Rutgers University Press, New Brunswick.

Root, William C. 1949. The Metallurgy of the Southern Coast of Peru. *American Antiquity* 15(1):10–37.

1962. Report on the Metal Objects from Mayapan. In *Mayapan, Yucatan, Mexico,* edited by H. E. D. Pollock, pp. 391–399. Carnegie Institution, Washington, D.C.

1969. Metal Artifacts from Chiapa de Corzo, Chiapas, Mexico. In *The Artifacts of Chiapa de Corzo, Chiapas, Mexico,* edited by T. Lee, pp. 203–207. New World Archaeological Foundation, Provo, Utah.

Rostoker, William, and James R. Dvorak. 1991. Some Experiments with Co-Smelting to Copper Alloys. *Archeomaterials* 5:5–20.

Rostworowski de Diez Canseco, María. 1970. Mercaderes del valle de Chincha en la época prehispánica: un documento y unos comentarios. *Revista Española de Antropología Americana* 5:135–177.

Rowe, Anne Pollard. 1980. Textiles from the Burial Platform of Las Avispas at Chan Chan. *Ñawpa Pacha* 18:81–148.

Rubín de la Borbolla, Daniel F. 1944. Orfebrería tarasca. *Cuadernos Americanos* 3(3):127–138.

Rutledge, John W., and Robert B. Gordon. 1987. The Work of Metallurgical Artificers at Machu Picchu, Peru. *American Antiquity* 52(3):578–594.

Ruvalcaba-Ruiz, Delfino C., and Tommy B. Thompson. 1988. Ore Deposits at the Fresnillo Mine, Zacatecas, Mexico. *Economic Geology* 83:1583–1596.

Sahagún, Fray Bernardino de. 1950–1982. *Florentine Codex: General History of the Things of New Spain.* 13 vols. (12 books). Translated and edited by A. J. O. Anderson and C. E. Dibble. School of American Research and University of Utah, Salt Lake City.

1993. *Psalmodia Christiana: Christian Psalmody.* Translated by A. J. O. Anderson. University of Utah Press, Salt Lake City.

Salas, Guillermo P. 1980. *Carta y provincias metalogenéticas de la República Mexicana.* Consejo de Recursos Minerales, Mexico.

1991a. Sierra Madre Occidental Metallogenic Province. In *Economic Geology, Mexico,* edited by G. P. Salas, pp. 197–198. The Geology of North America. Geological Society of America, Boulder.

1991b. Taxco Mining District, State of Guerrero. In *Economic Geology, Mexico,* edited by G. P. Salas, pp. 379–380. The Geology of North America. Geological Society of America, Boulder.

Salomon, Frank. 1978. Pochteca and Mindalá: A Comparison of Long-distance Traders in Ecuador and Mesoamerica. *Journal of the Steward Anthropolgical Society* 9:231–248.

——— 1986. *Native Lords of Quito in the Age of the Incas: The Political Economy of North-Andean Chiefdoms.* Cambridge University Press, Cambridge.

Sámano-Xerez. 1937. Relación. In *Cuadernos de historia del Perú,* vol. 2, edited by R. Porras B., pp. 63–68. Imprimeries Les Presses Modernes, Paris.

Santillán, Manuel. 1929. *Geología minera de las regiones norte, noroeste, y central del estado de Guerrero.* Boletín Instituto Geológico de México no. 48, pp. 47–102. Universidad Nacional Autónoma de México, Mexico.

Sauer, Carl O. 1948. *Colima of New Spain in the Sixteenth Century.* University of California Press, Berkeley

Schele, Linda, and Mary Ellen Miller. 1986. *The Blood of Kings.* Kimbell Art Museum, Fort Worth.

Scheubel, Frank R., Kenneth F. Clark, and Elise W. Porter. 1988. Geology, Tectonic Environment, and Structural Controls in the San Martin de Bolaños District, Jalisco, Mexico. *Economic Geology* 83:1703–1720.

Schiffer, Michael B. 1992. Archaeology and Behavioral Science: Manifesto for an Imperial Archaeology. In *Quandaries and Quests: Visions of Archaeology's Future,* edited by L. Wandsnider, pp. 225–238. Southern Illinois University Press, Carbondale.

Schiffer, Michael B., and James M. Skibo. 1987. Theory and Experiment in the Study of Technological Change. *Current Anthropology* 28:595–622.

Schöndube B., Otto. 1974. *Tamazula-Tuxpán-Zapotlán: pueblos de la frontera septentrional de la antigua Colima,* edited by J. M. Muriá, 2 vols. Tesis Profesional, Escuela Nacional de Antropología e Historia, Mexico.

——— 1980a. La nueva tradición. In *Historia de Jalisco,* vol. 1, pp. 213–258. Gobierno de Jalisco, Unidad Editorial, Guadalajara.

——— 1980b. La tradición de las tumbas de tiro. In *Historia de Jalisco,* vol. 1, pp. 171–212. Gobierno de Jalisco, Unidad Editorial, Guadalajara.

Scott, David A. 1986. Gold and Silver Alloy Coatings over Copper: An Examination of Some Artefacts from Ecuador and Colombia. *Archaeometry* 28:33–50.

——— 1988. Ancient Bronze, Copper-Arsenic and Tumbaga Alloys from South America: Some Ecuadorian Aspects. Manuscript on file, Getty Conservation Institute, Marina del Rey.

Shafer, Harry J., and Thomas R. Hester. 1991. Lithic Craft Specialization and Product Distribution at the Maya Site of Colha, Belize. *World Archaeology* 23:79–97.

Sharer, Robert J. 1985. Terminal Events in the Southeastern Lowlands: A View from Quirigua. In *The Lowland Maya Postclassic,* edited by A. F. Chase and P. M. Rice, pp. 245–253. University of Texas Press, Austin.

Shimada, Izumi. 1985. Perception, Procurement, and Management of Resources: Archaeological Perspective. In *Andean Ecology and Civilization,* edited by S. Masuda, I. Shimada, and C. Morris, pp. 357–399. University of Tokyo Press, Tokyo.

Shimada, Izumi, and John F. Merkel. 1991. Copper-Alloy Metallurgy in Ancient Peru. *Scientific American* 265(1):80–86.

Shook, Edwin M. 1965. Archaeological Survey of the Pacific Coast of Guatemala. In *Handbook of Middle American Indians,* vol. 2, edited by G. R. Willey, pp. 180–194. University of Texas Press, Austin.

Skibo, James M., and Michael B. Schiffer. 1987. The Effects of Water on Processes of Ceramic Abrasion. *Journal of Archaeological Science* 14:83–96.

Smith, Cyril S. 1965a. The Interpretation of Microstructures of Metallic Artifacts. In *Application of Science in Examination of Works of Art,* edited by W. J. Young, pp. 20–52. Museum of Fine Arts, Boston.

1965b. Materials and the Development of Civilization and Science. *Science* 148:908–917

1968a. Matter versus Materials: A Historical View. *Science* 162:637–644.

1968b. Metallographic Study of Early Artifacts Made from Native Copper. *Actes du XIe Congrès International d'Histoire des Sciences* 6:237–252.

1972. Metallurigal Footnotes to the History of Art. *Proceedings, American Philosophical Society* 116(2):97–135.

1975. Metallurgy as a Human Experience. *Metallurgical Transactions* 6A(4):603–623.

1977. *Metallurgy as a Human Experience: An Essay on Man's Relationship to His Materials in Science and Practice throughout History.* American Society for Metals, Metals Park.

1978. Structural Hierarchy in Science, Art, and History. In *On Aesthetics in Science,* edited by J. Wechsler, pp. 9–53. MIT Press, Cambridge.

1981. *A Search for Structure: Selected Essays on Science, Art, and History.* MIT Press, Cambridge.

Smith, Meritt R. 1977. *Harpers Ferry Armory and the New Technology: The Challenge of Change.* Cornell University Press, Ithaca.

Smith, Michael E. 1978. A Model for the Diffusion of the Shaft Tomb Complex from South America to West Mexico. *Journal of the Steward Anthropological Society* 9:179–204.

1990. Long-Distance Trade under the Aztec Empire: The Archaeological Evidence. *Ancient Mesoamerica* 1:153–169.

Smith, Michael E., and John F. Doershuk. 1991. Late Postclassic Chronology in Western Morelos, Mexico. *Latin American Antiquity* 2:291–310.

Smith, Michael E., and Cynthia M. Heath-Smith. 1980. Waves of Influence in Postclassic Mesoamerica? A Critique of the Mixteca-Puebla Concept. *Anthropology* 4(2):15–50.

1994. Rural Economy in Late Postclassic Morelos: An Archaeological Study. In *Economies and Polities in the Aztec Realm,* edited by M. G. Hodge and M. E. Smith, pp. 349–376. Institute for Mesoamerican Studies, Albany, N.Y.

Spicer, Edward H. 1964. Apuntes sobre el tipo de religión de los Uto-Aztecas centrales. In *35th International Congress of Americanists,* vol. 1, pp. 27–38. Mexico.

Stanford, E. Thomas. 1966. A Linguistic Analysis of Music and Dance Terms for Three Sixteenth-Century Dictionaries of Mexican Indian Languages. In *Yearbook II of the Inter-American Institute for Musical Research,* pp. 101–159. Tulane University Press, New Orleans.

Starbuck, David R. 1975. *Man-Animal Relationships in Pre-Columbian Central Mexico.* Ph.D. Dissertation, Yale University, New Haven.

Staudenmaier, John M. 1985. *Technology's Storytellers: Reweaving the Human Fabric.* MIT Press, Cambridge.

Steinberg, Arthur. 1973. Joining Methods on Large Bronze Statues: Some Experiments in Ancient Technology. In *Application of Science in Examination of Works of Art,* edited by W. J. Young, pp. 103–138. Museum of Fine Arts, Boston.

Steponaitis, Vincas P. 1984. Technological Studies of Prehistoric Pottery from Alabama: Physical Properties and Vessel Function. In *The Many Dimensions of Pottery: Ceramics in Archaeology and Anthropology,* edited by S. E. van der Leeuw and A. C. Pritchard, pp. 79–122. Universiteit van Amsterdam, Amsterdam.

Stevenson, Robert. 1968. *Music in Aztec and Inca Territory.* University of California Press, Berkeley.

Stimmell, Carole, and Richard L. Stromberg. 1986. A Reassessment of Thule Eskimo Ceramic Technology. In *Technology and Style,* edited by W. D. Kingery, pp. 237–250. Ceramics and Civilization, vol. 2. American Ceramic Society, Columbus.

Strong, William D. 1935. *Archaeological Investigations in the Bay Islands, Spanish Honduras.* Smithsonian Institution, Washington, D.C.

Strong, William D., Alfred V. Kidder II, and A. J. Drexel Paul, Jr. 1938. *Preliminary Report of the Smithsonian Institution–Harvard University Archaeological Expedition to Northwestern Honduras, 1936.* Smithsonian Institution, Washington, D.C.

Sullivan, Lawrence E. 1988. *Icanchu's Drum: An Orientation to Meaning in South American Religions.* Macmillan, New York.

Sutliff, Marie J. 1992. *El proceso productivo metalúrgico de la cultura Milagro: el caso de Peñón del Río.* Tesis de Licenciatura, Escuela Superior Politécnica del Litoral, Guayaquil.

Tedlock, Dennis (translator). 1985. *Popol Vuh: The Mayan Book of the Dawn of Life.* Simon and Schuster, New York.

Termer, Franz. 1928. Los bailes de culebra entre los Indios Quichés en Guatemala. In *23rd International Congress of Americanists,* pp. 661–667. New York.

Terrones, Alberto J. L. 1984. Overview of the Mineral Resource Potential of Latin America in Relation to Global Tectonic and Metallogenic Controls. *Global Tectonics and Metallogeny* 2:213–249.

Toor, Frances. 1947. *A Treasury of Mexican Folkways.* Crown Publishers, New York.

Torrence, Robin. 1984. Monopoly or Direct Access? Industrial Organization at the Melos Obsidian Quarries. In *Prehistoric Quarries and Lithic Production,* edited by J. E. Ericson and B. A. Purdy, pp. 49–64. Cambridge University Press, Cambridge.

Torres M., Luis, and Francisca Franco V. 1989. La orfebrería prehispánica en el Golfo de México y el tesoro del pescador. In *Orfebrería prehispánica,* pp. 217–270. Corporación Industrial Sanluis, Mexico.

Townsend, Richard F. 1992. *The Aztecs.* Thames and Hudson, London.

Tozzer, Alfred M. (editor). 1941. *Landa's Relación de las Cosas de Yucatán: A Translation.* Peabody Museum, Cambridge.

Tudela, José (transcriber). 1977. *Relación de las ceremonias y ritos y población y gobierno de los indios de la provincia de Michoacán (1541).* Balsal Editores, Morelia.

Turner, Terence S. 1980. The Social Skin. In *Not Work Alone: A Cross-Cultural View of Activities Superfluous to Survival,* edited by J. Cherfas and R. Lewin, pp. 112–140. Temple Smith, London.

Ubelaker, Douglas H. 1981. *The Ayalán Cemetery: A Late Integration Period Burial Site on the South Coast of Ecuador.* Smithsonian Institution, Washington, D.C.

———. 1983. Prehistoric Demography of Coastal Ecuador. *National Geographic Society Research Reports* 15:695–704.

van As, Abraham. 1989. Some Techniques Used by the Potters of Tell Hadidi During the Second Millennium B.C. In *Pottery Technology: Ideas and Approaches,* edited by G. Bronitsky, pp. 41–79. Westview Press, Boulder.

Vandiver, P. B., and C. G. Koehler. 1986. Structure, Processing, Properties, and Style of Corinthian Transport Amphoras. In *Technology and Style,* edited by W. D. Kingery, pp. 173–215. Ceramics and Civilization, vol. 2. American Ceramic Society, Columbus.

Varela R., Leticia T. 1986. *La música in la vida de los Yaquis.* Gobierno del Estado de Sonora, Sonora.

Velásquez G., Pablo. 1988. *Diccionario de la lengua phorhépecha.* Fondo de Cultura Económica, Mexico.

Ventura, Beatriz N. 1985. *Metalurgia: un aspecto poco conocido en la arqueología de las selvas occidentales.* Universidad de Buenos Aires, Buenos Aires.

Villalba O., Marcelo. 1988. *Cotocollao: una aldea formativa del valle de Quito.* Museos del Banco Central del Ecuador, Quito.

Wagley, Charles. 1940. World View of the Tapirape Indians. *Journal of American Folklore* 53:252–260.

Warren, J. Benedict. 1968. Minas de cobre de Michoacán, 1533. *Anales del Museo Michoacano* (2da época) 6:35–52.

———. 1985. *The Conquest of Michoacán: The Spanish Domination of the Tarascan Kingdom in Western Mexico, 1521–1530.* University of Oklahoma Press, Norman.

———. 1989. Información del licenciado Vasco de Quiroga sobre el cobre de Michoacán, 1533. *Anales del Museo Michoacano* (3ra época) 1:30–52.

Wauchope, Robert. 1948. *Excavations at Zacualpa, Guatemala.* Middle America Research Institute, Tulane University Press, New Orleans.

Weaver, Muriel Porter. 1981. *The Aztecs, Maya, and Their Predecessors: Archaeology of Mesoamerica.* 2nd ed. Academic Press, New York.

Weigand, Phillip C. 1985. Evidence for Complex Societies during the Western Mesoamerican Classic Period. In *The Archaeology of West and Northwest Mesoamerica,* edited by M. S. Foster and P. G. Weigand, pp. 47–91. Westview Press, Boulder.

———. 1992. Central Mexico's Influences in Jalisco and Nayarit during the Classic Period. In *Resources, Power, and Interregional Interaction,* edited by E. M. Schortman and P. A. Urban, pp. 221–232. Plenum Press, New York.

———. 1993. The Political Organization of the Trans-Tarascan Zone of Western Mesoamerica on the Eve of Spanish Conquest. In *Culture and Contact: Charles C. Di Peso's Gran Chichimeca,* edited by A. Woosley and J. Ravesloot, pp. 191–217. University of New Mexico Press, Albuquerque.

Weitlaner, Robert J. 1947. Exploración arqueológica en Guerrero. In *El occidente de México,* pp. 77–85. Sociedad Mexicana de Antropología, Cuarta Reunión de Mesa Redonda, Mexico.

Weitlaner J., Irmgard. 1964. Copper-Preserved Textiles from Michoacan and Guerrero. In *35th International Congress of Americanists,* vol. 1, pp. 525–536. Mexico.

Wenke, Robert J. 1981. Explaining the Evolution of Cultural Complexity: A Review. In *Advances in Archaeological*

Method and Theory, vol. 4, edited by M. B. Schiffer, pp. 79–128. Academic Press, New York.

Wertime, Theodore A. 1973. The Beginnings of Metallurgy: A New Look. *Science* 182:875–887.

West, Robert C. 1961. Aboriginal Sea Navigation between Middle and South America. *American Anthropologist* 63:133–135.

Wilbert, Johannes. 1974. The Calabash of the Ruffled Feathers. *Artscanada* 184–187:90–93.

Willey, Gordon R. 1962. The Early Great Styles and the Rise of the Precolumbian Civilizations. *American Anthropologist* 64:1–14.

1974. *Das alte Amerika*. Propyläen Verlag, Berlin.

Woodbury, Richard B., and Aubrey S. Trik. 1953. *The Ruins of Zaculeu, Guatemala*. United Fruit Company, New York.

Yerkes, Richard W. 1989. Lithic Analysis and Activity Patterns at Labras Lake. In *Alternative Approaches to Lithic Analysis*, edited by D. O. Henry and G. H. Odell, pp. 183–212. American Anthropological Association, Washington, D.C.

Young, David E., and Robson Bonnichsen. 1984. *Understanding Stone Tools: A Cognitive Approach*. Center for the Study of Early Man, University of Maine Press, Orono.

Index

Alloying. *See also* Color, altering through alloying
    cultural meanings of, 228–231
    development of, 16–17, 104, 106–112, 127–129, 171–180
Alloys, 106–121, 131–156, 158–167, 171–180
    copper-arsenic, 16, 21, 35–36, 38, 39, 41, 46, 51, 104, 106–121, 123, 127, 129, 134, 158, 166, 167, 169, 171–176, 180, 201, 222, 229
    copper-arsenic-lead, 220
    copper-arsenic-silver, 21, 139, 140
    copper-arsenic-tin, 21, 46, 127, 129, 145, 147, 158, 165, 175, 180, 193, 201
    copper-gold, 16, 17, 88, 107, 108, 112, 121, 123, 127, 152–153, 169, 180, 189, 219, 230, 235 (*see also* Tumbaga)
    copper-lead, 201, 206, 219–222, 224
    copper-silver, 16, 21, 35, 46, 52, 106–108, 111–120, 127, 131–132, 145, 147, 152–156, 169, 176–179, 229–233
    copper-silver-gold, 16, 108, 121, 127, 152–153, 176, 229
    copper-tin, 16, 21, 27, 36–37, 41, 46, 127, 129, 134–135, 138, 141, 145, 147, 152, 158–160, 162, 169, 179–180, 191, 193, 197, 199–201, 205, 219, 221, 224, 229
    gold-silver, 153, 229
Altun Ha, 47
Amapa, 14, 17, 49, 50, 51, 52, 53, 57, 59, 64, 66, 77, 80, 81, 92, 103, 104, 123, 249
Ancón, 181
Andean metallurgy, 16, 87–88, 120, 171–184, 187, 192
    copies of West Mexican artifacts, 183
Apatzingán, 127, 130, 132, 166, 184, 236
Arcelia, 38
Argentite. *See* Ore minerals
Arsenic, 22, 37. *See also under* Alloys; Ore minerals
    deposits, 27
    native, 21, 27, 39
Arsenopyrite. *See* Ore minerals
Autlán, 14

Awls
    Ecuadorian, 94, 108, 109, 110
    West Mexican, 52, 80–82, 161–163, 201, 207
Axayacatl, 243
Axe-monies
    Ecuadorian, 109, 110, 121, 167, 171–174
    Oaxacan, 166
    uses of, 167, 173
    West Mexican, 130–132, 156, 166–168, 171, 173
Axes
    Ecuadorian, 109, 100, 119
    West Mexican, 52, 73–75, 156–161
Ayalán, 91, 97, 109
Ayauhchicauaztli, 236–238, 245
Aztatlán ceramics, 49, 51, 131
Aztec state, 197, 202, 208, 223–224

Bahía culture period, 106
Balsas, Río, area, 50, 103, 130
Barra de Navidad, 64
Batanes del Tablazo, 187
Batán Grande, 171, 187
Bells
    Colombian, 99, 101
    copper-lead alloys in, 201
    cultural meanings of, 233–243
    Ecuadorian, 96, 109, 111, 112, 113, 121
    Huastec region, 217–219
    lost-wax-cast, 52–56, 101, 104, 122
    Maya region, 208, 209, 214, 219–220
    West Mexican, 52–58, 132–139, 202, 206, 207, 224
    wirework, 135–136, 201, 219
Bernard, 127, 131, 134, 140, 153, 162, 177
Bolivian metallurgy. *See* Andean metallurgy
Bonampak, murals at, 122, 233
Bottle gourd, 244

Bowls, 119

Bronze. *See* Alloys, copper-arsenic; Alloys, copper-tin

Bronze cannons, 39

Burkhart, Louise, 232

Buttons, 183, 184, 208–209

Calixlahuaca, 127, 130, 162

Cantares Mexicanos, 241–242

Capacha ceramics, 15–16, 103

Capilco, 127, 165, 193, 201, 202–208

Casas Grandes, 221

Cashibo, 232

Cassiterite. *See* Ore minerals

Cenote de Sacrificios, 127, 219–220, 224

Cerro Alto, 92, 93, 94, 106, 107, 108, 176

Cerro de Huistle, 47, 49, 50, 51, 52, 99, 104

Cervantes de Salazar, Francisco, 147

Chalchiuhtli Ycue, 238, 240

Chancay, 181

Chaquira, 101

Chiapa de Corzo, 127, 197, 221, 224

Chicauaztli, 236, 238, 239–240

Chimor, 102

Chimú, 171, 177

Chincha, 177, 181, 184–185, 191

Chipal, 221

Chisels, 201

Chorrera culture period, 106

Churumuco, 27, 38

Cinturón-sonaja, 57, 245

Classic Period, 10–12

Coalcomán, 38

Coamiles, 50

Cojumatlán, 47, 51, 52, 57, 83

Cold working, 75–83, 129, 139, 157, 162, 163, 167, 177
 and pitch in bells, 121, 122

Colombian metallurgy, 17, 183

artifact types, 98–99, 122–123, 175, 181–183, 209

fabrication methods, 88, 99, 101, 123, 183

metals and alloys, 98–99, 123, 180

Color, metallic
 altering through alloying, 129, 138–139, 144, 169, 180, 205, 229–230
 cultural significance of, 227, 228–233
 as emphasis of Period 2 metallurgy, 129, 156, 169
 as fundamental property, 3
 in sheet metal ornaments, 123

Compositional groups, of artifacts, 33–37

Contepec, 200

Copper, 21, 22. *See also under* Alloys; Ore minerals
 deposits, 22, 103
 native, 21, 34, 51, 52
 smelting regimes, 34–35
 zones of mineralization, 25–27, 106

Costa Rica, 99

Coyolxauhqui, 227

Creation, 246–248

Cuerauaperi, 240

Cuexcomate, 17, 127, 134, 162, 165–166, 174, 193, 197, 201, 202–208

Cuicuilco, 9, 10

Cuitzeo basin, 127

Culiacán, 131, 132, 140, 184, 221

"Cult of brilliance," 232

Curatame, 241, 244

Curicaueri, 240

Cyanocorax, 103–104, 186

Design, 18
 optimization through alloying, 127

Díaz del Castillo, Bernal, 200

Diffusion of technology, 101, 184, 197–225. *See also* Distribution (circulation) of artifacts; Transmission of technology

Digging sticks, 156, 192
Disks, 153, 179, 183
Distribution (circulation) of artifacts, 197–199, 201, 207–208, 209, 214, 219–222, 224

Ecuador, West Mexican contacts with, 12, 15–17, 99–105, 184–186
Ecuadorian metallurgy, 16, 87–124, 171–180
    artifact types, 91–98, 104, 107, 123–124, 171–176, 180, 192
    cultural emphasis of, 121–122
    development of, 105–112
    fabrication methods, 87–88, 91–98, 100, 107, 112, 123, 173–175, 179
    introduction to West Mexico, 99–105, 122–123, 184–186
    metals and alloys, 101, 104, 106–119, 123, 134, 171–180, 190
    ore minerals, 120, 179, 187
El Azúcar, 91, 94, 106, 107–108, 177
El Chanal, 127, 131, 153, 177, 184
El Paraíso, 122
El Talar, 180
Enargite. *See* Ore minerals
Enrichment effects and techniques, 154, 189
Esmeraldas, 97

Finite element analysis and tweezer function, 67–69
Fishhooks
    Ecuadorian, 95–96, 106, 108, 111, 112
    West Mexican, 82–83, 104, 156, 168
Formal design, 45, 46, 54, 131
Formative (Preclassic) Period, 8–10, 14

Gold, 21, 52, 105, 106. *See also under* Alloys; Ore minerals
    cultural significance, 155, 228–233
    documentary accounts of, 155

Guangala culture period, 94, 107–108, 177
Guaraxo, 38
Guasave, 131, 132, 184, 221
Guayas basin, 108
Guzmán, Nuño de, 155

Hachuelas, 166
Helms, Mary, 124
Hilltop House, 221
Hiripan, 229
Hoes, 139, 156, 192
Hot working, 129, 139, 147
Huandacareo, 127, 130, 134, 140, 145, 166
Huastec region, 57, 127, 168, 179, 201–202, 216–219, 224, 241. *See also* Platanito; Vista Hermosa
Huayapán, 221
Huichol, 104, 244
Huitzilopochtli, 227, 241
Huy, 220

Ica, 185
Indium, 29, 37
Industrial Minera México (IMMSA), 25
Infiernillo, 49, 50, 80, 91, 92, 94, 97, 103, 127, 130, 140, 145, 153
Ingapirca, 106, 112
Ingots, 199, 200, 214, 217
Inguarán, 38
Inka state, 102, 179, 182–183
Integration Period, 108–112
Intermediate processing material, 217
Introduction of metallurgy to West Mexico. *See* Maritime exchange system; Transmission of technology
Ixtlán del Río, 14

Jambelí culture period, 108
Jauja, 183

Knives, 109, 110

La Compañía, 95, 96, 106, 110, 112, 177
La Florida, 107
La Huacana, 38
La Libertad, 94, 95
Lamanai, 17, 101, 127, 145, 184, 201, 207–213, 214
Láminas, 169
Landa, Diego de, 199–200, 222
La Tolita, 97, 98, 104, 112
La Verde, 27
La Villita, 50, 52, 83, 127, 130, 132, 166
Leach, Edmund, 232
Lead, 21, 204, 222. *See also under* Alloys; Ore minerals
League of Merchants, 101, 105, 106, 108
Lechtman, Heather, 5, 230
Legajo 1204, 39
Lemonnier, Pierre, 5
Leone, Mark, 5
Lidchi Bigu, 221
Lienzo de Jucutacato, 38
Lo Arado, 127, 131, 153
Loma de los Cangrejitos, 95, 106, 109–110, 113, 176
Los Cabires, 200
Lost-wax casting, 52–56, 122, 208, 209
    in Colombia and lower Central America, 101, 104
    and pitch, 122, 127–128, 136–137
Lower Central American metallurgy, 17, 88, 101, 104, 185

Madero, 197, 220
Manabí, 92
Manta, 101
Manteño culture period, 106, 184
Manuel Elordi, 180
Mapa Metalogénico de México, 22–27, 200
Maracas, 246

Maritime exchange system, 100–104, 184–186
Matlazinca, 127, 223–224
Maya culture, 11–12, 208–216, 219–221
Mayahuel, 238
Mayapán, 220
Mechanical properties of metals, 18, 55
Meliponidie, 123
Mesoamerica, geographical limits of, 6
Metal, cultural definitions of
    as marker of status, 155
    and the sacred, 156, 227
    as standard of value or medium of exchange, 121
Mictlantecuhtli, 157
Milagro, 110, 111
Milpillas, 17, 127, 130, 134, 140, 177, 205, 207
Mines, preconquest, 39–41
Mist rattleboard. *See Ayauhchicauaztli*
Mixteca-Puebla ceramic tradition, 104
Moche, 177
Molino, 221
Mollendo, 185
Monte Albán, 11, 197, 221
Moquegua, 185
Mountjoy, Joseph, 49, 103, 104
Museo Antropológico de Guayaquil (MAG), 89, 106, 109, 112–116, 192, 256
Museo Regional de Guadalajara (RMG), 131, 132–136, 192, 254–256

Nahuatl language, 227, 228–229, 233, 236, 241–243
Naipes, 120
Napa Tecutli, 237
Naranjo, 166
Navacoyan, 221
Needles
    Ecuadorian, 92–93, 108, 109–110, 111, 112, 174–176, 180

West Mexican, 52, 79–80, 104, 163–166, 174, 180, 201, 207
Nicoya, Gulf of, 104
Nochixtlán, 221
Nohmul, 220
Nose rings, 109, 111, 112, 119

Oaxaca, 152, 169, 197, 221–222
OGSE-Ma-172 site, 106, 108
Olid, Cristóbal de, 155
Olmec culture, 8–9
Opochtli, 238, 240
Ore minerals
    Ecuador, 105, 120, 187
        arsenopyrite, 120, 186, 187, 189
        atacamite, 187
        chalcopyrite, 187
        enargite, 105, 120, 186, 187, 189
        tetrahedrite-tennantite series, 105, 120, 186, 187
    Mexico, 21–41, 186–189
        argentite, 21, 27, 31
        arsenopyrite, 21, 27, 29, 31, 37, 39, 186, 187, 189, 207
        azurite, 27, 34, 51
        bismuthinite, 36, 37
        bornite, 39
        cassiterite, 21, 27, 29, 39, 138, 179, 186, 188, 191, 197, 200–201, 206, 207
        chalcocite, 27
        chalcopyrite, 22, 25, 27, 29, 34, 35, 36, 38, 39, 123, 186, 187, 197, 207
        enargite, 27, 29, 35, 38, 186, 187
        freibergite, 35, 41
        galena, 201
        malachite, 22, 25, 27, 34, 36, 186
        pentlandite, 36, 37
        polybasite, 21, 27, 31
        proustite, 27, 29, 31
        pyragyrite, 29
        stephanite, 29
        stibnite, 36
        tennantite, 27, 31, 35, 38
        tetrahedrite, 31, 186, 187
    Peru, 120, 187
        arsenopyrite, 189
        chalcopyrite, 187, 188
        enargite, 187
        tetrahedrite-tennantite series, 27
    stannite (southern Andes), 188
Otomí, 223
Oxtuma, 223

Painted jay. *See Cyanocorax*
Palmar, 92
Palos Blancos, 158
Panama, 99
Pátzcuaro, 38, 49, 127, 130, 155
Pendants, 109
Peñitas, 50, 52, 83
Peñon del Río, 106, 111
Period 1 metallurgy
    artifact types, 52, 59, 63–64, 73, 75, 80, 82, 89–98
    Ecuadorian elements in, 89–98
    fabrication methods, 51–66, 79–83, 91, 93, 95–96
    introduction of, 99–105
    lower Central American and Colombian elements in, 98–99, 123
    metals and alloys, 50, 186
    technical and cultural emphases, 83–85
Period 2 metallurgy
    artifact types, 127, 132–134, 139, 155, 168, 180–184, 189, 193

Period 2 metallurgy *(continued)*
    fabrication methods, 140–141, 147–149, 161, 163–165, 174–175
    introduction of, 184–186
    metals and alloys, 127, 129, 134–139, 140, 144, 152, 155–156, 158, 171–180, 186–191, 193
    technical and cultural emphases, 127–129, 168–170, 189–193
Peruvian metallurgy. *See* Andean metallurgy
Petersen, Ulrich, 25, 33
Peyote, 103, 104, 244
Pirincay, 107
Pitch, 56–57. *See also* Sound
    cultural interest in, 127, 136–138, 228
    in Ecuadorian bells, 121, 122
Platanito, 127, 201, 216–219, 224
Pochteca traders, 224
Popol Vuh, 157, 246–248
Postclassic Period, 12–13
Preclassic Period. *See* Formative Period
Psalmodia Christiana, 241

Quemistlan, 220
Quetzalcoatl, 235, 238

Rattleboards, 139. *See also Ayauhchicauaztli*
Rattles, 122, 240, 243, 244
Rattlesnakes, 244
Rattlesticks, 39, 139, 230, 235, 236. *See also Chicauaztli*
Rattling sounds, cultural significance of, 243–246
Recycling of metal stock, 206–207, 214, 222
Reflectivity, in sheet metal ornaments, 123
Regional Developmental Period, 106
Reichel-Dolmatoff, Gerardo, 230
Relación de Ajuchitlán, 227, 230, 246
Relación de Michoacan, 59, 145, 154, 155, 156, 158, 229, 241, 243

Relación de Sinagua, 38
Relación de Sultepec, 201
Relaciones Geográficas, 38–39
Rings
    Ecuadorian, 91–92, 109, 111, 112, 119, 121, 177, 180
    hollow, 112
    use of, 123
    West Mexican, 59–63, 139–145, 177, 180
Río Tambo, 94
Rodela, 134, 154
Ruiz de Estrada, Bartolomé, 103, 184, 190

Saavedra, Alvaro, 155
Sailing rafts, 101, 103
Salango, 101, 106, 107, 108–109, 110, 174, 176, 192
Salangone, 106
San Lorenzo (Ecuador), 91, 108
San Lorenzo (Mesoamerica), 8–9
Santa Rita, 220
Sauer, Carl, 41
Schöndube, Otto, 15
Shaft tombs, 15–16
Sheet metal ornaments, 112, 152–155, 176–177
Shipibo, 232
Sicán, 171
Silver, 21, 22, 27, 29, 39, 52, 105, 106. *See also under* Alloys; Ore minerals
    cultural significance of, 155, 228–233
Simulation studies, 70–73, 231
Sinagua, 38, 39
Sipán, 97
Slag, 127, 131
Smelting regimes, 123, 138, 157, 187–189
Smith, Cyril, 5, 251
Sound, 3, 121, 127, 169. *See also* Pitch
    cultural significance of, 235–241, 246–248

Spiral ornaments, 97
Spondylus, 101–103, 105, 184, 185, 189, 249
Standardization of alloys, 138, 159, 168, 250
Sultepec, 223–224

Tabasco, 197, 199
Tajumulco, 221
Tamaulipas, 197
Tamazula, 14
Tamtok, 206
Tamulté Las Sabanas, 220, 224
Tarascan language, 229, 231, 233, 240, 241
Tarascan region, 123, 130, 145, 176, 183, 192
Tarascan state, 74, 130, 155, 197, 223–224
Tariacuri, 241
Tarquea, 139
Taxco, 39
Technical choice, 3–4, 6, 18, 71, 85, 122, 124, 145, 169, 189–194, 225, 227, 248, 250
Technical diversity, 129
Technological chronology, 45–46. See also Period 1 metallurgy; Period 2 metallurgy
Technology, as cultural activity, 3–6
Teloloapan, 200, 206, 223–224
Temascaltepec, 223
Tennantite. *See* Ore minerals
Tenochtitlan, 39, 202, 224, 250
Tenosique, 220
Teocalitche, 200
Teotihuacan, 9, 10, 11, 14, 103, 235, 248–249
Tepecoacuilco, 39
Tepulan, 38
Tetela del Río, 39
Teuchitlán, 14, 47, 249
Tezcatlipoca, 241, 243
Tezcatzoncatl, 238
Tikal, 11

Tin, 22, 37, 38. *See also* Zacatecas tin province; *see also under* Alloys; Ore minerals
  control of deposits, 223–224
  disseminated in chalcopyrite ores, 52, 123, 197
  indium concentrations in, 29, 37, 206
  locations of deposits, 22, 101, 179, 200, 206
  nature of deposits, 29
Tingambato, 14, 47
Titlacauan, 241
Tiwanaku, 185
Tizapán el Alto, 47, 51, 77
Tlacolula, 221
Tlaloc, 157, 235–237, 244
Tlatilco, 233
Tohil, 247
Tomatlán, 17, 47, 49, 52, 59, 61, 63, 64, 73, 80, 81, 83, 91, 92, 94, 97, 103, 104, 123
Tonatiuh, 228
Topia, 221
Transmission of technology, 100–105, 122–124, 184–186, 189. *See also* Diffusion of technology; League of Merchants; Maritime exchange system
Tres Cerritos, 127, 145, 166
Tukano, 231
Tumbaga, 8, 17, 88, 107, 112, 123, 177, 189, 214, 230, 235. *See also* Alloys, copper-gold
Turner, Terence, 231
Tuxcacuesco, 130–131, 132, 140
Tweezers
  in central and southern Andes, 181–182
  Colombian, 145, 181–182
  cultural significance, 145, 191–192, 231
  design requirements, 69–70, 72, 147
  Ecuadorian, 93–94, 98, 108, 109, 110, 111, 112, 180–183
  fabrication methods, 63–66, 145–152, 231
  function, determination of, 66, 70–73, 149–151, 231

shell design, 145–152, 180–183

significance of removal of body and beard hair, 231–232

West Mexican, 52, 104, 145–152, 183, 201, 207

Tzapotlan Tenan, 240

Tzintzuntzan, 14, 127, 130, 134, 140, 145, 155, 158, 166

Uixtociuatl, 238

Urichu, 17, 91, 127, 130, 140, 144, 145, 162, 166, 177

Utilitarian aspects of West Mexican metallurgy, 129, 192–193

Vista Hermosa, 127, 201, 216–219

Warren, J. Benedict, 39

West Mexican metalworking zone, 13–15

influence of Teotihuacan in, 14, 127, 199, 248

Wirework. *See* Bells, wirework

Xaratanga, 229, 240

Xilonen, 240

Xipe Totec, 230, 239

Xochipala, 166

Yucatán Peninsula, 197

Zacapu, 127

Zacatecas tin province, 22, 27, 197, 200, 217, 222

metallic minerals in, 12

Zacatula, 103, 186

Zacualpa, 221

Zaculapan, 201

Zaculeu, 221